Foundations of EMERGENCY MANAGEMENT

David McElreath • Adrian Doss • Carl Jenson • Michael Wigginton
Robert Nations • Jeffrey Van Slyke • Julie Nations

Kendall Hunt
publishing company

Cover image © Shutterstock, Inc.

Kendall Hunt
publishing company

www.kendallhunt.com
Send all inquiries to:
4050 Westmark Drive
Dubuque, IA 52004-1840

Copyright © 2014 by Kendall Hunt Publishing Company

ISBN 978-1-4652-3488-9

All rights reserved. No part of this publication may be reproduced, stored in a retrieval system, or transmitted, in any form or by any means, electronic, mechanical, photocopying, recording, or otherwise, without the prior written permission of the copyright owner.

Printed in the United States of America

CONTENTS

About the Authors v

CHAPTER 1: Foundations of Emergency Management 1

CHAPTER 2: History of Emergency Management in the United States 23

CHAPTER 3: Principle Hazards Facing the United States 69

CHAPTER 4: Power and Policy of Stakeholders and First Responders 111

CHAPTER 5: Roles in Emergency Management 145

CHAPTER 6: Strategy 167

CHAPTER 7: Roles in Emergency Management: Volunteer Organizations and NGOs 193

CHAPTER 8: Role of the Emergency Manager and Developing an Effective Emergency Management Organization 215

CHAPTER 9: Preparing for the Inevitable 241

CHAPTER 10: The Emergency Management Process: Risk Perception and Analysis, Prevention, and Hazard Mitigation 267

CHAPTER 11: The Emergency Management Process: Disaster Recovery 309

CHAPTER 12: Professionalization of Emergency Management 341

CHAPTER 13: International Emergency Management 391

CHAPTER 14: Future of Emergency Management 413

DEDICATION

This text is dedicated to both those homeland security and emergency management professional who put their lives on the line every day to ensure the safety of our communities and nation and to those whose love and support has proven so important to each of us. Leisa, Ashley, Emily, Bethany, Liam and Hughes; Margaret and Caleigh; Brenda, Genevieve, and Elyse; Lisa; Renee, Jessica, Matthew, Jacob and Cannon; our Colleagues in Public Safety and the NPS/CHDS class of 2010 without whose love, support and encouragement this text would not have been possible. The authors 2014.

ABOUT THE AUTHORS

David H. McElreath, PhD: Dr. McElreath's background includes Professor and Chair, Department of Legal Studies, University of Mississippi; Professor and Chair, Department of Criminal Justice, Washburn University; Associate Professor, Southeast Missouri State University; Colonel, United States Marine Corps; and Law Enforcement and Corrections positions with the Oxford (Mississippi) Police and Forrest County (Mississippi) Sheriff's Department. His education and training include a PhD in adult education and criminal justice, University of Southern Mississippi; MSS, United States Army War College; MCJ, University of Mississippi; BPA, University of Mississippi; graduate of the United States Army War College. He is also the author of numerous publications on the criminal justice system. He and his wife Leisa live in Oxford, Mississippi.

Daniel Adrian Doss, PhD: Dr. Doss's background includes tenured Assistant Professor, College of Business Administration, University of West Alabama; Assistant Professor, Belhaven College; Adjunct Assistant Professor, Embry-Riddle Aeronautical University; and Chair of Graduate Business and Management, University of Phoenix (Memphis). His professional career consisted of software engineering and analytical positions in both the defense and commercial industries. Corporate entities included full-time and contract positions with Federal Express and uMonitor.com, and contract positions, via Data Management Consultants (Biloxi, MS), with Loral Corporation (formerly IBM Federal Systems) and Lockheed-Martin. Additional credentials include Colonel, Mississippi State Guard; graduate of the Lafayette County Law Enforcement Academy; and graduate of the Law Enforcement Mobile Video Institute. He has also co-authored a variety of peer-reviewed journal articles and conference proceedings. His education consists of a PhD in business administration, Northcentral University; MCJ in homeland security, University of Mississippi; MA in computer resources and information management, Webster University; MBA, Embry-Riddle Aeronautical University; graduate certificate in forensic criminology, University of Massachusetts (Lowell); graduate certificate in nonprofit financial management, University of Maryland (Adelphi); and a BS in computer science with a mathematics minor, Mississippi State University. He is currently pursuing a second doctorate, in police science, with the University of South Africa.

Carl J. Jensen, PhD: Carl J. Jensen is the Director of the University of Mississippi's (UM) Center for Intelligence and Security Studies. He is also a member of UM's Legal Studies department and serves in an adjunct capacity as a Senior Behavioral Scientist with the RAND Corporation. Dr. Jensen served as a Special Agent with the Federal Bureau of Investigation (FBI) for 22 years; his FBI career included service as a field agent, a Forensic Examiner in the FBI Laboratory, and an Instructor and Assistant Chief of the Behavioral Science Unit. He has published extensively and lectured throughout the world. Dr. Jensen received a BS degree from the U.S. Naval Academy, an MA from Kent State University, and a PhD from the University of Maryland. He and his family reside in Oxford, Mississippi.

Michael Wigginton, PhD: Mike Wigginton's background includes Assistant Professor of criminal justice and Director of the University of Mississippi Master of Criminal Justice Executive Cohort Program, Department of Legal Studies, with the University of Mississippi; former Assistant Professor, Southeast Louisiana University, and an adjunct professor with Tulane University; Special Agent, United States Customs Service; Special Agent, United States Drug Enforcement Administration; detective and State Trooper, Louisiana State Police; police officer, New Orleans Police Department; and a United States Air Force Security Police Dog Handler with service in Vietnam. His education and training include a PhD in criminal justice, University of Southern Mississippi; MS, the University of New Orleans; MS, the University of Alabama; BA, Loyola University of New Orleans. He is also the author of numerous publications on the criminal justice system.

Robert Nations, MA: Director Robert (Bob) Nations, Jr., is currently the Director of the Shelby County Office of Preparedness, which represents the coordination of both Homeland Security and the Emergency Management Agency for Memphis/Shelby County, Tennessee and the Memphis/Shelby Urban Area Security Initiative (UASI). Bob is a career law enforcement officer and police administrator serving in those related fields from 1972 through his current position. Prior to coming to Shelby County, he served as State Director of Homeland Security for Mississippi, and was Chief of Operations for Homeland security during Hurricane Katrina. Bob is in Project Management with the University of Mississippi in the legal studies department and is a part-time faculty member at the University of Memphis in the Department of Criminal Justice. Bob is a graduate of the Naval Post Graduate School's Center for Homeland Security and Defense—Executive Leadership

Program and holds a master's degree from the University of Memphis. He currently serves on the Homeland Security Advisory Board at Daniel Webster College in Nashua, New Hampshire. Bob is a recognized public speaker/writer at regional and national conferences.

Jeffrey Van Slyke, PhD: Dr. Jeff Van Slyke is a retired Chief of Police with an inclusive background of emergency management/crisis response, threat assessment, law enforcement services, and special security details. During his career as a Chief of Police, Dr. Van Slyke has experientially managed and responded to such emergencies as tornados, bomb threats, Hurricane Katrina, nor'easter storms, plane crash, suicides/homicides, chemical spills, residence hall fires, and mitigated two credible active-shooter scenarios. Dr. Van Slyke also assisted with facilitating 21 presidential visits, and was responsible for maintaining the security of a presidential library, a nuclear laboratory, a university airport, athletic events, concerts, and movie sets. While serving as Chief of Police at the University of Texas at Austin, Dr. Van Slyke assisted the United States Secret Service with the protective responsibilities of President George W. Bush's daughter (Jenna), and was awarded the U.S. Secret Service Certificate of Appreciation by President Bush in recognition of his efforts. Dr. Van Slyke earned a bachelor's degree in criminal justice, Auburn University; master's degree, Western Carolina University in public administration; doctorate in higher education from the University of Texas. Dr. Van Slyke also attended the F.B.I. National Academy and the U.S. Secret Service National Threat Assessment Center.

Julie Nations: Julie Nations, Memphis Police Department (MPD) Grants Manager, has been employed with the City of Memphis since 2001 and with the Memphis Police Department in her current position since 2006. She is responsible for MPD grants development and administration. Julie graduated Magna Cum Laude from Arkansas State University with a bachelor's degree in education; attended Birmingham School of Law in Birmingham, Alabama; and received her master's degree in Security Studies (Homeland Security and Defense) from the Naval Postgraduate School in Monterey, California. Julie's career concentrations include research, strategy development, public safety project planning and coordination, and policy analysis.

FOUNDATIONS OF EMERGENCY MANAGEMENT

"Sometimes it's not about trying to correct the wrong thing; but rather, it's about trying to get through the wrong thing the correct way."

Jeffrey M. Van Slyke`

Figure 1.1 San Francisco City Hall after the Great San Francisco Earthquake of 1906.

OBJECTIVES

The objectives of this chapter are to:
- define the terminology of emergency management;
- identify principles of emergency management;
- introduce man-made and natural disasters;
- discuss the four phases of emergency management;
- discuss the concepts of civil defense; and
- discuss various historical incident necessitating emergency management.

Introduction

On a daily basis, the United States of America, and the entire world in general, faces the reality and responsibility of managing hazards, emergencies, and disasters. This chapter will serve as an introduction to the text and highlight the importance of effective emergency management.

Specifically, this chapter defines significant terms and phrases relating to emergency management, discusses the importance of emergency management, and also describes the comprehensive emergency management (CEM) concepts of hazard mitigation, disaster preparedness, emergency response, and disaster recovery. It also includes a brief description of the present emergency management system, and concludes with identifying effective principles of emergency management that are meant to provide fundamental best practices during an event.

Several recent natural and man-made disasters will be mentioned in this chapter along with the associated loss of life and economic cost of each. A succinct case will be made to demonstrate that effective emergency management, utilizing the framework mapped out in later chapters, can and will reduce the lives lost and dollars spent during future natural and man-made emergencies.

Definition of Emergency Management

Over the years, the concepts and terminology of emergency management have seemingly evolved and continue to be a work-in-progress based upon efforts to survive, cope, and deal with conditions and circumstances that affect the general safety and welfare of humanity. Invariably, significant events caused by nature or man both continue to frequent the damage and disruption to communities. Such consistently recurring incidents have served as the impetus to create the discipline and profession commonly known as emergency management.

To truly understand what emergency management is, it is first helpful to decompose and understand the individual meanings of emergency and management.

According to the *Merriam-Webster Dictionary*, emergency is defined as: "1) an unforeseen combination of circumstances or the resulting state that calls for immediate action; and 2) an urgent need for assistance or relief."[2] *Merriam-Webster* further defines management as: "1) the act or art of managing; the conducting or supervising of something; 2) judicious use of means to accomplish an end; and 3) the collective body of those who manage or direct an enterprise."[3]

While one can seemingly understand the simplistic literary explanation of emergency management, with respect to a practitioner-based explanation, there have been numerous characterizations of this subject matter. Over the years, and as a result of numerous disasters, emergencies, and critical incidents, the term emergency management has become a profoundly interchangeable moniker. The following are examples of how the definition of emergency management has evolved over the past several decades:

1. Emergency Management (1995): "Organized analysis, planning, decision-making, and assignment of available resources to mitigate (lessen the effect of or prevent) prepare for, respond to, and recover from the effects of all hazards. The goal of emergency management is to save lives, prevent injuries, and protect property and the environment if an emergency occurs" (Blanchard 2008, p. 344; FEMA, EMI, 1995).[4]
2. Emergency Management (1997): "The process by which the uncertainties that exist in potentially hazardous situations can be minimized and public safety maximized. The goal is to limit the costs of emergencies or disasters through the implementation of a series of strategies and tactics reflecting the full life cycle of disaster, i.e., preparedness, response, recovery, and mitigation" (Blanchard, 2008, p. 344; FEMA, EMI, 1997).[5]
3. Emergency Management (2001): "The process through which America prepares for emergencies and disasters, responds to them, recovers from them, rebuilds, and mitigates their future effects" (Blanchard, 2008, p. 345; *FEMA, Disaster Dictionary*, 2001, p. 40, citing FEMA Strategic Plan).[6]
4. Emergency Management (2002): "The process through which the Nation prepares for emergencies and disasters, mitigates their effects, and responds to and recovers from them" (Blanchard, 2008, p. 345; *A Nation Prepared, FEMA Strategic Plan: Fiscal Years, 2003–2008*, 2002, p. 57).[7]
5. Emergency Management (2003): "Activities that include prevention, preparedness, response, recovery, rehabilitation, advocacy, and legislation, of emergencies irrespective of their type, size, and location, and whose purpose is reduction in death, disability, damage, and destruction" (Blanchard, 2008, p. 344; *Toward an International System Model in Emergency Management*, Dykstra 2003).[8]
6. Emergency Management (2006): ". . . the term 'emergency management' means the governmental function that coordinates and integrates all activities to build, sustain, and improve the capability to prepare for, protect against, respond to,

recover from, or mitigate against threatened or actual natural disasters, acts of terrorism or other man-made disasters; ..."⁹

7. Emergency Management (2007): "An ongoing process to prevent, mitigate, prepare for, respond to, and recover from an incident that threatens life, property, operations, or the environment."¹⁰
8. Emergency Management (2007): "Emergency management is the managerial function charged with creating the framework within which communities reduce vulnerability to hazards and cope with disasters."¹¹

The primary objective of emergency management is to sustain life, minimize damage to property, and maintain the safety and welfare of a community. Fundamentally, emergency management is a twofold comprehensive process. The first part necessitates knowing what to do, which involves the coordination of personal, procedures, and provisions that are essential for the prevention of, the preparation for, the response to, and the recovery from the effects of natural disasters, acts of terrorism, man-made hazards, disasters, emergencies, and critical incidents.

The second part requires doing what you know, which involves an inclusive collaboration among all key-stakeholders to include the government, the private

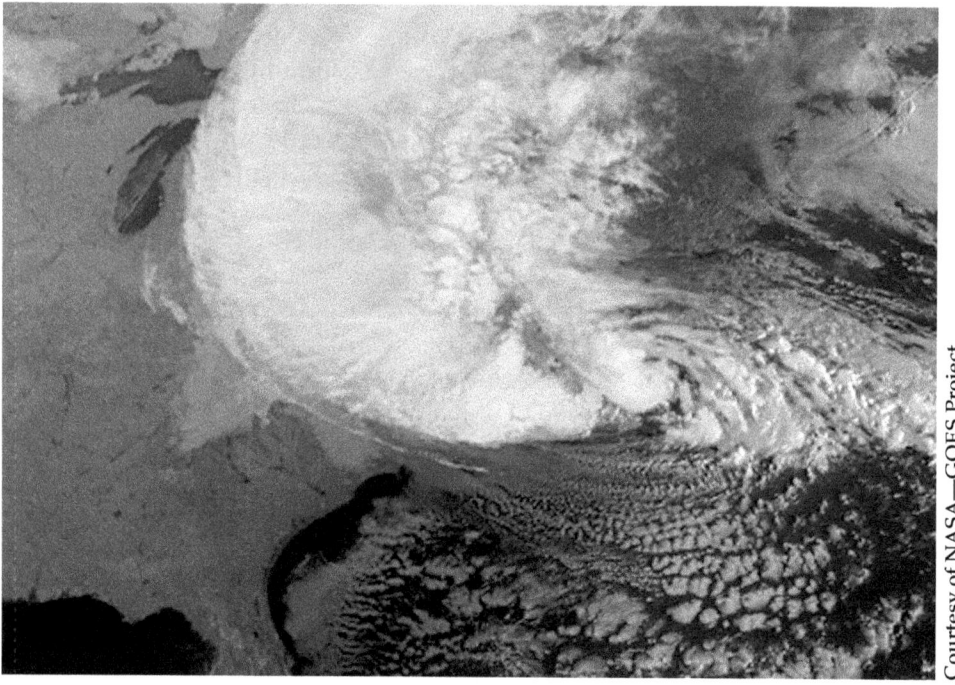

Figure 1.2 Hurricane Sandy as it impacts the United States.

Source: National Aeronautics and Space Administration

sector, the public, and the media. The collaboration aspect must be designed to identify functions, roles, and responsibilities, and requires a commitment to promote the communication and trust components, which are essential to sustain the functional effectiveness of all service deliverables during a disaster event. Comfort and Cahill acknowledge the essential nature of collaboration within the emergency management function: "In environments of high uncertainty, this quality of inter-personal trust is essential for collective action. Building that trust in a multi-organizational operating environment is a complex process, perhaps the most difficult task involved in creating an integrated emergency management system."[12] The importance of collaboration is a critical dynamic of emergency management. In essence, collaboration involves the development of mutual cooperation within a community that creates the environment in which coordination can function effectively.

Disaster Response

The United States has evolved from a group of European colonies into an international superpower. Over the centuries, the growth of the United States has been an incredible testament to the benefits of democracy; modern America is a nation whose boundaries have extended beyond the continental coastlines, with highly populated cities, and a well-developed national infrastructure. Although the United States has focused primarily on international terrorism as the major challenge facing the nation during the last decade, there exists a much broader range of threats. This "all-hazards" approach is based upon a progressive process of risk assessment and vulnerability awareness.

For much of the history of the nation, the major threats have been natural events. Over the last three centuries, storms, earthquakes, floods, tornadoes, and fires have impacted many U.S. cities. The damage inflicted by many of these "disasters" has been significant, but the damage has been relatively localized. For the most part, the recovery was performed by both state and local responders.

During the early history of the United States, the federal government had limited abilities to respond to such events; nevertheless, efforts were made to ensure the national government had a formal responsibility during disaster response. The beginning of federal disaster relief originated with the Congressional Act of 1803, which was passed following a devastating fire in Portsmouth, New Hampshire. The Congressional Act of 1803 can be viewed as a precursor to the modern Federal Emergency Management Agency (FEMA). From 1803 forward, the United States experienced many events that overwhelmed the response capability of local government. The Chicago fire of 1871, which devastated much of the city; the various Yellow Fever epidemics of the 1870s; the Jamestown Flood in 1889; the 1900 Galveston, Texas hurricane; the San Francisco earthquake of 1906; and the 1918 Spanish influenza epidemic are examples of several of the major events that created casualties and property damage far in excess of the local community's ability to respond without external assistance.

Foundations of Civil Defense and Emergency Management

Since independence, the United States has been in a constant state of evolution. Born in the flames of conflict, the nation has faced international and domestic security threats throughout its history. During the first century and a half of its existence, the United States was not a major power. During this period, European nations competed to create global empires (e.g., Britain controlled the seas). The United States experienced several conflicts including the War of 1812, the Mexican War of the 1840s, the War Between the States, and the Spanish-American War. Over this time, the United States expanded its boundaries reaching from ocean to ocean. The nation increased international trade and created a small navy and army to protect American interests abroad.

The protection of the nation and its interests has always remained one of the primary responsibilities of the national government. The Monroe Doctrine of 1823 informed the European Powers that the United States assumed a responsibility for the Western Hemisphere, although the nation had little ability to actually enforce its intent. From its earliest days, the nation continued its expansion. Alaska was purchased from Russia in 1867; Hawaii became a U.S. possession in 1898; and the nation gained control of areas including Cuba, Puerto Rico, and the Philippines as a result of the Spanish-American War. By the beginning of the twentieth century, the United States itself had become an imperialistic power and on the threshold of international importance.

U.S. presidents were historically content to avoid international entanglements or push the nation onto the world stage, although enforcement of federal laws related to customs and the border became an increasing concern as the United States evolved. During 1853, the U.S. Customs Border Patrol was established when the secretary of the treasury authorized the collector of customs to hire customs mounted inspectors for patrol duty along the borders of the nation.

The border shared with Mexico had traditionally been an area of unrest. For example, one of the driving forces behind the formation of the famous Texas Rangers was the enforcement of border security. Border concerns led to many states passing their own border protection and immigration legislation. As a result, an 1875 U.S. Supreme Court decision ruled that creation and enforcement of immigration regulations was a federal responsibility. During 1882, the Chinese Exclusion Act and Immigration Act of 1882 targeted immigration issues putting increased emphasis upon enforcement.

By the end of the nineteenth century, border-related issues had become so significant that, in 1915, the United States authorized the Bureau of Immigration to deploy mounted guards along the U.S. border. Even this effort proved ineffective; U.S. military forces were deployed to the Mexican border from 1916 to 1917. Finally, in 1924, Congress created the U.S. Border Patrol to consolidate border protection efforts in a single agency.

As the twentieth century dawned, the vast majority of European and Asian powers were governed by monarchies. The 1901 assassination of President McKinley by anarchist Leon Czolgosz led to the presidency of Theodore Roosevelt. Roosevelt, who had once served as the New York City Police Commissioner, visualized the United States as an emerging world power, not only with international possessions but also with a growing global responsibility.

During 1941, with the World War II unfolding in Europe and Asia, the Office of Civilian Defense was established within the Office for Emergency Management to assure effective coordination of federal relations with state and local governments engaged in furtherance of war programs; to provide for necessary cooperation with state and local governments with respect to measures for adequate protection of civilian population in war emergencies; and to facilitate participation by all persons in war programs. New York Mayor Fiorello La Guardia was named as director, serving on a volunteer basis without compensation.

During August 1941, the U.S. Citizens Defense Corps was established. It gave the first complete and coordinated plan for local organization of civilian defense and was the prototype of all following civil defense (CD) organizations. With the nation engaged in World War II, the concern for the security of the continental United States increased. The civil defense program increased in popularity and scope. During 1942, the Office of Civil Defense conducted training for Civil Defense police volunteers among 46 cities with the support of J. Edgar Hoover and the Federal Bureau of Investigation.

The end of World War II found the United States as one of two remaining global superpowers. In 1945, the Office of Civilian Defense was disbanded; but soon the new threats of the cold war, including the fear of a nuclear attack, led to renewed discussion as to the need for local level "civil defense." In 1950, President Harry Truman signed Executive Order 10186 creating the Federal Civil Defense Administration (FCDA) within the Office for Emergency Management, Executive Office of the President. This action authorized the federal civil defense program until 1994 (when it was repealed by Public Law 93-337). Thus, in the 1950s, as the nation developed a civil defense system to address the threat of nuclear attack by the Soviet Union, the government officials who staffed these programs at all levels of government became the nation's first emergency managers.

During this period, the world was dominated by the United States and the Soviet Union, which maintained an uneasy peace in what was called the cold war.

From the earliest days of the cold war, preparedness focused upon events such as a nuclear attack from the Soviets. Bomb drills were conducted in schools, many Americans constructed bomb shelters, and it was very common to see buildings that had been designated "fall-out" shelters. Feeding, evacuation, communications, and medical response plans were also developed as safeguards against potential attacks.

When the cold war ended, it was hoped that the world would become a safer place and that international threats would reduce, but that proved not to be the case. New threats challenged the nation. With the 1994 repeal of the civil defense statute, parts were retained and incorporated into the Stafford Disaster Relief and Emergency Assistance Act including using the term "Emergency Preparedness" wherever the term "Civil Defense" previously appeared in the statutory language.

An emergency management program encompasses potential emergencies and disasters based on the risks posed by likely hazards; it also develops and implements programs aimed toward reducing the impact of these events on the community, prepares for those risks that cannot be eliminated, and prescribes the actions required to deal with the consequences of actual events and to recover from those events. Emergency management includes participants from all governmental levels and within the private sector. Activities are geared according to phases before, during, and after emergencies. The effectiveness of emergency management rests on a network of relationships among partners in the system.

Principles of Emergency Management

The profession of emergency management has become an inclusive triad of process, responsibility, and discipline. First, the process of emergency management can be found in the actual response, or "boots on the ground" to an emergency, disaster, or critical incident after it happens. For the most part, every incident or event is different, and will result in a profound learning moment based upon mistakes that are made and/or achievements arrived at. Because every incident/event seemingly has a life of its own, this often requires different response and recovery approaches that either were not documented in standard protocol, were unforeseen, or demanded immediate modification of "best practice."

Invariably, the process or "game plan" results in the often-used paradigm of "adapt, improvise, and overcome." The Four Ds can serve as a basic foundation for the emergency management process: (1) discover—lessons learned, achievements, deficiencies, effectiveness; (2) determine—what needs to be done, necessary resources, relationships, readiness; (3) develop—mitigation, preparedness, operational response/recovery methods; and (4) deploy—when, where, set-out, set-up, step-up, stand-down. The ever-changing dynamics associated with an adverse event will continue to develop and influence the emergency management process to the extent that there will never be a panacea or "one method fits all" approach.

The responsibility of emergency management involves key stakeholders to include local/state/federal government agencies, first responders, and organizational entities (community business, volunteers, etc.). For example, during the response and recovery phases, the leadership involved must take the responsibility to ensure that the necessary provisions such as personnel and supplies are adequate to the extent that the front-line responders and the public can depend upon and trust those tasked with the readiness to react when needed. Emergency managers often have few resources under their immediate control, but have access to an incredible amount of assistance from government, nongovernmental partners, and the private sector.

Specifically, the emergency manager must act responsibly by being able to engage with those partners, understand how these other agencies work in an effort to make the best use possible of those partner-agency resources during an emergency, and then demonstrate responsible leadership. In addition, communication is the responsibility of everyone involved prior to, during, or after an incident. The absence of, or deficiency in, communication may completely hinder the responsibility of those involved with the emergency management process to include the propensity to compromise the safety and welfare of the public, responders, property, or the environment.

The profession and practices of emergency management have continued to evolve as a valid discipline and necessary subject matter that has earned its rightful place in the national headlines as well as within the mindset of the public. Emergency management has evolved into the discipline of ensuring that communities, businesses, and organizations are able to successfully endure through all aspects of an emergency or significant critical incident. Thus, the field of emergency management has become in itself a science that does more than just "deal with" extreme events. The discipline employs methods, means, and mindset to ensure the protection of people, property, and the environment. This goal is accomplished through mitigation, prevention, response, and recovery.

The study of previous emergency responses can help to assess areas of improvement as well as the ability to determine the causation of disasters that, if possible, may enable opportunities for prevention in the future. As disasters become more numerous and begin to encompass people from all nationalities, socioeconomical levels, races, and geographical locations, the knowledge from both the experiences of a disaster, and the attempt to negotiate our way through such disasters, will help improve and develop the distinctive phases of emergency management.

Four Phases of Emergency Management

The most recognized benchmark for describing the four phases of emergency management is the terminology provided by the Federal Emergency Management Agency (FEMA): (1) mitigation, (2) preparedness, (3) response, and (4) recovery.

10 Foundations of Emergency Management

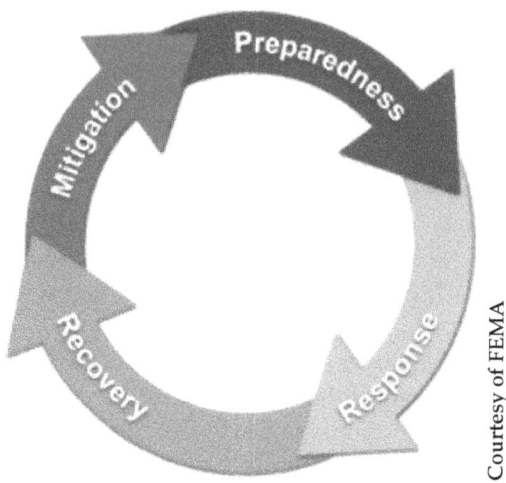

Figure 1.3 Four phases of emergency management—NEHRP, 2009.

The emergency management cycle is a description of how people, in general, respond to a major disaster. Understanding the phases is the first step in planning out a detailed response plan for any major disaster. This process follows a predictable pattern, and understanding how each component affects the other is essential in developing a well-crafted response plan. The interdependency of each functional activity of emergency management remains its most significant characteristic. For example, the value of mitigation and preparation should not be underestimated and are extremely constructive aspects of the entire process.

Essentially, all communities are in one of the established phases. Mitigation includes measures taken to prevent or subdue the effects of a disaster (securing homes, buying insurance, building levees). Preparedness includes the establishment of education and training measures (planning evacuation routes, performing emergency drills, compiling a supply list). Response occurs after the disaster and maintains the safety of the public (search and rescue missions, disaster response teams). The recovery phase occurs while the affected area is getting back on its feet (reducing financial stress, rebuilding, taking measures to prevent the same devastating result from future disasters). The following are definitions of each phase of emergency management:

1. Mitigation (prevention): Involves taking proactive measures to prevent or reduce the possibility of disaster or emergency from occurring, or marginalizing its impact within the community, by identifying and assessing risk, and implementing preventative measures that may reduce the risk.

2. Preparedness: Involves organizing and preparing a community's response to emergencies with the foremost objective to save lives and to help response and rescue operations; includes aspects such as the development of emergency response plans and the procurement of supplies, to educating the community about procedures for disaster response.
3. Response: Involves how a community reacts to a disaster, hazard, or emergency situation, including crisis communication and the treatment and protection of key assets, as well as, managing and protecting life, property, the environment, and critical infrastructure.
4. Recovery: Involves the timely resumption of "normalcy" within the community; in essence, the process and measures necessary for moving from the "disaster" mode to the "normal" mode through treatment/therapy, rebuilding, reorganization, and revitalization.

Phase I: Mitigation/Prevention

In preparing for a disaster, it is important to start with mitigation, or prevention, as it is sometimes known. In this phase, households, businesses, cities, counties or larger organizations review the type of disaster, hazard, or emergency they expect to face. In thinking of the emergency management cycle, it is often useful to use scenarios. In this case, we will use a hurricane on the eastern coast of New York.

If a family were to move to the East Coast, they are likely to already be aware of previous hurricanes and the threat posed by them. Mitigation begins the moment

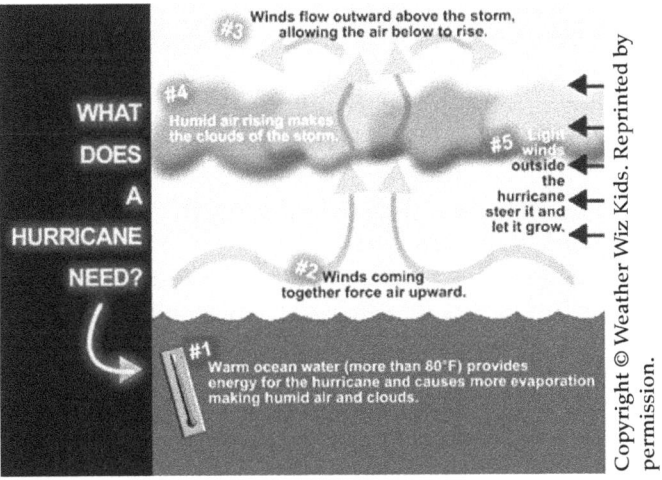

Figure 1.4 The sequence of weather conditions needed to form a hurricane.

they decide to make the move. The news coverage of Hurricane Sandy and the damaging winds, which were magnified by the tidal surge of 15–20 ft, would have been hard to miss. Thus, a family has to determine whether *where* they settle and reside may eventually compromise their safety and welfare due to prior history and the future propensity to be affected by a hurricane. Bottom line—how willing does a family desire to settle in harm's way?

Specifically, in an effort to avoid the potential nightmare of enduring the likelihood of a hurricane, a family that is considering moving into a hurricane alley would need to determine whether the home on their radar to purchase is strong enough to endure the strike of a hurricane, and/or if they would be willing to possibly evacuate if necessary. The family must decide what would best protect their safety and welfare from a storm's impact and determine their comfort level regarding whether to evacuate or shelter-in-place based upon home construction and location.

For example, factors to be considered regarding a hurricane's storm surge or damaging winds should include: (1) the elevation of the house, (2) the house construction (brick, wood, or vinyl siding), and (3) whether their home may possibly be subject to a lower risk of the storm's impact based upon the proximity of the home relative to the time the hurricane strikes the area. These are not all inclusive reasons, but considering the construction of a home and its location may well marginalize the impact, and risk, of a hurricane.

Phase II: Preparedness

The second phase of emergency management is the preparedness phase. Preparedness comes in many forms, but most of the forms have a similar goal. The knowledge of how to plan for disasters is critical in emergency management. Planning can make a difference when attempting to mitigate the effects of a disaster, including saving lives and protecting property, and helping a community recover more quickly. During the preparedness phase, within the organization of an emergency management agency, plans would have been devised, employees would have been trained in procedures, and then they actually practice the procedural steps of a plan by conducting table-top exercises.

In the previous scenario, the family in question would need to develop an overall plan. With a pending storm due to strike, maintaining a state of readiness is a must. The family might attend a seminar about disaster preparedness or view Internet resources that delineate preparedness materials. Either way, preparing for the worst and hoping for the best should be the mindset for developing a plan. This plan should include things like what family essentials to pack other than clothing (legal documents, heirlooms, etc.), how far away from the coast to drive, which road(s) to take, and obtaining an emergency supply kit.

Figure 1.5 Military assistance occurring after Hurricane Sandy.

Finally, the family should run through their plan. Conducting a brief pre-disaster exercise will enable family members to determine the approximate amount of what all can actually be stored in the vehicle, map-out at least three travel option destinations, and determine where to obtain gas and how much will be needed to reach final recovery point. Generally speaking, the preparedness process may be the last time a family will be able to modify their plan before the disaster occurs.

Phase III: Response

The third phase of the emergency management cycle is known as the response phase. The response phase is characterized by the initial response of first responders, agencies like FEMA field reps and local law enforcement, as well as the initial actions taken by those affected by the disaster. It is during the response phase that those adversely affected by the disaster are provided emergency assistance from first responders. The nature and severity of the event/incident will determine the service deliverables necessary and expected by those who may be in need of medical care, rescue, or evacuation.

14 Foundations of Emergency Management

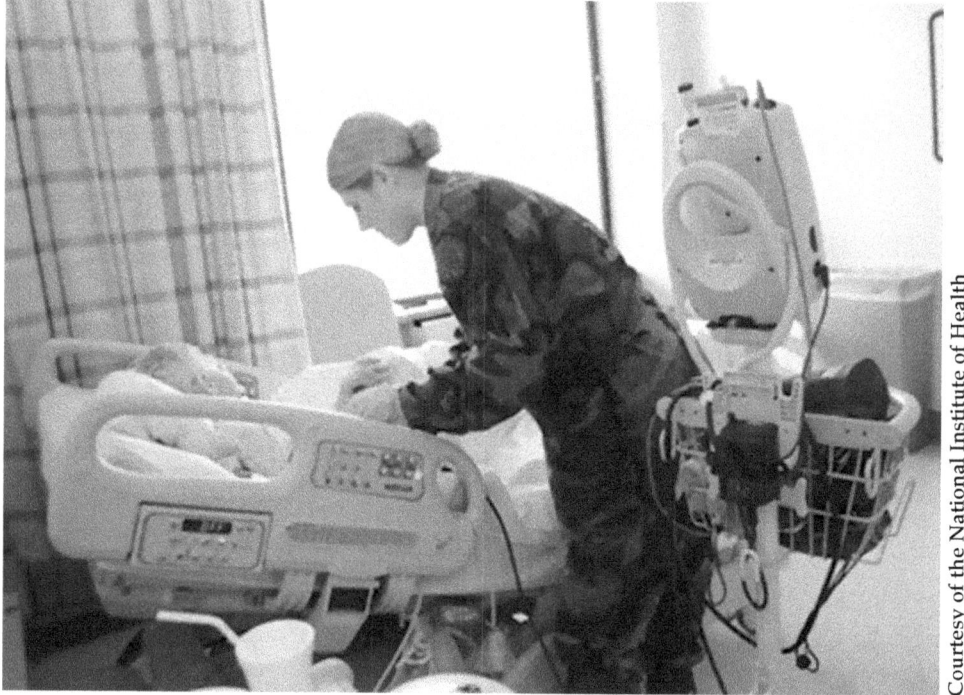

Figure 1.6 "LT Kelly Kerr, CC oncology nurse and Commissioned Corps officer, attended to a patient at the Brookdale University Hospital and Medical Center during the aftermath of Hurricane Sandy."[1]

The timely response will begin once the scene and initial damage assessments are conducted to ensure that the assistance to the public by first responders is not compromised and that there is minimal risk for all involved. Once these assessments are made, additional support, known as emergency responders, is called to the scene or to support facilities, like hospitals. The resources available during this phase directly impact the success of the recovery phase.

In addition, what is commonly referred to as a Community Emergency Response Team (CERT) are also created within neighborhoods. A CERT can be an invaluable asset to local first responders. The CERT consists of trained volunteers who provide additional resources in times of need during a disaster, and can be diligently organized during an incident. The CERT is well prepared to assist with first-aid, search and rescue, triage, suppression, fire, and cost estimates of damage.

---

[1]Clinical Center—National Institute of Health, "Commissioned Corps Officers from the CC Provide Care to Hurricane Survivors," 2013, http://clinicalcenter.nih.gov/about/news/newsletter/2013/jan2013/newsletter.html (accessed June 28, 2013).

CHAPTER 1 Foundations of Emergency Management **15**

The CERT is also provided with training assistance from the local emergency management agency, which is an important aspect of being prepared to play an integral part in the response process. For example, one operable training component for a CERT includes an introduction to the process of the Incident Command System (ICS) and the Incident Management System (IMS)—all of which provides a CERT with an organizational framework, roles and responsibilities, and the radio communication dynamics necessary to operate during a critical incident. In addition, other components of a CERT training program will include exposure to disaster preparedness, fire safety, disaster medical operations (first-aid), light search and rescue, disaster psychology, and the dynamics of terrorism.

Generally, prior to when a disaster is actually expected to strike the at-risk area, the local/state emergency response plan will be initiated. This will include an initial public communication process regarding life-safety, evacuation, etc. Likewise, along with local/state emergency managers and first responders, FEMA personnel may arrive to assist, and the Emergency Operations Center (EOC) becomes fully staffed and operational. The family in our example will activate their evacuation

Figure 1.7 "The close-knit community of Breezy Point lost more than one hundred homes to fire during Hurricane Sandy."[2]

---

[2]Federal Emergency Management Agency, "Damaged Caused by Hurricane Sandy," 2012, http://www.fema.gov/photolibrary/photo_details.do?id=62609 (accessed June 28, 2103).

plan, packing essential items and documents while preparing to move away from the hurricane area. It is also common, and strongly urged, that private business organizations administer evacuation plans, like airlines moving aircraft to safer airfields, or business moving machinery out of possible storm surge areas. It is important to note that response is handled at the local and state levels, and the federal government cannot intervene unless officially requested by the local or state governments.

Phase IV: Recovery

The recovery phase often overlaps efforts during the response phase of a disaster. In most disasters, the recovery phase is characterized by two closely related parts of the recovery, namely, the short-term and long-term recovery efforts. The basic concept underlying recovery involves a return to a sense of normalcy following an incident.

Figure 1.8 "A resident of Ortley Beach, NJ takes a break from cleaning up the remainder of his damaged home which occurred when Hurricane Sandy swept through the area."[3]

---

[3]Federal Emergency Management Agency, "Recovering from Hurricane Sandy," 2012, http://www.fema.gov/photolibrary/photo_details.do?id=61761 (accessed June 28, 2013).

Short-term recovery will often be localized, such as the return of power to neighborhoods. The size and scale of the short-term recovery will differ from disaster to disaster, but the focus during the short term is the return or resolution of issues pertaining to basic survival necessities, such as food, water, and shelter. Short-term recovery is also the focus for emergency responders, to treat the injured, restore emergency services, and begin removing debris.

For the family, this would involve returning to their home and beginning the clean-up process. During this time, they will cover broken windows, make immediate repairs, boil their water, and secure their home as best they can. Basic survival for the family is the most important part of the short-term recovery. For businesses, the short-term recovery is muted by the emphasis on the employees' part to secure their homes and families. Businesses will focus on securing any goods that they may have from looting as well as putting up tarps or canopies to protect their building from the elements.

Education facilities (childcare, primary, secondary, colleges/universities) must also begin the arduous process of assessing damage, and then determine the prognosis for repair and restoration to normalcy prior to engaging students. In an effort to ensure that kids/students are not subjected to anything that may pose a risk to their health and safety, this process may be very labor intensive, but is extremely necessary.

Health-care facilities are also a significant function that is essential for a community to have restored during the recovery process. Without the assurance of operable health care, families will be at risk should an injury or illness occur, or that may in need of remedial therapy, or awaiting a newborn, all of which can create a high degree of anxiety. Many hospitals in the New York City area were evacuated during Hurricane Sandy, as the result of underground transformers becoming inoperable or generators losing power due to massive flooding.

Long-term recovery, though, is the return to normalcy. This is characterized by the return of electricity and water, and other services. Street repairs are completed, power lines replaced, and structures repaired. Normal police and fire responses return and employees return to work. An essential part of this phase is the lessons learned portion. During the long-term recovery, those who felt the impact evaluate their response to the disaster, make adjustments as needed, and transition to the mitigation (preparedness) process for planning a future response, which is the first phase of the emergency management cycle.

Typically, the priorities during a family recovery process include a type of homecoming, whereby the return to normalcy will include learning the probable time for returning to work and school, and to also determine whether the local grocery stores have sustainable food items, and that there will be sufficient fuel at the gas pumps. Eventually, at some point, families that have experienced the disaster elements should take the time to assess the before, during, and after dynamics of

Figure 1.9 "Homes damaged by Hurricane Sandy."[4]

their journey, and determine what proactive steps will be necessary in preparation for a response in the future.

Specifically, upon sizing up their learning curve, the family may come to a better understanding of what nonessential items were brought versus other items that were needed, yet left behind. Also, check on obtaining insurance or making proper adjustment to their current policy to ensure full coverage. For businesses, the goal of the long-term recovery process would be a return to functional operability and a normal work schedule. Subsequently, employees can begin to concentrate more on returning to the routine of a normal operation.

Commentary Regarding Emergency Management

Emergencies arise from both man-made and natural origins. They also vary in size, scope, length, and magnitude. Some may incur the ultimate tragedy of human deaths, whereas others may result in no casualties. Certainly, thousands died during the event of September 11, 2001. The effects of the terrorist attacks against the United States were felt globally. However, other emergencies may be confined either regionally or locally. For instance, during January 2014, an explosion occurred at a

---

[4] Federal Bureau of Investigation, "Aftermath of Hurricane Sandy Raises Potential for Disaster Fraud," 2012, http://www.fbi.gov/news/news_blog/aftermath-of-hurricane-sandy-raises-potential-for-disaster-fraud (accessed June 28, 2013).

bio-diesel plant in New Albany, Mississippi.[5] No losses of human life were reported during the aftermath of the event.[6] In both cases, emergencies occurred, but they exhibited different situational attributes.

These contrasting scenarios also show the uniqueness of emergencies. No two incidents are alike; they are unique. Despite the differences that are associated with emergencies, there is some commonness that permeates such incidents. Emergency management must also incorporate some attributes of managerial activities regarding the functions of controlling, leading, coordinating, organizing, and planning.

Controlling and leading must exist throughout the emergency management cycle to enhance its efficiency and effectiveness. For instance, some established chain-of-command must exist with respect to organizational and individual responsibilities. If uncontrolled, the emergency management cycle may exhibit some characteristics of chaos and may involve different amounts of confusion among first-responder entities. Leaders may be faced with a variety of scenarios in which information quality and availability may vary. Thus, leaders may exercise control with respect to conditions that involve structured, semi-structured, or unstructured decisions.

Coordination also is a consideration of emergency management. For instance, the timing and processing of materials and resources logistically is also critical to bolstering the efficiency and effectiveness regarding emergency incidents. Communication must exist to facilitate the routing of emergency resources.

Organizing and planning are also essential regarding the entirety of the emergency management cycle. Any organization, whether commercial or government, must plan for disaster. Any organization, regardless of its size, must plan for disaster. Although it is impossible to plan for any and all potential threats and contingencies that may affect an organization, some thoughtfulness must be devoted to potential emergencies. After all, having some plan is much better than not having any plan whatsoever. Through planning, the basic characteristics affecting organization within the emergency management cycle are identified and established. In other words, an organization understands what resources are necessary, when they are needed, who represents positions in the chain-of-command, where resources are located, and how they are to be utilized. Organizing also facilitates an understanding of what quantities of resources may be necessary.

Regardless of the type of incident considered, it involves every phase of the emergency management cycle. Within the context of the emergency management cycle, the presences of the managerial functions of controlling, coordinating, leading, organizing, and planning are immutable. All incidents, regardless of their origin or scope, must be managed functionally and with respect to the phases of the emergency management cycle.

---

[5]Edgemon, Erin, "No Injuries Reported in Explosion, Fire at New Albany, Miss. Bio-Diesel Facility; Firefighters Can't Get Close Enough," 2014, *Alabama Media Group*, http://blog.al.com/wire/2014/01/no_injuries_reported_in_explos.html (accessed March 31, 2014).

[6]Ibid.

Chapter Comments and Summary

The maturation of the emergency management profession will continue to be a work-in-progress based upon consistent evaluation, constant speculation, and proactive integration of continually transformed principles, preparation, protocols, and performance-based best practices. In hindsight, we may definitively discuss the "what-ifs," as well as speculate on future "what-ifs." Yet, irrespective of such reality and post-incident experience, there will always be a degree of uncertainty and sometimes no definitive answers with respect to the outcome of a disaster, emergency, or hazardous event.

Principally, however, the four phases of emergency management remain a significant systematic representation regarding the adversarial evolution of a disaster, hazard, or emergency situation. Practically speaking, these phases can be both immediate moments in action, or years in length. Moreover, each phase may not have a specific start or stop point, but their intentions and related objectives are representative of our instinctive reaction to the consequences of life experience, which attempts to maintain the status quo in spite of a natural disaster, a technological crisis, or an intentional hazard. Understanding the dynamics of each phase affords a better understanding of the expectations involving the responsibility, the process, and the discipline that is—emergency management.

Terminology

- All-Hazards
- CERT
- Civil Defense
- Disaster
- Emergency
- Emergency Management
- First Responder
- Man-Made Disaster
- Management
- Mitigation
- Natural Disaster
- Preparedness
- Prevention
- Recovery
- Response

Thought and Discussion Questions

1. Perform some research regarding the history of your locality. Identify two man-made and two natural disasters that affected your community. Write a brief essay that highlights your findings.
2. Perform some research regarding significant disasters of national scope that affect your locality or region. Identify two of these events. Write a brief essay that highlights your findings.
3. Consult your local emergency services organizations regarding their perceptions of emergency management. For instance, discuss with your local fire department

their activities regarding preparedness, mitigation, response, and recovery. Write a brief essay that highlights your findings.
4. Consult some local businesses regarding their perceptions of emergency management. What impacts do you believe either a man-made or natural disaster would have regarding their activities? Write a brief essay that highlights your findings.

References

1. *Pasha Bulker* grounded after storm off the coast of Nobbys Beach in New Castle, South Wales, Australia. http://www.thiscrazyweb.com/marine-accidents-19-pics/.
2. *Merriam-Webster Dictionary*. Definition of *emergency*. http://www.merriam-websters.com/dictionary/emergency (accessed January 11, 2013).
3. *Merriam-Webster Dictionary*. Definition of *management*. http://www.merriam-webster.com/dictionary/emergency (accessed January 11, 2013).
4. Blanchard, B. Wayne (2008, p. 344). *A tutorial on emergency management, broadly defined, past and present.* Federal Emergency Management Agency. Introduction to Emergency Management (1995). Emmitsburg, MD: Emergency Management Institute, FEMA.
5. Blanchard, B. Wayne (2008, p. 344). *A tutorial on emergency management, broadly defined, past and present.* Federal Emergency Management Agency—Emergency Planning Workshop Instructor Guide (Drabek, Thomas, 1997). Emmitsburg, MD: Emergency Management Institute, FEMA.
6. Blanchard, B. Wayne (2008, p. 344). *A tutorial on emergency management, broadly defined, past and present.* FEMA, Disaster Dictionary: Common Terms and Definitions Used in Disaster Operations (9071.1-JA Job Aid). Washington, DC: FEMA, May, 2001.
7. Blanchard, B. Wayne (2008, p. 345). *A tutorial on emergency management, broadly defined, past and pPresent.* A Nation Prepared, FEMA Strategic Plan: Fiscal Years 2003-2008, (2002, p. 57). http://www.fema.gov/pdf/library/fema_strat_plan_fy03-08(no_append).pdf.
8. Blanchard, B. Wayne (2008, p. 344). *A tutorial on emergency management, broadly defined, past and present.* Toward an International System Model in Emergency Management (Eelco H. Dykstra, 2003). Call for Papers—Public Entity Risk Institute Symposium on www.riskinstitute.org.
9. Blanchard, B. Wayne (2008, p. 346). *A tutorial on emergency management, broadly defined, past and present.* Post-Katrina Emergency Management Reform Act of 2006, pp. 1394-1433, Title VI of Public Law 109-295 (120 Stat. 1394), Department of Homeland Security Appropriations Act, 2007, Washington, DC, October 4, 2006.
10. Blanchard, B. Wayne (2008, p. 346). *A tutorial on emergency management, broadly defined, past and present.* National Fire Protection Association (NFPA) 1600:

Standard on Disaster/Emergency Management and Business Continuity Programs, 2007 Edition, Quincy, MA. http://www.nfpa.org/assets/files//PDF/CodesStandards/1600–2007.pdf.
11. Blanchard, B. Wayne (September 11, 2007, p. 3). *Principles of emergency management.* FEMA Emergency Management Higher Education Learning Project 2007) 11. https://training.fema.gov/EMIWeb/edu/08conf/Conference.
12. Edgemon, Erin. 2014. No injuries reported in explosion, fire at New Albany, Miss. Bio-Diesel Facility; firefighters can't get close enough. *Alabama Media Group.* http://blog.al.com/wire/2014/01/no_injuries_reported_in_explos.html (accessed March 31, 2014).
13. Comfort, Louise K. and Anthony G. Cahill. 1988. *Managing disaster, strategies and policy perspectives.* Durham, NC: Duke Univ. Press.
14. *Images of Hurricane Sandy.* http://www.gettyimages.com/Search/Search.aspx?contractUrl=2&language=en-US&family=editorial&p=hurricane+sandy&assetType=image# (accessed January 13, 2013).
15. *Hurricane Sandy Satellite Image Courtesy of NASA.* http://news.discovery.com/earth/hurricane-sandy-damage-photos-121030.htm (accessed January 11, 2013).
16. National Earthquake Hazards Reduction Program (NEHRP). Introduction to emergency management (November 1, 2009). http://training.fema.gov/EMIWeb/EarthQuake/NEH0101220.htm (accessed January 21, 2013).
17. The sequence of weather conditions needed to form a hurricane. http://www.weatherwizkids.com/ (accessed January 12, 2013).

HISTORY OF EMERGENCY MANAGEMENT IN THE UNITED STATES

"I respectfully suggest the propriety of having stationed at the arsenal a full company of U. S. troops, that they may be made available in any emergency, from fire, insurrection, or anything else." [1]

Thomas L. Smith

Figure 2.1 Three Mile Island—site of the 1979 nuclear incident. "On March 28, 1979, an accident at the Three Mile Island (TMI) nuclear plant in Pennsylvania transforms the hypothetical into reality. The TMI accident ends without the need for a general evacuation, but it is made clear that existing emergency planning requirements are unsatisfactory. On December 7, 1979, in accordance with the Kemeny Commission (which investigated the TMI accident), the Federal Emergency Management Agency (FEMA) is designated as the lead agency for dealing with offsite nuclear power plant emergencies." [2]

---

[1] Thomas Smith, "Emergency Quotes," *Brainyquote.com*, http://www.brainyquote.com/quotes/keywords/emergency_3.html (accessed December 18, 2012).

[2] U.S. Nuclear Regulatory Commission, "History," 2013, http://www.nrc.gov/about-nrc/emerg-preparedness/history.html (accessed June 9, 2013).

OBJECTIVES

The objectives of this chapter are to:
- define emergency management;
- discuss the historical maturing of American emergency management;
- explain the major historical legislation affecting emergency management;
- discuss various events that impacted emergency management; and
- discuss the history of American emergency management.

Introduction

An emergency is defined as "an unforeseen combination of circumstances or the resulting state that calls for immediate action" and as "an urgent need for assistance or relief."[3] Any number of events and situations may be classified as emergencies ranging from localized instances of bank robbery and hostage negotiation to national calamities necessitating the engaging of an enemy state during periods of war. Figure 2.1 shows a classic example of an American emergency involving the Three-Mile Island incident. Although no visible signs of disaster were readily viewable, the incident represented a nuclear emergency. Emergency management is defined as the "organized analysis, planning, decision-making, and assignment of available resources to mitigate (lessen the effect of or prevent) prepare for, respond to, and recover from the effects of all hazards. The goal of emergency management is to save lives, prevent injuries, and protect property and the environment if an emergency occurs."[4]

Since its origin in the colonial period, the United States has experienced a variety of emergencies representing national, regional, state, and local prominences. Each incident tested the mettle and resolve of governmental administrations, and contributed toward the fashioning of policies, technologies, and resources that have strengthened the ability to manage emergency situations. Each incident was unique, and demonstrated a different threat to which national reactions and responses became seasoned in due time. Essentially, these situations contributed both individually and collectively toward the defining and crafting of American emergency management paradigms.

---

[3]"Emergency," 2012, *Merriam-Webster Dictionary*, http://www.merriam-webster.com/dictionary/emergency (accessed December 19, 2012).

[4]Wayne Blanchard, *Guide to Emergency Management and Related Terms, Definitions, Concepts, Acronyms, Organizations, Programs, Guidance, Executive Orders and Legislation, Federal Emergency Management Agency*, 2008, http://training.fema.gov/EMIWeb/edu/termdef.asp (accessed December 19, 2012), p. 344.

The historical maturing of emergency management is viewed from two perspectives: natural disasters and man-made threats. Both categories have the potential of affecting human life internationally, nationally, regionally, and locally. The characteristics of natural disasters are unchanging and involve varying amounts of danger and damage. The dangerousness of floods, hurricanes, storms, tornadoes, droughts, fires, blizzards, volcanoes, and earthquakes pervades American history. Such events resulted in the bettering of the attributes of preparedness, reaction, response, and recovery involving natural disasters.

Man-made emergencies also have posed threats. One of the earliest situations that necessitated attention from the federal government was the Whiskey Rebellion of 1794. This event involved the military use of approximately 13,000 troops to suppress an insurrection that occurred regarding the implementing of taxation.[5] The outcome of the Whiskey Rebellion showed that the federal government was "prepared, willing, and capable of using physical force to suppress violent opposition to its laws."[6] This event was the first use of federal power to quash internal insurrection nationally. Being the initial use of such force, the Whiskey Rebellion established a precedent from which succeeding government administrations could fashion policies to accommodate emergency situations domestically.

Throughout American history, many other emergency situations honed the acumen and knowledge of government administrations and emergency personnel. Examples of such instances include the Great Fire of Portsmouth in 1802 that destroyed 132 buildings;[7] the Cholera epidemic of 1832 that devastated New York;[8] the Galveston, Texas hurricane of 1900;[9] the 1927 flooding of the Mississippi River;[10] the 1964 Alaskan earthquake;[11] the erupting of Mount St. Helens in 1980;[12]

---

[5]Larry Schweikart and Michael Allen, *A Patriot's History of the United States: From Columbus's Great Discovery to the War on Terror* (New York, NY: Penguin Group, 2004).

[6]Deanna Spingola, *The Ruling Elite: A Study in Imperialism, Genocide, and Emancipation* (Bloomington, IN: Trafford Publishing, 2011), 273.

[7]Portsmouth New Hampshire Fire Department, "Portsmouth Fire: A Chronological History of a New England Seaport Fire Department," 2012, http://www.cityofportsmouth.com/fires/history.htm (accessed December 19, 2012).

[8]Charles Rosenberg, *The Cholera Years: The United States in 1832, 1849, and 1866* (Chicago, IL: Univ. of Chicago Press, 1987).

[9]Casey Greene and Shelly Kelly, *Through a Night of Horror: Voices from the 1900 Galveston Storm.* (Galveston, TX: Rosenberg Library, 2000).

[10]John Barry, *Rising Tide: The Great Mississippi Flood of 1927 and How it Changed America* (New York, NY: Touchstone Publishing, 1997).

[11]David Dowrick, *Earthquake Resistant Design and Risk Reduction,* 2nd ed. (Hoboken, NJ: Wiley and Sons Publishing, 2009).

[12]Patricia Lauber, *Volcano: The Eruption and Healing of Mount St. Helens* (New York, NY: Aladdin, 1993).

the Tucson, Arizona, flood of 1983;[13] the 1996 blizzard that devastated New York;[14] the events and aftermath of September 11, 2001;[15] the powerfulness of Hurricane Katrina in 2005;[16] and the 2012 mass shooting in Aurora, Colorado.[17] Certainly, many more incidents may be identified that show the dangerousness of both man-made emergencies and natural disasters and the associated necessity of emergency management.

Both natural disasters and man-made events are within the domain of emergency management. Accommodating both categories represents an all-hazards approach to emergency management. Although American history does not exhibit this approach throughout its entirety, it now represents a foundational basis for mitigating the dangerousness of perilous situations. During 1979, because of fragmentation and inadequacies regarding local, state, and federal organizations to manage disasters and to mitigate their effects, President Jimmy Carter created the Federal Emergency Management Agency (FEMA).[18] The creation of FEMA ushered the all-hazards approach to managing disasters.[19]

Contemporary emergency management now emphasizes an all-hazards paradigm regarding the phases of preparedness, reaction, response, and recovery for both man-made emergencies and natural disasters.[20] This current all-hazards paradigm exists because of the culmination of previous experiences with emergencies throughout history. It reflects the knowledge gained from each experience that allowed leaders to understand shortcomings in the abilities of managing emergencies, and realizing how these shortcomings could be rectified. Throughout American history, the methods of emergency management and government policies have respectively improved and increased through time. Each incident is unique, and much can be learned from the experiences of individual incidents. This chapter explores some of the salient incidents and policies that affected the maturing of American emergency management throughout the history of the nation.

---

[13]Thomas Saarinen, Victor Baker, Robert Durrenberger, and Thomas Maddock, *The Tucson, Arizona Flood of 1983* (Washington, D.C.: National Academy Press, 1984).

[14]Christopher Burt, *Extreme Weather: A Guide and Record Book* (New York, NY: W.W. Norton & Company, Inc, 2007).

[15]David Simpson, *9/11: The Culture of Commemoration* (Chicago, IL: Univ. of Chicago Press, 2006).

[16]Judith Fradinand and Dennis Fradin, *Hurricane Katrina* (Tarrytown, NY: Marshall Cavendish Benchmark, 2010).

[17]Glenn Muschert and Johanna Sumiala, *School Shootings: Mediatized Violence in a Global Age* (Bingley, UK: Emerald Group Publishing, 2012).

[18]Philip Purpura, *Security: An Introduction* (Boca Raton, FL: CRC Press, 2011), 75.

[19]Ibid.

[20]Jane Bullock, George Haddow, and Damon Coppola, *Introduction to Homeland Security: Principles of all-Hazards Risk Management*, 4th ed. (Waltham, MA: Butterworth-Heinemann Elsevier Publishing, 2013).

The Eighteenth Century

The United States originated in the eighteenth century following the American Revolution. Emerging from the Colonies was a fledgling nation whose infrastructure and capacity for managing a wide array of emergencies was poor. Communication was primarily written or oral. Travel depended upon horses or ships. Any knowledge of the biological basis of disease was virtually nonexistent. Government was newly formed and did not manifest a numerous collection of agencies and programs through which emergencies could be managed. Complex law enforcement organizations, fire departments, and hospitals, as are known today, were nonexistent. Benevolence during emergencies depended upon the goodwill of individuals or groups.

When emergencies occurred, no formal paradigm for managing the incident and its aftermath existed. Regardless, early Americans experienced a variety of emergency situations and conditions that affected their lives and the functioning of cities, towns, and communities. Both natural and man-made emergencies affected the new nation.

Earthquakes

Emergencies have been a part of American national history since before the founding of the United States. During the 1700s, the state of Massachusetts was affected by earthquakes.[21] Accounts of these events are summarized as follows:

> . . . in 1727 (November 9) an earthquake described as 'tremendous' in one report and 'violent' in another caused much damage at Newbury. The shock was felt from the Keenebec River to the Delaware River and from ships at sea to the extreme western settlements. Several strong aftershocks were reported from the area through February 1728. Eastern Massachusetts was shaken moderately on February 17, 1737, and June 24, 1741. Then on June 14, 1744, large numbers of bricks were shaken from tops of chimneys at Boston and other towns and stone walls were shaken down. Many persons in Newbury and Ipswich were alarmed. The earthquake was reported felt severely at Falmouth, Maine.[22]

Similar disasters occurred in the colonies. Earthquakes and their associated shocks impacted other communities within the colonies and were also felt among

---

[21]Thomas Birkland, *Lessons of Disaster: Policy Change After Catastrophic Events* (Washington, D.C.: Georgetown University Press, 2006), 105.
[22]U.S. Geological Survey, "Massachusetts Earthquake History," 2012, http://earthquake.usgs.gov/earthquakes/states/massachusetts/history.php (accessed November 19, 2012).

maritime vessels at sea.[23] On December 18, 1737, an earthquake "toppled chimneys at New York City and was reported felt at Boston, Massachusetts, Philadelphia, Pennsylvania, and New Castle, Delaware."[24] Additional earthquake shocks, originating from outside Pennsylvania, occurred in "1758, 1783, and 1791."[25]

Revolutionary Hurricane Weather

Weather also affected the fledgling American nation. During the closure of the American Revolution, a hurricane occurred in Yorktown that prevented the resupplying and reinforcing of British military forces that were commanded by General Cornwallis.[26] This event ushered the surrender of British forces and facilitated an American victory.[27] During this period, no early warning systems existed through which Colonial or British forces could have anticipated the severity of the coming storm. Although devastating, the storm was a catalyst that impeded the functioning of the British military thereby providing an advantageous situation for the American Colonies.

Disease

Disease also constituted emergency situations within the colonies. Yellow fever and smallpox were deadly diseases that killed much of the population. The yellow fever death rate "in Colonial America varied between 12% and 80%" and the death rate from smallpox ranged between "10 and 50%" among those "who contracted the disease naturally."[28] Managing the emergencies of disease was impeded by "a lack of understanding regarding germs and the spread of disease."[29] The failure to manage such diseases resulted in the deaths of many within the American colonies and devastated the Indian population.[30]

---

[23]Ibid.

[24]U.S. Geological Survey, "Pennsylvania Earthquake History," 2012, http://earthquake.usgs.gov/earthquakes/states/pennsylvania/history.php.

[25]Ibid.

[26]Jennifer Wilson, "The State of Emergency Management 2000: The Process of Emergency Management Professionalization in the United States and Florida" (Doctoral Dissertation, Florida International University, Miami, FL), 47.

[27]Ibid.

[28]Clifton Bryant, *Handbook of Death and Dying* (Thousand Oaks, CA: Sage Publishing, 2003), 186.

[29]Dorothy Mays, *Women in Early America: Struggle, Survival, and Freedom in a New World* (Santa Barbara, CA: ABC-CLIO Publishing, 2004), 106.

[30]Roland Robertson, *Rotting Face: Smallpox and the American Indian* (Caldwell, ID: Caxton Press, 2001).

The Whiskey Rebellion

Man-made emergencies also affected the early nation. The Whiskey Rebellion represented an emergency of man-made origin. It had the potential of impacting the United States economically, financially, and politically. Its occurrence ushered a new paradigm for countering dissent within the early American nation. President George Washington deployed the militias of several states to quash the dissent thereby "establishing a precedent for the federal government to enter individual states with a military force to provide security."[31] This precedent was followed by succeeding government administrations in the coming centuries.

Emergency Management Implications

Formal methods of emergency management were practically nonexistent during this period. Emergencies affected various communities within the colonies separately and were often localized. Emergency management practices and paradigms were not complex, and exhibited an "incipient or rudimentary, ad hoc organization."[32] Communication and travel modalities were also rudimentary and took much time depending upon the distance considered. Few, if any, resources existed through which emergency management could be conducted efficiently and effectively. With the exception of military force used during the Whiskey Rebellion, there was practically no federal oversight for managing disasters and emergencies during the formative years of the American nation.

The Nineteenth Century

Although the eighteenth century closed with no consideration of organized emergency management paradigms, the nineteenth century exhibited frameworks for establishing the methods of managing emergencies. The nineteenth century opened with modest concerns of both natural and man-made emergencies by the federal government. However, by the end of the century, significant events occurred that necessitated the crafting of organizations, legislation, policies, and technologies to counter the effects of emergencies. This century witnessed the primary origins of the emergency management domain.

---

[31]Jeffrey Dorwart, *Invasion and Insurrection: Security, Defense, and War in the Delaware Valley 1621–1815* (Danvers, MA: Rosewood Publishing, 2008), 188.
[32]Wilson, "The State of Emergency Management 2000," p. 47.

The Burning of Portsmouth

During 1802, Portsmouth, New Jersey, was gutted by a fire that destroyed 132 buildings.[33] This catastrophe "laid the town in ashes at its conclusion."[34] The community raised and distributed approximately "$45,000 to the poor" who suffered losses.[35] These funds were obtained from benevolent individuals.[36] Although the goodwill of fellow citizens provided some measure of relief, the following year was significant because federal attention was given to the ravaged community.

The Portsmouth incident resulted in the first offering of federal assistance by the American government to assist the ravaged community. During 1803, the U.S. Congress drafted legislation to assist the Portsmouth community in the form of near-term financial assistance and relief. The text of the bill is given as follows:[37]

> Be it enacted, by the Senate and House of Representatives of the United States of America, in Congress assembles, that the Secretary of the Treasury be, and he hereby is authorized and directed to cause to be suspended for months, the collection of bonds due to the United States by merchants of Portsmouth, in new Hampshire, who have suffered by the late conflagration of that town.[38]

This legislation represents the initial use of federal assistance during the aftermath of an emergency incident. It is known as the Fire Disaster Relief Act of 1803. The federal government did not provide direct financial assistance to the community. Instead, the legislation facilitated the suspending of bond payments that would have been rendered to the government. Allowing the community to retain its monies provided a source of financial and monetary capital through which it could begin to recover from the incident.

This one piece of legislation also established a precedent. It represented an actionable instance of federal community assistance that now may occur during the aftermath of an emergency incident. The establishing of financial relief provided a foundational concept that slowly increased in both scope and magnitude through time. Between the years 1803 and 1950, the U.S. Congress passed a total of

---

[33]Portsmouth New Hampshire Fire Department, "Portsmouth Fire: A Chronological History of a New England Seaport Fire Department," 2012, http://www.cityofportsmouth.com/fires/history.htm (accessed December 19, 2012).

[34]Russell Lawson, *Portsmouth: An Old Town by the Sea* (Charleston, SC: Arcadia Publishing, 2003), 109.

[35]Ibid.

[36]Ibid.

[37]Rebecca Katz, *Essentials of Public Health Preparedness* (Burlington, MA: Jones and Bartlett Publishing, 2013), 48.

[38]Ibid.

128 separate "laws dealing with disaster relief."[39] Although the 1802 fire that devastated Portsmouth was tragic, its resulting legislation in 1803 provided a basis for future emergency management legislation and policies.

Fire Suppression Systems

The dangerousness of fire was acknowledged in larger metropolitan areas. Early American water systems to suppress fires were designed in the 1800s.[40] During 1817, a New York City fireman George Smith developed the first valve-system fire hydrants (i.e., fire-plugs) that allowed emergency responders to access the piping of the municipal water system when fires occurred.[41] These fire hydrants were used in conjunction with both below ground and above ground water "piping systems."[42] The crafting of this system represented a systematic approach to fire suppression in a municipal setting.

Westward Expansion of the United States

The American nation and its population increased during the first half of the nineteenth century. During this period, the American economy slowly changed from an agrarian foundation to that of an industrialized nation. Westward expansion and the gold rush facilitated population shifts that crossed the Mississippi River. Migrating from the cities and lands of the east toward the promises of the west entailed journeys that were fraught with danger. People depended "only on each other to survive," and were "self-reliant and isolated."[43]

During this expansion, both individuals and the federal government provided services, goods, and protection for the migrants.[44] Both government entities and migrants expected that it was the duty of the federal government to ensure the "safe passage" of travelers.[45] Accomplishing this goal occurred in conjunction with military outposts where travelers could obtain "accurate information" about the "best routes and current conditions," could receive both "emergency medical aid

---

[39]Federal Emergency Management Agency, *A Citizen's Guide to Disaster Assistance* (Washington, D.C.: U.S. Government Emergency Management Institute, 2003), 3–2.

[40]National Fire Protection Association, *Fire Officer: Principles and Practice* (Sudbury, MA: Jones and Bartlett Publishers, 2010), 8.

[41]Ibid.

[42]Ibid.

[43]Richard White, *It's Your Misfortune and None of My Own: A New History of the American West* (Norman, OK: The University of Oklahoma Press, 1991), 199.

[44]Ibid, p. 204.

[45]Ibid.

and emergency supplies," and could use "blacksmith shops."[46] Western migration increased the scope and presence of the federal government. It necessitated the building of trails and "military roads" through which migrants accessed western lands.[47]

Reporting Emergencies

The technological capabilities of reporting emergencies also advanced during this period. During the 1850s, public call boxes were introduced in Washington, D.C., as a method of reporting fires.[48] These devices facilitated communication between citizens and emergency dispatchers through the use of a "coded telegraph signal."[49] The coded signal identified the location of the call box.[50] After emergency responders arrived at the location of the call box, citizens could further direct them to the location of the emergency incident.[51] Although the use of call boxes did not facilitate a direct routing of emergency personnel to the scene of the reported emergency, their use improved the efficiency and effectiveness of emergency communication.

The War Between the States

The 1860s manifested the War Between the States. This conflict was the bloodiest in American history, and claimed the lives of approximately 600,000 Confederate and Union soldiers.[52] It also claimed the lives of approximately 250,000 civilians.[53] This conflict necessitated the use of field hospitals and rudimentary medical treatments. Within American history, it represented the origin of "military physicians and volunteer nurses in roles related to emergency management."[54] It also introduced the concept of triage in which patients are examined and treated according to the severity of their medical condition.

The "military model" of managing and evaluating patients that was crafted during the War Between the States remains a foundational component of modern

---

[46]Ibid.

[47]Ibid.

[48]National Fire Protection Association, *Fire Officer: Principles and Practice* (Sudbury, MA: Jones and Bartlett Publishers, 2010), 8.

[49]Ibid.

[50]Ibid.

[51]Ibid.

[52]Henry Hobhouse, *Seeds of Change: Six Plants that Transformed Mankind* (Washington, D.C.: Shoemaker and Hoard, 2005), 232.

[53]Ibid.

[54]DeMond Miller and Jason Rivera, *Comparative Emergency Management: Examining Global and Regional Responses to Disasters* (Boca Raton, FL: CRC Press, 2011), 6.

emergency management.⁵⁵ Specifically, this model is a systematic and "graded response system" involving the "identification of ground zero, triage, and rescue of the injured that proceeds rearward to an acute care site, and, last, retrieval of the dead."⁵⁶ During modern times, this basic concept is continued to be used to manage "mass casualty events involving private citizens."⁵⁷

The War Between the States devastated the American nation, its infrastructure, and its people. During the aftermath of the war, the Freedmen's Bureau was created by the U.S. Congress in 1865 as a method of providing "food, clothing, medical care, and education both to freedmen and to white refugees."⁵⁸ Administratively, the Freedman's Bureau was housed within the War Department, thereby establishing the U.S. Army as the federal resource that was responsible for "providing disaster relief."⁵⁹ Although tragic, the War Between the States improved the ability to manage emergency situations.

Dillon's Rule

During 1868, Dillon's Rule stated the notion that "local governments may exercise only those powers explicitly granted to them by the state, those clearly implied by these explicit powers, and those absolutely essential to the declared objectives and purposes of the local government."⁶⁰ Essentially, Dillon's Rule facilitates the expressing of "statutes that grant authority to local governments," thereby granting power only if it is "necessarily implied or essential to the exercise of powers that are expressly granted."⁶¹ Simply explained, localities can only derive powers from the state.

Although it originated in 1868, Dillon's Rule has various implications for emergency management during modern times. Because Dillon's Rule constrains the power of localities, any power exhibited by a locality must be allocated by the constitution of a state, the statutes of a state, a "home rule charter," implied powers that complement expressed powers, and any powers that are "essential to the declared purposes of the municipality."⁶² This notion represents a constraint regarding the

---

⁵⁵Ibid.

⁵⁶Ibid.

⁵⁷Ibid.

⁵⁸David Kennedy and Lizbeth Cohen, *The American Pageant: Since 1865*, vol. 2 (Boston, MA: Wadsworth, 2013) 469.

⁵⁹Katz, *Essentials of Public Health Preparedness*, p. 48.

⁶⁰Ann Bowman and Richard Kearney, *State and Local Government: The Essentials* (Boston, MA: Wadsworth, 2012), 234.

⁶¹Sammis White, Richard Bingham, and Edward Hill, *Financing Economic Development in the 21ˢᵗ Century* (Armonk, NY: M.E. Sharpe, 2003), 34.

⁶²Clifton Smoke, *Company Officer*, 2nd ed. (Clifton Park, NY: Thomson-Delmar, 2005), 231.

self-governance capability of a locality involving emergency events. For example, some states limit the ability of localities regarding decisions that are affiliated with the "adoption and enforcement of fire and building codes."[63] In this instance, the ability of the locality to render decisions, during phases of emergency management preparedness, are severely limited. State-level authorities may not understand the need of the locality or may be slow in responding to the needs of the locality in such instances. As a result, the preparedness efforts of the locality may be delayed and hampered or altered because of imposed constraints mandated by Dillon's Rule.

Reconstruction and Home Rule

Reconstruction was implemented within the ravaged southern states following the War Between the States, and strongly influenced the peoples of the affected land between the years 1865 and 1877.[64] The period of reconstruction politically "forced the South" into one party as a method of maintaining home rule.[65] The notion of home rule is generally "associated with local autonomy and referred to the capacity of local governments to act in accordance with local communities' needs free of the interference of higher tiers of the state."[66] Two perspectives of home rule existed during Reconstruction: state and local.

The state perspective viewed home rule as the right of the state to "self-determination" without any interference federally regarding the resolving of problems and the determining of its future.[67] Reconstruction involved the use of military rule and influence in economic and political decisions among the southern states. For example, the state of North Carolina endured military rule during the years 1867 and 1868.[68] Officers of the U.S. army were stationed in North Carolina, and "intervened in political and economic decisions" at both the local and state levels of government.[69] The political ideologies of these officers varied, and many were aligned with the "traditionalist Southern white elites."[70]

The local perspective of home rule was viewed as a form of localized self-government involving the "popular election of county officials such as magistrates

---

[63]Ibid.

[64]Merton Coulter, *The South During Reconstruction: 1865–1877*, vol. 8 (Baton Rouge, LA: Louisiana State University Press, 1975).

[65]Ibid., p. 377.

[66]Gordon Clark, *Judges and the Cities: Interpreting Local Autonomy* (Chicago, IL: Univ. of Chicago Press, 1985), 172.

[67]Deborah Beckel, *Radical Reform: Interracial Politics in Post-Emancipation North Carolina* (Charlottesville, VA: Univ. of Virginia Press, 2011), 54.

[68]Ibid.

[69]Ibid.

[70]Ibid., p. 55.

and judges."[71] The governmental structure of antebellum period often exhibited the appointing of local officials by the "ruling elites."[72] Therefore, individuals whom occupied these positions were unelected by the populace. The act of appointing officials was often unpopular with the populace because the decisions of appointed officials influenced practically every aspect of its existence. The influence of appointed officials "affected voting procedures, judicial and police powers, citizenship rights and responsibilities, and countless other aspects of people's lives."[73]

Despite the allusions to self-direction and self-government implied by home rule, the extensiveness of federal power, presence, and authority permeated the South. These characteristics of the federal government affected economic and political decisions of both states and localities with varying amounts of bias. Such influences and biases either quashed or diminished the foundational elements of the home rule concept. After the presidential election of 1876, a compromise was reached in 1877 that diminished the federal influence that affected home rule.

The election of 1876 produced no clear winner because neither of the candidates, Rutherford Hayes and Samuel Tilden, had established a clear majority of Electoral College votes. Disputes arose regarding the voting outcomes of three states: South Carolina, Florida, and Louisiana.[74] Within Louisiana and South Carolina, Democrats interrupted forcefully Republican meetings, threatened Republican voters and candidates, and campaigned in "armed, uniformed groups."[75] Similar actions occurred in Florida.[76] Although the "raw returns" indicated outcomes for Tilden, election boards, consisting of Republicans, were capable of counteracting the "fraud and violence."[77] The voting returns from counties in which allegations of violence and fraud existed were rejected by Republican officials.[78] Democratic state officials disagreed, made allegations of fraud, and "certified the votes of the Democratic electors."[79] These scenarios resulted in the existences of "two sets of conflicting electoral votes" in Louisiana, Florida, and South Carolina.[80]

During 1877, a special election commission was organized and formed in the event that the two houses of Congress disagreed regarding the counting of electoral

---

[71] Ibid.

[72] Ibid.

[73] Ibid.

[74] Ballard Campbell, *Disasters, Accidents, and Crises in American History, A Reference Guide to the Nation's Most Catastrophic Events* (New York, NY: Facts on File, 2008), 145.

[75] Ibid.

[76] Ibid.

[77] Ibid., p. 146.

[78] Ibid.

[79] Ibid.

[80] Ibid.

votes.[81] The commission determined that Hayes won the election.[82] Continuing disputes occurred between Democrats and Republicans concerning the election necessitating compromise. Essentially, the Republican faction agreed to the termination of Reconstruction provided that Democrats and the South "agreed to accept Hayes's victory as president."[83]

The Compromise of 1877, between Republicans and Democrats, represented the ending of Reconstruction with the departing of federal troops from the "Southern states."[84] This compromise terminated the interfering of federal entities in southern matters, and "ushered a period of Redemption."[85] It also facilitated various concessions politically and economically, and the appointing "of a Southerner to the cabinet" of President Rutherford Hayes.[86] The promises made within the compromise also impacted the South with respect to governance without interference from federal factions, thereby instantiating home rule.[87]

Epidemic Disease

During 1878, a Yellow Fever epidemic plagued the nation. It caused "approximately 20,000 deaths" along rivers and waterways in Vicksburg, Mississippi; Memphis, Tennessee; and New Orleans, Louisiana, and in communities along the Tennessee, Mississippi, and Ohio rivers.[88] The economic and financial devastation generated by this illness and deaths were approximately $100 million in "relief efforts" and "trade losses."[89] Mitigating the effects of this incident resulted in the passing of the Quarantine Act of 1878 and the forming of the National Board of Health in 1879.[90]

Although quarantine measures were primarily conducted by states, the passing of the Quarantine Act of 1878 interjected federal involvement by "giving the Marine Hospital Service responsibility to stop disease from coming ashore via

---

[81]Ibid.

[82]Ibid.

[83]Ibid.

[84]Judith Peacock, *Reconstruction: Rebuilding After the Civil War* (Mankato, MN: Capstone Press, 2003), 36.

[85]Hans Trefousse, *Historical Dictionary of Reconstruction* (Westport, CT: Greenwood Press, 1991), 49.

[86]Ibid., p. 69.

[87]David Lincove, *Reconstruction in the United States: An Annotated Bibliography* (Westport, CT: Greenwood Press, 2000), 113.

[88]Ballard Campbell, *Disasters, Accidents, and Crises in American History, A Reference Guide to the Nation's Most Catastrophic Events* (New York, NY: Facts on File, 2008), 150.

[89]Ibid.

[90]Ibid.

sailors from ships."[91] The emergency responses to the illness varied among locales and states. The city of Memphis, Tennessee, implemented quarantine measures and "blocked railroad lines," thereby impacting the flow of goods to and from the city, but reversed its decision upon the threat of a lawsuit by businessmen.[92] Somewhere between 25,000 and 27,000 people fled the city to escape the illness.[93] The use of "shotgun barricades" was necessary to prevent the refugees from entered uninfected communities, and the madam of a Memphis brothel, Annie Cook, assisted during the epidemic by converting her place of business to a hospital, where she nursed the stricken. She died from the disease in September.[94] In Memphis, alone, approximately 5,000 individuals perished from Yellow Fever, and the epidemic disease bankrupted the city.[95]

The losses resulting from the Yellow Fever epidemic were tremendous economically, politically, and socially. Although the epidemic was horrific, it instigated much change among municipal infrastructures. Renovations of sanitary conditions and hygiene were necessary to diminish further instances of the disease. Municipalities facilitated such improvements by "building sewer and drainage systems."[96] During the aftermath of the epidemic, the National Board of Health was established in 1879 as an initial attempt to investigate seriously the characteristics and cause of the disease.

During the remainder of the nineteenth century, much attention was devoted to Yellow Fever. Its effects were recognized as a significant danger to human life and the economic performance of the nation. Therefore, it was acknowledged that some methods of preventing, diminishing the effects of, or eradicating the disease should be discovered. These considerations are expressed as follows:[97]

> Yellow fever should be dealt with as an enemy which imperils life and cripples commerce and industry. To no other great nation of the earth is yellow fever as calamitous as to the United States of America. In a single season, more than a hundred thousand of our people were stricken in their homes, and twenty thousand lives sacrificed by this preventable disease. Systematic, scientific study should be unceasingly directed against this

---

[91]"The Great Fever," Mississippi Public Broadcasting—Public Broadcasting Service, 2012, http://www.pbs.org/wgbh/amex/fever/peopleevents/e_1878.html (accessed December 17, 2012).
[92]Ibid.
[93]Ibid.
[94]Ibid.
[95]Ibid.
[96]Campbell, *Disasters, Accidents, and Crises in American History*, p. 150.
[97]Ernest Hardenstein, *The Epidemic of 1878 and its Homeopathic Treatment: A General History of the Origin, Progress, and End of the Plague in the Mississippi Valley* (New Orleans, LA: J.S. Rivers Publishing, 1879), 53–54.

subtle enemy until our weapons are so perfected as to destroy or to surely hold it in check. In the benefits flowing from scientific research, America has received from European nations more than she has bestowed, but the opportunity is now offered to pay a part of the debt by continuing to completion, so far as human skill will permit, the work which has been begun.

In the light of scientific experience, and of such facts as have been obtained, the Board has given careful consideration to the outlines of a system of observation of yellow fever and cholera; and, as connected therewith, to shipping in foreign ports, to the interchange of information, and to the inspection and sanitation of infected vessels and persons outside of our ports and after their arrival; together with the supervision of inter-state travel and traffic in times of epidemic within our borders. The objective aimed at, is to present the outlines of a system of quarantine, which may afford the greatest attainable degree of protection against the introduction and spread of infectious epidemic diseases; and at the same time inflict only a minimum of injury and inconvenience upon commerce.

This passage is significant regarding the maturing of emergency management within American society. It shows that the federal government was acutely aware and concerned with the effects of epidemic diseases and their social and economic consequences. This acknowledging of epidemic disease represents its prominence as a national threat which could neither be ignored nor discounted. The actions described within the passage show a consideration of strategic thinking with respect to managing instances of epidemic illnesses.

The passage also shows the realization of the outcomes regarding the incident of yellow fever and represents an unequivocal dedication to countering epidemic diseases through a continuous investment in scientific inquiry. The passage implies that American scientific research was previously uncharacteristic when compared with European efforts and accomplishments regarding beneficial outcomes and achievements. However, this status and performance of American scientific inquiry changed because of the yellow fever epidemic.

During the late nineteenth century, an American physician, Walter Reed, led a group of individuals into Cuba to investigate the cause of yellow fever.[98] Their research yielded the conclusion that yellow fever was carried and spread by mosquitoes.[99] Eventually, yellow fever was eradicated in Cuba by the "systematic destruction of mosquitoes on the island."[100] From the perspective of emergency management, this

---

[98]Cynthia Northrup, *The American Economy: A Historical Encyclopedia* (Santa Barbara, CA: ABC-CLIO Publishing, 2003), 84.
[99]Ibid.
[100]Ibid.

research initiative represents an early effort of investigating scientifically the cause of disease and crafting measures through which its effects are managed and countered.

The American Red Cross

Benevolence also characterized the nineteenth century. The American Red Cross was founded on May 21, 1881, in Washington, D.C.[101] This organization conducted numerous disaster relief initiatives (both foreign and domestic), provided aid to the American military during the Spanish-American War, and also "campaigned successfully for the inclusion of peacetime relief work" in conjunction with the "global Red Cross network."[102] The American Red Cross continues to provide benevolence and to facilitate emergency management initiatives during modern times.

The Year Without Summer

Although the yellow fever epidemic was a form of natural disaster that involved the carrying and transmitting of epidemic disease by mosquitoes, other forms of disaster imperiled human life and economic functioning. The Great Blizzard of 1888 represented a force of nature that instigated emergency conditions throughout the northeastern United States. The blizzard was so drastic that 1888 became known as the "year without summer" because it contributed to the producing of snowfall during the summer months.[103]

The blizzard deposited approximately 50 inches of snow within the northeast, and brought to a standstill the functioning of cities, towns, and communities.[104] Mass travel and communication were practically impossible because roadways and railways were covered with impassable levels of snow and telegraph lines were destroyed.[105] Businesses were closed and were unable to distribute goods and services ranging from food and beverage to heating coal.[106]

Despite its destructiveness and dangerousness, the blizzard incited a radical change in the infrastructure of northeastern cities. Communication telegraph lines were moved underground to protect them from such future calamities.[107] The above-ground railway infrastructure of New York was deemed susceptible to future calamities. Because of its status as a national "commercial and financial

---

[101]American Red Cross, "Our History," 2012, http://www.redcross.org/about-us/history (accessed December 18, 2012).

[102]Ibid.

[103]Niles Eldredge, *Life on Earth: An Encyclopedia of Biodiversity, Ecology, and Evolution* (Santa Barbara, CA: ABC-CLIO, Inc., 2002), 354.

[104]Ibid.

[105]Ibid.

[106]Ibid.

[107]Ibid.

center," construction began on underground subway systems.[108] Little advance warning preceded the blizzard to warn people of the impending storm. Therefore, weather-monitoring stations were situated in the "Atlantic region" from "Nassau, the Bahamas, and Bermuda in the south the New Hampshire and Newfoundland in the north."[109]

From the perspective of emergency management, despite its catastrophic consequences, the blizzard produced beneficial outcomes in due time. The communications and logistics systems of the northeast were vastly improved, thereby facilitating the ability to communicate and to travel during future emergencies. Placing these systems underground provided a measure of protectiveness from the ravages of nature. The situating of an array of weather-monitoring stations in the Atlantic region provided both a method of tracking storms and of providing advance warning regarding inclement conditions that could impact human life and physical infrastructure.

The Galveston Hurricane

The nineteenth century closed with a vengeance in the Gulf of Mexico with the depositing of a hurricane over Galveston, Texas, on September 8, 1900. Before this event, the city had experienced at least 11 hurricanes during the nineteenth century.[110] Approximately 6,000 people perished in the city itself whereas approximately between 4,000 and 6,000 people perished "elsewhere on the island and on the nearby mainland."[111] The financial cost of property damage was approximately $30 million, and included the destruction of "3,600 homes."[112] The storm destroyed roadway infrastructure leaving only railways as the accessible method of traveling to the mainland.[113]

The response and recovery activities that occurred during the aftermath of the storm represented an organized approach to emergency management. The American Red Cross participated in the activities by distributing food and clothing.[114] The Central Relief Committee was formed to conduct activities regarding finance, correspondence, sheltering of refugees, new house construction, and for

---

[108]Ibid.

[109]Ibid.

[110]Texas State Historical Association, "Galveston's Response to the Hurricane of 1900," 2012, http://www.texasalmanac.com/topics/history/galvestons-response-hurricane-1900 (accessed December 14, 2012).

[111]Ibid.

[112]Ibid.

[113]Ibid.

[114]Ibid.

repairing damaged homes.[115] A Deep Water Committee was formed to consider the methods of governance through which the recovery period could be managed.[116] Strategically, a plan was crafted through which the city could be possibly protected from future calamities. This plan accommodated the erecting of a "concrete seawall three miles long" and "raising the entire city" to a higher elevation.[117] These improvements protected the city significantly from the effects of future storms. During 1915, when another hurricane impacted the area, it "destroyed 90 percent of the buildings outside the seawall and flooded the downtown area."[118] Despite the damage, only eight people perished in the city whereas 304 were killed in other locations.[119]

From the perspective of emergency management, the Galveston incident shows an improvement in the overall characteristics of emergency management that existed at the beginning of the nineteenth century. The response and recovery activities exhibited the characteristics of a thoughtful paradigm through which emergency management operations were performed both efficiently and effectively. Coordination and cooperation occurred among multiple organizations to conduct operations ranging from the sheltering of refugees to the crafting of engineering plans to raise the elevation of the city and to erect a seawall. Although devastating, the Galveston hurricane showed an improvement of American emergency management abilities to respond to and recover from a natural calamity.

Emergency Management Implications

The beginning of the nineteenth century exhibited little governmental involvement with the managing of emergencies. However, the ending of the century exhibited a vastly different perspective. Throughout the nineteenth century, numerous incidents and endeavors occurred that necessitated federal attention and involvement regarding emergency management. The initial piece of legislation was passed by the U.S. Congress that afforded relief for Portsmouth during the aftermath of its fire. The westward expansion necessitated protecting federally travelers and rendering medical assistance during their journeys.

The second half of the nineteenth century was no less tumultuous. The War Between the States advanced emergency medical patient evaluation and field treatment and demonstrated the perils of political emergencies. Instances of epidemic disease facilitated improvements in hygiene and sanitation among municipalities, necessitated quarantine, and instigated a meaningful, serious investment

---

[115] Ibid.
[116] Ibid.
[117] Ibid.
[118] Ibid.
[119] Ibid.

in scientific research to deter and eradicate threatening diseases. Infrastructure improvements caused the construction of underground communication and logistics systems in the northeast. Various technological advancements improved the reporting of emergency incidents, thereby improving the effectiveness and efficiency of emergency management. Engineering improvements contributed to the erecting of the Galveston seawalls and the elevating of the city.

The nineteenth century was marked by change regarding the perspectives of locales, states, and federal administrations regarding emergencies of varying scopes and magnitudes. The nineteenth century witnessed the changing and transforming of the federal government from an uninvolved entity to one that committed unequivocally resources toward protecting American society from the effects of both man-made and natural emergencies. The origin of emergency management legislation occurred in this century as well as the initial forming of organizations to counter the effects of both natural and man-made emergencies. Because of the growing population and maturing of the American nation, coupled with the knowledge and experience of both man-made and natural emergencies, the federal government could neither ignore nor discount the importance of its role in emergency management.

The Twentieth Century

Both natural and man-made emergencies permeated the twentieth century. The ability to manage such emergencies continued to mature federally, regionally, and locally through numerous technological improvements, continued scientific research, organization, legislation, and policy. The twentieth century demonstrated numerous emergencies ranging from floods and wars to hurricanes and domestic terrorism. Certainly, the experiencing of both man-made and natural emergencies contributed toward the maturing of emergency management paradigms.

The Changing Paradigms of Civil Defense

The origins of twentieth-century civil defense occurred during World War I. President Woodrow Wilson petitioned Congress for a declaration of war against Germany on April 6, 1917.[120] The initial landing of the American Expeditionary Force occurred in France on July 3, 1917.[121] Approximately a year before America's entry into the conflict, the Council of National Defense was created and organized as the first civil defense program within the United States, and was "established on

---

[120] "WWI Timeline," Mississippi Public Broadcasting—Public Broadcasting Service, 2012, http://www.pbs.org/greatwar/timeline/time_1917.html (accessed December 19, 2012).
[121] Ibid.

August 29, 1916."[122] It represented a "non-military effort to prepare American civilians for possible military attack."[123] The Council represented an advisory board to the president, and included the "Secretaries of War, Navy, Interior, Agriculture, Commerce, and Labor."[124]

During World War II, this entity was replaced by the Office of Civilian Defense. This organization was created in 1941 by executive order for the purpose of coordinating local, state, and federal defense interactions concerning the protection of civilians during air raids and various emergencies.[125] It also facilitated the participation of civilians within programs that were necessary to support the war effort.[126] This entity was responsible for "morale maintenance, promotion of volunteer involvement, and nutrition and physical education."[127] In 1945, it was dissolved by executive order.[128]

Perilous times continued during the cold war. With the passing of the National Security Act of 1947, the National Security Resources Board was created. It was responsible for the mobilization of military and civilian support, and the maintaining of "adequate reserves and effective resource use in the event of war."[129] The successful testing of a Soviet nuclear weapon provided the United States with a formidable, deadly opponent. It heightened tremendously the tensions of the cold war. Across the nation, localities petitioned the federal government to craft plans regarding what must be done locally should an emergency event arise. This petitioning resulted in the creation of the *Blue Book* which provided recommendations concerning civil defense.[130]

Given the recommendations of the *Blue Book* and the opinion of President Harry Truman that civil defense was a primary responsibility of state and local governments, Congress passed the Federal Civil Defense Act of 1950 which assigned much of civil defense responsibilities to the individual states.[131] It also created the Federal Civil Defense Administration which provided guidance to the individual states

---

[122]"History + Evolution of Emergency Management," Providence Rhode Island Emergency Management Agency and Office of Homeland Security, 2012, http://www.providenceri.com/PEMA/about/history-evolution-of-emergency-management (accessed December 19, 2012).

[123]Ball State University, "MSS 268—World War II Government Publications, 1941–1945," 2012, http://www.bsu.edu/libraries/archives/findingaids/MSS268.pdf (accessed December 20, 2012).

[124]Homeland Security National Preparedness Task Force, *Security: A Short History of National Preparedness Efforts* (Washington, D.C.: U.S. Government, 2006), 5.

[125]Ibid.

[126]Ibid.

[127]Ibid.

[128]Ibid.

[129]Ibid, p. 6–7.

[130]Ibid.

[131]Ibid.

Figure 2.2 Civil defense poster.

when crafting civilian defense programs.[132] Many beneficial programs resulted from this legislation, including the creation of shelter facilities.[133] Civil defense also depended upon the assistance of volunteers from the citizenry. Figure 2.2 shows a typical recruiting poster designed to lure volunteers from the populace.

The 1960s ushered the presidential administration of John F. Kennedy. By executive order, Kennedy divided the Office of Civil Defense and Mobilization into two

---

[132]Ibid.

[133]Ibid.

separate components: Office of Civil Defense and Office of Emergency Planning.[134] During 1961, President Kennedy became the first president to publically address the nation regarding the necessity of protecting the American population.[135] This event heralded a demand for fallout shelters within the populace.[136] During the Cuban Missile Crisis, additional emphasis regarding sheltering was advocated by the Kennedy administration.[137]

During the 1960s, the administration of President Lyndon Johnson witnessed a diminishing in the emphasis regarding civil defense. Because of obligations in the Vietnam War coupled with anti-war sentiments that existed within the population, less emphasis was placed upon civil defense.[138] Further, because of the concept of mutually assured destruction, which held "the United States population hostage to Soviet nuclear attack as a reassurance that the United States would never attack the Soviet Union," facilitated a "strategic environment in which civil defense was seen as destabilizing to a dangerous degree."[139]

The 1970s involved a different approach to civil defense. The administration of President Richard Nixon demonstrated various changes in civil defense paradigms, but witnessed a "downward slide of actual capability to protect the population."[140] The diminished attention regarding civil defense was facilitated by the Strategic Arms Limitation Treaty (i.e., SALT I).[141] Given the tenets of this treaty, both the United States and the Soviet Union perceivably agreed "that they would take no steps to limit the other side's ability to cause catastrophic damage in retaliation to a first strike."[142] Any advocating of civil defense could possibly be perceived as a method through which vulnerability could be reduced.[143]

The administration of President Gerald Ford again changed the civil defense paradigm, and returned it to its original emphasis of "nuclear attack preparedness."[144] After a five-year suppressing of "intelligence observations" of the Soviet civil defense capabilities, new intelligence reports indicated that the Soviets had made

---

[134]Ibid.

[135]"Civil Defense: The Kennedy Administration," University of Richmond, 2003, https://facultystaff.richmond.edu/~wgreen/Ecdkennedy.htm (accessed December 20, 2012).

[136]Ibid.

[137]Ibid.

[138]"Civil Defense: The Johnson Administration," University of Richmond, 2003, https://facultystaff.richmond.edu/~wgreen/Ecdjohnson.htm (accessed December 20, 2012).

[139]Ibid.

[140]"Civil Defense: The Nixon Administration," University of Richmond, 2003, https://facultystaff.richmond.edu/~wgreen/Ecdnixon.htm (accessed December 20, 2012).

[141]Ibid.

[142]Ibid.

[143]Ibid.

[144]Homeland Security National Preparedness Task Force, *Security: A Short History*, p. 16.

great progress toward protecting strategically their own population.¹⁴⁵ Although the Ford administration attempted to make significant progress in reviving civil defense, the efforts were relatively ineffective.¹⁴⁶

Significant change to civil defense occurred during the administration of President Jimmy Carter. Upon assuming the presidency, Carter issued Presidential Directive 41 (PD41) which clarified the civil defense perspective. According to PD41, civil defense was a strategic tool to embellish stability and deterrence regarding the potential of nuclear conflict.¹⁴⁷ The Carter administration deemed also that it was unnecessary to pursue any level of "equivalent survivability" with the Soviets.¹⁴⁸ Further, the administration pursued evacuation as the dominant theme of civil defense policy regarding the potential of nuclear attack.¹⁴⁹

However, President Carter radically changed American civil defense after the nuclear incident of Three Mile Island. Although the event itself was not the primary catalyst for change, it contributed toward the forming of the FEMA in which numerous, independent government agencies and organizations were conglomerated into a solitary entity.¹⁵⁰ During this period, the creating of FEMA was the "single largest consolidation of civil defense efforts in U.S. history."¹⁵¹ President Carter emphasized the need for FEMA to address peacetime emergencies.¹⁵²

The administration of President Ronald Reagan witnessed the amending of the 1950 Civil Defense Act with respect to the duality of natural emergencies and attacks against the United States.¹⁵³ President Reagan's approach to civil defense differed from the approaches of his predecessors because his stance was much more aggressive and proactive.¹⁵⁴ Essentially, this approach emphasized the notion that civil defense was a "necessary complement to U.S. war-fighting" capability, and that a strong civil defense demonstrated the American resolve to "fight and survive a nuclear war."¹⁵⁵

The perspectives of the Reagan administration were critical aspects of emphasizing a truly all-hazards approach to civil defense. This period witnessed the crafting of the Integrated Emergency Management System (IEMS) which represented

---

¹⁴⁵Ibid.
¹⁴⁶Ibid.
¹⁴⁷Ibid., p. 18.
¹⁴⁸Ibid.
¹⁴⁹Ibid., p. 19.
¹⁵⁰Ibid,. p. 18
¹⁵¹Ibid.
¹⁵²Ibid.
¹⁵³Ibid., p. 20.
¹⁵⁴Ibid., p. 21.
¹⁵⁵Center for Defense Information, "President Reagan's Civil Defense Program," *The Defense Monitor* 11, no. 5 (1982): 1–8.

CHAPTER 2   History of Emergency Management in the United States   47

Figure 2.3   Exxon Valdez.

the "first statement of a true all-hazards approach including the complete spectrum of natural and man-made disasters and national security threats, as well as the identification of functions, such as communications, warning, direction and control, feeding, sheltering, and others, that were critical to any response to any disaster."[156] This perspective represented "substantive changes to existing civil defense programs and the first movement toward an emergency management focus."[157]

President George H. Bush witnessed the downfall of the Soviet Union and the resulting changes in American civil defense that necessitated less emphasis regarding nuclear attack. After the dissolving of the Soviet Union, the nuclear threat against the United States significantly abated. However, his tenure within the presidency witnessed numerous natural emergencies that further defined American civil defense. During his administration, the *Exxon Valdez* oil spill was recognized as the worst in American history.[158] Figure 2.3 shows the Exxon Valdez cargo ship.

Numerous hurricanes affected the east coast and Atlantic region, and an earthquake shattered California.[159] The Bush administration and FEMA were criticized for the poor quality of emergency management regarding these incidents.

---

[156]"Civil Defense: The Reagan Administration," University of Richmond, 2003, https://facultystaff.richmond.edu/~wgreen/Ecdreagan.htm (accessed December 20, 2012).
[157]Ibid.
[158]Homeland Security National Preparedness Task Force, *Security: A Short History*, p. 21.
[159]Ibid.

The administration of President William Clinton crafted numerous changes to civil defense. The FEMA was elevated to "cabinet-level status," thereby improving executive communication with the agency; the Federal Civil Defense Act of 1950 was repealed; and a new emphasis regarding the ability to detect hazards of numerous varieties emerged.[160] Numerous calamities occurred that further defined the civil defense function. The Murrah Building was bombed in Oklahoma City, Oklahoma; several terrorist attacks occurred against American targets overseas; the calamities of other nations (e.g., sarin gas attack in a Tokyo subway); and an evaluating of U.S. security policies contributed to the redefining of American civil defense and emergency management.

In many respects, the Clinton years signaled the closure of a civil defense instantiation that endured over half of a century. With no Soviet threat and the continued maturing of an all-hazards approach that accommodated both natural and man-made emergencies, the Clinton administration represented an initial period of change and transition regarding the approaches to civil defense. During its origin, governmental administrations approached civil defense with respect to the potential dangers that were posed by a nation state. During the closure years of the Clinton administration, the threats posed by nation states were significantly less severe than had been witnessed during the cold war.

Despite the dynamics of change regarding the civil defense paradigm, new man-made threats were emerging that posed danger for American society (e.g., domestic terrorism). The dangers of natural disasters were unchanged because the nation remained susceptible to the ravages of hurricanes, floods, blizzards, earthquakes, and so forth. After the fall of the Soviet Union, the advent and proliferation of increased globalism slowly began changing the American economy, thereby interjecting new concerns of security. Protecting the American populace demanded new insights, methods, resources, and strategies. This period of change represented one in which conditions were right for a new paradigm and approach to emergency management: homeland security.

Flooding Legislation

During 1927, the Mississippi River flooded numerous cities, towns, and rural areas along its course. The incident was deemed as the "most destructive river flood" in American history, and resulted in the deaths of 500 people and caused approximately 600,000 people to become homeless.[161] The flooding impacted the states of "Illinois, Missouri, Kentucky, Tennessee, Arkansas, Mississippi and Louisiana," and encompassed "some 16 million acres of land (26,000 square miles)."[162] In the

---

[160]Ibid., p. 23.
[161]Kayla Webley, "Mississippi River, 1927." *Time Magazine*, 2011, http://www.time.com/time/specials/packages/article/0,28804,2070796_2070798_2070780,00.html (accessed December 20, 2012).
[162]Ibid.

terms of 1927 dollars, the financial costs of the resulting damages were approximately $400 million.[163]

This flood resulted in the passing of the Flood Control Act of 1928. It represented the "first comprehensive flood control act."[164] Before the enacting of this legislation, flood control was primarily the responsibility of individual localities and states.[165] This legislation interjected federal responsibility for flood control along the Mississippi River through a "system of levees in conjunction with floodways, outlet channels, and improved tributary basins."[166]

During the 1930s, coupled with the New Deal, the Roosevelt administration undertook numerous public projects through which the effects of natural disasters could either be averted or reduced.[167] These endeavors involved the passing of additional legislation. The Flood Control Act of 1936 and succeeding legislation provided the authorization for the "U.S. Army Corps of Engineers to design and construct control projects" across the nation to counter the effects of flooding.[168] Managing nature involved the use of "levees, floodways, and dams."[169] Since the passing of these two original pieces of legislation regarding flooding, numerous other flood control acts were passed that further facilitated government involvement in emergency management regarding flooding events.

Defense Production Act of 1950 and the War Power Act of 1973

The Defense Production Act (DPA) of 1950 was enacted after the commencing of the Korean War. It provided a means of ensuring that the United States had available resources to satisfy its military needs concerning resource requirements.[170] It also provided a capacity for supplying the resources that were necessary for facilitating civil defense.[171] Essentially, the DPA authorizes the federal government to "create, maintain or expand a domestic production capability needed for national

---

[163]Jeffrey Bumgarner, *Emergency Management: A Reference Handbook* (Santa Barbara, CA: ABC-CLIO Publishing, 2008), p. 5.
[164]Ibid.
[165]Ibid.
[166]Ibid.
[167]Ibid.
[168]Ibid.
[169]Gareth Pender and Hazel Faulkner, *Flood Risk Science and Management* (West Sussex, UK: Blackwell Publishing, 2011), 500.
[170]Linda Burns, *FEMA (Federal Emergency Management Agency): An Organization in the Crosshairs* (New York, NY: Nova Science Publishers, 2007), 8.
[171]Ibid.

defense."[172] It is a temporary law that necessitates reauthorization approximately "every two-to-three years."[173]

The War Power Act of 1973 also accommodates a military perspective during periods of national emergency. It allows the president the authority to deploy the military for a period not to exceed 60 days if a national emergency arises involving an attack upon the U.S. territory before requesting congressional approval for the use of such force.[174] However, Congress has the "right to demand that the president withdraw the troops before the sixty-day limit."[175]

The emergency management implications of these acts are serious. Together, they facilitate a quick, effective response to emergency situations and provide a legal basis for obtaining any necessary resources to support the initiative. For example, if an attack occurs against U.S. territory, the military may be mobilized quickly to repel the aggressor force. If resources are required to support the mobilization and initiative, then they may be procured. For example, if a supply of bullets is needed, then Revlon lipstick production lines could be converted to produce bullet and shell casings. By having such legislation actively in place, a foundational, a legal basis exists to both mobilize and deploy military forces and to equip them during the period of activation. During the period of activation, a presidential request may be made to Congress to prolong the initiative or for a declaration of war.

Stafford Disaster Relief and Emergency Assistance Act of 1988

The Stafford Disaster Relief and Emergency Assistance Act of 1988 provides a foundational basis for providing federal assistance to local and state governments as a form of assistance when managing "major disasters and emergencies."[176] Under the provisions of the legislation, assistance may consist of "federal resources, medicine, food and other consumables, work assistance and financial relief."[177] This legislation broadened the scope of the 1974 Disaster Relief Act regarding the distribution of relief assistance including "debris removal, temporary housing,

---

[172]Rich Mirsky, "Trekking Through that Valley of Death—The Defense Production Act," *Innovation: America's Journal of Technology Commercialization*, 2005, http://www.innovation-america.org/trekking-through-valley-death-defense-production-act (accessed December 21, 2012).

[173]Daniel Else, *Defense Production Act: Purpose and Scope* (Washington, D.C.: Congressional Research Service, 2009), 5.

[174]Karl Rogers, *Debunking Glenn Beck: How to Save America from Media Pundits and Propagandists* (Santa Barbara, CA: ABC-CLIO Publishing, 2011), 38–39.

[175]Ibid.

[176]Tener Veenema, *Disaster Nursing and Emergency Preparedness for Chemical, Biological, and Radiological Terrorism and Other Hazards* (New York, NY: Springer, 2013), 4.

[177]Ibid.

individual and household financial assistance, infrastructure repair, emergency communications, military support for the preservation of life and property, and crisis counseling" through the establishment of cost sharing between state and federal governments.[178]

Through the Stafford Act, the president maintains a permanent authority to direct assistance to affected regions of the United States and its territories.[179] Governors of areas in which emergencies occur are required to "declare a state of emergency, activate the state emergency response plan, and furnish information to FEMA regarding the availability of state resources" that are dedicated to the incident including specification of cost-sharing requirements between the state and federal governments.[180] When disasters strike federal properties, the president may declare the state of emergency without gubernatorial consent.[181] Depending on the severity of the situation, state governors can declare a state of emergency before an incident occurs provided that the "consequences of a disaster are clear and imminent" and necessitate action to either avert or reduce the potency of an impending threat.[182]

The Stafford Act improved upon previous legislation by eliminating ambiguities that existed within the expressed law. It expressed a clarification of declarations of emergency, expressed responsibilities for relief assistance and public entities, emphasized the relevancy and significance of preparedness and mitigation, and expressed the cooperative process of obtaining relief assistance. These specifications improved the understanding of processes, procedures, and incident characteristics that are necessary to obtain federal assistance when managing emergencies. As a result, the efficiency and effectiveness of timely responses are enhanced when managing emergencies.

Disaster Mitigation Act of 2000

The Disaster Mitigation Act (DMA) of 2000 was an amendment to the Stafford Disaster Relief and Emergency Assistance Act of 1988. It mandated the existence of "hazard mitigation planning" as a prerequisite for obtaining disaster assistance.[183] This legislation replaced the former requirements of mitigation with the "intent to emphasize the role of state, tribal, and local entities in the coordination of

---

[178]Julie Framingham and Martell Teasley, *Behavioral Health Response to Disasters* (Boca Raton, FL: CRC Press, 2012), 39.

[179]Ibid.

[180]Ibid.

[181]Ibid.

[182]Tener Veenema, *Disaster Nursing and Emergency Preparedness*, p. 4

[183]Donald Menzel and Harvey White, *The State of Public Administration: Issues, Challenges, and Opportunities* (Armonk, NY: M.E. Sharpe, 2011), 215.

mitigation planning and implementation efforts."[184] Essentially, through the DMA, pre-disaster mitigation programs were established.

Program funding is available through the National Pre-Disaster Mitigation Fund as a method of assisting local and state governments regarding the implementation of "cost-effective hazard mitigation activities that complement a comprehensive mitigation program."[185] If candidates for such funding are deemed to be susceptible to the risks of floods, then they must participate in the National Flood Insurance Program and have implemented an active mitigation plan.[186] Maintaining a mitigation plan is necessary to remain eligible for DMA funding assistance.

Executive Order 10427

During the cold war, Executive Order 10427 was a 1950s attempt to further define and establish a federal role in emergency management. This order was signed by President Harry Truman as a method of delineating federal responsibilities versus those of state and local governments. The order clarified the notion that federal responsibility was to be a supplementary construct to the functions of state and local governments.[187] It was not to replace the functions of either state or local government functions.[188] These notions are expressed within Section 6 of the order as follows:

> Federal disaster relief provided under this act shall be deemed to be supplementary to relief afforded by State, local, or private agencies and not in substitution therefore; Federal financial contributions for disaster relief shall be conditioned upon reasonable State and local expenditures for such relief; the limited responsibility of the federal Government for disaster relief shall be made clear to state and local agencies concerned; and the States shall be encouraged to provide funds which will be available for disaster relief purposes.[189]

---

[184] ABA Center on Children and the Law, *Children, Law, and Disasters: What have We Learned from the Hurricanes of 2005?* (Chicago, IL: American Bar Association, 2009), 204.

[185] Jane Bullock, George Haddow, Damon Coppola, and Sarp Yeletaysi, *Introduction to Homeland Security: Principles of All-Hazards Risk Management*, 3rd ed. (Burlington, MA: Butterworth-Heinemann, 2009), 319.

[186] Ibid.

[187] National Academy of Public Administration, *Coping with Catastrophe: Building an Emergency Management System to Meet People's Needs in Natural and Manmade Disasters*, National Academy of Public Administration (Washington, D.C.: U.S. Government, 1993), 11.

[188] Ibid.

[189] "Executive Order 10427," Harry S. Truman Library and Museum, 1953, http://trumanlibrary.org/executiveorders/index.php?pid=669 (accessed December 21, 2012).

The order also facilitated preparedness activities through which federal entities could better respond to emergency events. Within its contents, the order states, "In order to further the most effective utilization of the personnel, equipment, supplies, facilities, and other resources of Federal agencies pursuant to the act during a major disaster, such agencies shall from time to time make suitable plans and preparations in anticipation of their responsibilities in the event of a major disaster."[190] This passage shows an attempt to maximize the benefits of planning activities to anticipate any federal actions and resources that would be necessary during any periods of emergency. Through establishing a foundation of anticipative planning, the order represents a method of establishing a preparedness mindset among government entities.

Federal Emergency Management Agency

The FEMA originated during the administration of President Carter in 1979. Modern disaster legislation authorizing a disaster response during periods of emergency is in accordance with the Stafford Disaster Relief and Emergency Assistance Act of 1988. The Stafford Act is the "statutory authority for most federal disaster response activities especially as they pertain to FEMA and FEMA programs."[191]

Within its time period, the forming of the FEMA organization represented one of the largest actions of integrating numerous government agencies into a single entity within U.S. history. Its formation signaled the formal acknowledging of the federal government regarding its role concerning emergency and disaster management. Its formation and existence exhibited the seriousness with which federal authorities viewed the impacts of emergencies and disasters.

Before the creation of FEMA, there was no solitary government organization of such breadth and depth whose primary devotion was emergency and disaster management. During the twentieth century, federal involvement gradually increased regarding disaster and emergency management through various forms of assistance and oversight, and culminated in the passing of approximately 100 pieces of legislation involving responses to incidents of "hurricanes, earthquakes, floods and other natural disasters."[192]

Certainly, disasters occurred throughout the twentieth century before the forming of FEMA. However, during the 1960s and the 1970s, disasters of mass proportions occurred that necessitated significant federal involvement. Hurricane Carla (1962), Hurricane Betsy (1965), Hurricane Camille (1969), Hurricane Agnes (1972),

---

[190]Ibid.

[191]Federal Emergency Management Agency, "About the Agency," 2012, http://www.fema.gov/about (accessed December 22, 2012).

[192]Ibid.

an Alaskan earthquake (1964), and a California earthquake (1971) were all devastating events that necessitated significant federal involvement.[193]

Although federal involvement occurred concerning these incidents, they involved a "fragmented" approach to disaster and emergency management.[194] Numerous "programs and policies" existed concurrently at both the local and state levels, thereby increasing the "complexity of federal disaster relief efforts."[195] Despite attempts by the National Governor's Association to reduce the quantity of agencies that were involved with such incidents, an appeal to President Jimmy Carter was necessary to generate improvements in emergency and disaster management federally.[196]

During 1979, President Carter issued Executive Order 12127 that created FEMA. This action integrated "many of the separate disaster-related responsibilities" into a solitary entity.[197] After it was formed, FEMA rendered assistance during the nuclear incident that involved Three Mile Island, the Love Canal contamination, and the "Cuban refugee crisis."[198] For the remainder of the twentieth century, FEMA rendered various forms of assistance during a variety of emergencies ranging from hurricanes to floods.

Although FEMA matured organizationally throughout the remainder of the century, its focus changed significantly upon the commencing of the twenty-first century. The terrorist attacks against the United States on September 11, 2001, ushered a new, deadly threat that demonstrated an ability to surprisingly strike American society and inflict both mass casualties and mass damage. After this incident, a new paradigm of "national preparedness and homeland security" was tasked within FEMA.[199] Coordination occurred between FEMA and the Office of Homeland Security to both train and to equip "first responders" regarding incidents involving "weapons of mass destruction."[200]

This altered paradigm necessitated focusing the all-hazards approach of emergency and disaster management toward issues involving homeland security.[201] The creation of the Department of Homeland Security (DHS) reorganized federal agencies, and placed FEMA within the DHS hierarchy. This reorganization facilitated "a coordinated approach to national security from emergencies and disasters—both natural and man-made."[202]

---

[193]Ibid.
[194]Ibid.
[195]Ibid.
[196]Ibid.
[197]Ibid.
[198]Ibid.
[199]Ibid.
[200]Ibid.
[201]Ibid.
[202]Ibid.

FEMA expresses its mission as follows: "to support our citizens and first responders to ensure that as a nation we work together to build, sustain and improve our capability to prepare for, protect against, respond to, recover from and mitigate all hazards."[203] Throughout its history, FEMA has endured many disasters and their aftermaths. The organization has provided assistance to numerous communities throughout the nation. During modern times, FEMA provides resources that support every phase of emergency management including preparedness, reaction, response, and recovery. FEMA resources are available to mitigate the effects of emergency situations. Both natural and man-made emergencies are accommodated within the mission and operational scope of FEMA. Although FEMA represents an entity that accommodates all phases of emergency management, its efficiency and effectiveness were questioned on a variety of occasions despite many successes. Regardless, the establishing of FEMA remains a significant event in American history through which the abilities of emergency and disaster management were improved greatly.

Homeland Security Presidential Directives

The terrorist attacks of September 11, 2001, profoundly illustrated that managing such an incident was beyond the capability of one entity. The aftermath of 9/11 generated a profound awareness for the necessity of the emergency management process to expand and become a nationwide collaborative effort. As such, significant efforts were made to improve the emergency management process to include all levels of government, the private sector, and nongovernmental agencies.

Subsequently, the Homeland Security Act of 2002 created the DHS and assigned the Secretary of Homeland Security responsibility for coordinating federal emergency operations within the United States. Federal emergency operations include preparing for, responding to, and recovering from terrorist attacks, major disasters, and other emergencies.

Thereafter, a series of Homeland Security Presidential Directives (HSPDs) were developed with the intention of identifying, developing, and implementing necessary emergency planning requirements. The following HSPDs HSPD-5 and HSPD-7 both played an integral part in the development of common approaches for emergency management:

HSPD-5: Issued by President George W. Bush, February 28, 2003, and titled *Management of Domestic Incidents*. The purpose of HSPD-5 was to prevent, prepare for, respond to, and recover from terrorist attacks, major disasters, and other emergencies, and establish a single comprehensive approach to domestic incident management.[204] Basically, there are three significant aspects of HSPD-5:

---

[203]Ibid.
[204]George W. Bush, *Homeland Security Presidential Directive/HSPD-5: Management of Domestic Incidents*, Office of the Press Secretary, White House, February 28, 2003. http://www.whitehouse.gov/news/releases/2003/02/20030228-9.html (accessed January 25, 2013).

1. HSPD-5 authorized the secretary for the DHS as the principle administrator for domestic incident management and responsible for coordinating federal operations within the United States to prepare for, respond to, and recover from terrorist attacks, major disasters, and other emergencies.[205]
2. HSPD-5 authorized the secretary of the DHS to develop and administer a National Incident Management System (NIMS), which provides interoperability and compatibility among federal, state, and local capabilities. NIMS builds on existing incident management and emergency response systems by providing concepts, principles, terminology, and technologies covering the Incident Command System (ICS), Unified Command, training, and establishes concepts for multi-agency coordination and public information systems.[205]
3. HSPD-5 authorized the Secretary of the DHS to develop and administer a National Response Plan (NRP), which integrated the federal government's domestic prevention, preparedness, response, and recovery plans into one system. The NRP is also the operational mandate for federal support to state, tribal, and local emergency managers. The NRP includes protocols for operating under different threat levels, and incorporated all existing federal emergency management plans.[206]

HSPD-7: Issued by President George W. Bush, December 17, 2003, and titled *Critical Infrastructure Identification, Prioritization, and Protection*. This directive superseded the earlier Presidential Decision Directive No. 63 (PDD-63), which was issued by President Clinton in May of 1998.[207] The purpose of HSPD-7 was to create a comprehensive strategic plan that would enhance the ability of all federal agencies to protect the Nation's Critical Infrastructure and Key Resources (CIKR).[208] Basically, there are three significant aspects of HSPD-7:

1. HSPD-7 authorized the president to designate the Secretary of Homeland Security as the principal federal official to lead CIKR protection efforts among federal departments and state and local governments with the objective being to protect against terrorism.[209]
2. HSPD-7 directed the DHS to identify all CIKR within the United States, determine security vulnerabilities, and to prioritize and develop new protective safety

---

[205]Ibid.

[206]Ibid.

[207]George W. Bush, *Homeland Security Presidential Directive/HSPD-7: Critical Infrastructure Identification, Prioritization, and Protection*, Office of the Press Secretary, White House, December 17, 2003, http://www.whitehouse.gov/news/releases/2003/12/20031217-5.html (accessed January 26, 2013).

[208]Ibid.

[209]Ibid.

measures.²¹⁰ Critical infrastructures and key resources are, ". . . those assets, systems, and networks whether physical or virtual, so vital that their failure or destruction would have a debilitating impact upon security (law enforcement, military), continuity of government, continuity of operations, public health and safety (fire, EMS, hospitals, and public health care centers), public confidence, or any combination of the effects" (PATRIOT ACT, Section 1016(e)).²¹¹ Subsequently, since the inception of HSPD-7, a total of eighteen critical infrastructure sectors have been identified by the DHS.²¹²

3. HSPD-7 improved critical infrastructure protection by establishing a framework for the DHS partners to identify, prioritize, and protect the critical infrastructures (identified above) known as the National Infrastructure Protection Plan (*NIPP*).²¹³, ²¹⁵ The NIPP provides our nation with guidance to achieve the complex task of protecting our nation's CIKR.

The goal of the NIPP is to "Build a safer, more secure, and more resilient America by preventing, deterring, neutralizing, or mitigating the effects of deliberate efforts by terrorists to destroy, incapacitate, or exploit elements of our nation's CIKR and to strengthen national preparedness, timely response, and rapid recovery of CIKR in the event of attack, natural disaster, or other emergency." ²¹⁴

The purpose of the NIPP is to provide guidance for implementing protection of CIKR in one plan that can be applied throughout multiple departments, entities, and organizations. The goals of the NIPP not only exemplify the importance of protection against terrorist attacks but also recognize the danger that CIKR face when affected by natural disasters and other emergencies.²¹⁵ The reason for protection of the CIKR is obvious. Destruction or deprivation of any part of the nation's CIKR would cripple the nation, which serves as the impetus to prevent this from occurring.

The Twenty-First Century

The closure of the twentieth century and the period of change demonstrated by the 1990s facilitated a fresh perspective for managing emergencies: homeland security. Accommodating the needs of modern emergency management necessitates

---

²¹⁰Ibid.

²¹¹*USA PATRIOT Act* (H.R. 3162), EPIC—Electronic Privacy Information Center, October 24, 2001, http://epic.org/privacy/terrorism/hr3162.html (accessed January 29, 2013).

²¹²Ibid.

²¹³Department of Homeland Security, *National Infrastructure Protection Plan*, 2009, http://www.dhs.gov/national-infrastructure-protection-plan (accessed January 29, 2013).

²¹⁴Ibid.

²¹⁵Ibid.

a continuation of the all-hazards paradigm. This concept is certainly applicable within the context of national emergencies and is also relevant for regions, states, and locales. Emergencies often occur unexpectedly without warning (e.g., terrorism) or they may have a sufficient period of notice in which to warn residents and to marshal resources (e.g., blizzards). In any case, managing emergencies during the twenty-first century necessitates a continued emphasis regarding both natural and man-made emergencies.

Post-Katrina Emergency Management Reform Act of 2006

During 2005, Hurricane Katrina devastated the Gulf Coast and impacted the nation. When compared and contrasted with storms that had occurred during the preceding 100 years, it was "one of the strongest storms to impact the coast of the United States."[218] Hurricane Katrina resulted in approximately 1,800 deaths, "billions of dollars" in financial costs, and the displacing of "between 700,000 and 1.2 million people."[219] Despite issuing public warnings that advised people to evacuate the affected region, many remained and endured the full force of natural disaster.

The severity of this incident necessitated federal involvement. Although FEMA responded to the incident, its efficiency and effectiveness were questioned and debated within American society and among government factions. The FEMA response exhibited failures in communication, coordination, and preparation.[216] For example, FEMA demonstrated an unawareness of the actions of other federal agencies regarding the rescuing of "storm victims."[217] The Department of the Interior had "offered to send boats, planes, trucks and personnel to rescue Katrina's victims immediately after the Aug. 29 storm hit."[218] Such characteristics of FEMA during the response to Hurricane Katrina caused much animosity and generated poor perceptions of the organization in the minds of many Americans and government entities.

Significant changes and improvements were necessary federally to improve the effectiveness and efficiency of future emergency incidents. The Post-Katrina Emergency Management Reform Act of 2006 resulted from the failures that were observed during Hurricane Katrina. This legislation facilitated numerous alterations within FEMA. Specifically, it revised organizational positions and requirements of leadership, introduced "new missions" and restored some missions that were previously abandoned, and mandated an array of "activities" both before an incident and afterwards.[219]

---

[216] Associated Press, "FEMA Official Acknowledges Poor Response to Hurricane Katrina," 2006, http://www.foxnews.com/story/0,2933,183178,00.html (accessed December 19, 2012).
[217] Ibid.
[218] Ibid.
[219] "Post Katrina Emergency Management Reform Act of 2006 Law & Legal Definition," U.S. Legal, 2012, http://definitions.uslegal.com/p/post-katrina-emergency-management-reform-act-of-2006/ (accessed December 19, 2012).

This legislation modified the organizational structuring of FEMA to accommodate a stronger capacity of facilitating preparedness. Notable organizational changes included transferring the following organizations underneath the FEMA hierarchy:

- U.S. Fire Administration
- Office of Grants and Training
- Chemical Stockpile Emergency Preparedness Division
- Radiological Emergency Preparedness Program
- Office of National Capital Region Coordination

The Post-Katrina Emergency Management Reform Act of 2006 represents the acknowledging and expressing of enhancements to federal capacities and capabilities for preparedness. Although Hurricane Katrina was devastating, it was the catalyst that showed various weaknesses of FEMA. Both people and organizations learned from experience. In this case, experiencing Hurricane Katrina heralded significant improvements federally through which enhancements were crafted regarding preparedness activities for both natural and man-made emergencies.

Chapter Comments and Summary

American emergency management had humble beginnings. Its origins are derived from the founding of the nation, and often incorporated the goodwill of others to lessen the local impacts of emergencies. The new government of the United States had little or no experiential precedents from which to conduct emergency management operations and functions. Any knowledge of a biological basis of disease was in its infancy. Technology was rudimentary and basic. However, both natural and man-made emergencies tested the mettle of the new nation, and established precedents from which future emergency management initiatives were either crafted or improved.

The role of federal involvement in emergency management increased through time. The Whisky Rebellion was the first use of military force to suppress domestic insurrection regarding issues of taxation. Since then, government involvement during emergencies increased through the centuries to represent a highly complex arrangement of organizations, politics, and resources during modern times.

Throughout much of American history, the approach to emergency management involved fragmented paradigms federally. Numerous organizations, functions, and policies were paralleled within other factions of government. Various tensions existed among local, state, and federal governments that impeded and diminished the quality of emergency management. However, with the maturing of the nation and the experiencing of numerous calamities, the approaches to emergency management became methodical and systematic.

Throughout American history, legislation regarding emergency management has accommodated a wide array of threats ranging from its origin involving relief after the Portsmouth fire to the modern tenets contained within the Post-Katrina Emergency Management Reform Act of 2006. Experience is often the teacher of hard lessons regarding emergency management. In many cases throughout American history, legislation occurred as the result of experiencing both man-made and natural emergencies.

The changes of governmental administrations also affect emergency management. For example, during the cold war, the administration of President Reagan exhibited one of the most stringent civil defense policies regarding the ability to fight, survive, and win a nuclear conflict, whereas the policies of his predecessors were less emphatic. Numerous civil defense policy changes occurred when new presidents were elected, thereby influencing the methods and paradigms of emergency management.

In essence, American emergency management has matured fluidly and experientially. Although the nation has a much better capacity to effectively and efficiently conduct federal emergency management functions than it did at its origin, some aspects of the emergency management domain are unchanged since the founding of the nation. Human nature is static. Any instantiation of emergency management is susceptible to the strengths and weaknesses of humans. The basic premise of emergency origins also is unchanged. Emergencies arise from either man-made or natural origins. Further, the maturing of emergency management has exhibited change in one context: technology.

Calamities will always affect the nation through time. Regardless of the situation, each event represents a learning experience through which improvements may be achieved and from which precedent may possibly be established. In any case, emergency management is a serious aspect of societal functioning and government involvement. Each succeeding generation of Americans will continue to witness events and incorporate technological advancements that impact and influence emergency management.

Terminology

- All-hazards
- Autonomy
- Call box
- Cold war
- Compromise of 1877
- Dillon's Rule
- Disaster
- Disaster, anagement
- Disaster relief
- Effectiveness
- Efficiency
- Emergency
- Emergency management
- Epidemic
- Executive Order 10427
- Executive Order 12127

FEMA
Field hospital
Financial capital
Fire Disaster Relief Act of 1803
Fire hydrant
Flood Control Acts
Great Blizzard of 1888
Home rule
Infrastructure
Legislation
Man-made emergency
Management
Mitigation
Mutually assured destruction
Natural disaster
National Security Act of 1947
Quarantine Act of 1878
Policy
Post-Katrina Emergency Management Reform Act of 2006
Reaction
Recovery
Response
Self-governance
Stafford Act
Technology
Triage
Yellow Fever

Thought and Discussion Questions

1. The primary functions of traditional management consist of controlling, coordinating, leading, organizing, and planning. During 2012, Hurricane Sandy impacted the United States along the east coast. Do some research, and compare and contrast basic characteristics of Hurricane Sandy and Hurricane Katrina from the perspective of emergency management involving FEMA. Given your observations, write a substantive, critical essay that addresses the FEMA issues of emergency management preparedness and response between these two storms.

2. The Whiskey Rebellion represented the initial use of military force to quash insurrection domestically regarding issues of taxation. Obviously, it established a precedent within the context of emergency management. Since then, other presidential administrations have employed military forces in conjunction with domestic incidents (e.g., Kent State University incident, integration of the University of Mississippi, Hurricane Katrina). Do some research, and determine what other historical events have established precedent within the context of emergency management. Based upon your research, list and briefly discuss each event.

3. From a historical perspective, this chapter introduced a variety of both man-made and natural emergencies that affected the United States. Select an event, and consider the characteristics of efficiency and effectiveness that were exhibited during each of the phases of emergency management. Write an essay that substantively and critically analyzes the efficiency and effectiveness of the response that is associated with your selected event.

4. This chapter introduced only a limited array of historical incidents that contributed toward the maturing of American emergency management. However, countless others exist among locales, states, regions, and the nation. Do some historical research, and identify two events (not contained herein) that contributed toward the maturing of American emergency management. Write an essay that substantively and critically analyzes the merits and significances of your chosen events with respect to the maturing of American emergency management.

References

1. ABA Center on Children and the Law. 2009. *Children, law, and disasters: What have we learned from the hurricanes of 2005?* 204. Chicago, IL: American Bar Association.
2. American Red Cross. 2012. Our History. http://www.redcross.org/about-us/history (accessed December 18, 2012).
3. Associated Press. 2006. FEMA official acknowledges poor response to hurricane Katrina. http://www.foxnews.com/story/0,2933,183178,00.html (accessed December 19, 2012).
4. Ball State University. 2012. MSS 268—World War II Government Publications, 1941–1945. http://www.bsu.edu/libraries/archives/findingaids/MSS268.pdf (accessed December 20, 2012).
5. Barry, J. 1997. *Rising tide: The great Mississippi flood of 1927 and how it changed America*. New York, NY: Touchstone Publishing.
6. Beckel, D. 2011. *Radical reform: Interracial politics in post-emancipation North Carolina*, 54. Charlottesville, VA: University of Virginia Press.
7. Birkland, T. 2006. *Lessons of disaster: Policy change after catastrophic events*, 105. Washington, DC: Georgetown University Press.
8. Bowman, A. and R. Kearney. 2012. *State and local government: The essentials*, 234. Boston, MA: Wadsworth.
9. Bryant, C. 2003. *Handbook of death and dying*, 186. Thousand Oaks, CA: Sage Publishing.
10. Bullock, J., G. Haddow, D. Coppola, and S. Yeletaysi. 2009. *Introduction to Homeland Security: Principles of all-hazards risk management*. 3rd ed., 319. Burlington, MA: Butterworth-Heinemann.
11. Bullock, J., G. Haddow, and D. Coppola. 2013. *Introduction to Homeland Security: Principles of all-hazards risk management*. 4th ed. Waltham, MA: Butterworth-Heinemann Elsevier Publishing.
12. Bumgarner, J. 2008. *Emergency management: A reference handbook*, 5. Santa Barbara, CA: ABC-CLIO Publishing.

13. Burns, L. 2007. *FEMA (Federal Emergency Management Agency): An organization in the crosshairs*, 8. New York, NY: Nova Science Publishers.
14. Burt, C. 2007. *Extreme weather: A guide and record book*. New York, NY: W.W. Norton & Company, Inc.
15. Bush, G. W. 2003. Homeland Security Presidential Directive/HSPD-5: Management of domestic incidents. *Office of the Press Secretary, White House, February 28, 2003*. http://www.whitehouse.gov/news/releases/2003/02/20030228-9.html (accessed January 25, 2013).
16. Bush, G. W. 2003. Homeland Security Presidential Directive/HSPD-7: Critical infrastructure identification, prioritization, and protection. *Office of the Press Secretary, White House, December 17, 2003*. http://www.whitehouse.gov/news/releases/2003/12/20031217-5.html (accessed January 26, 2013).
17. Bush, G. W. 2003. Homeland Security Presidential Directive/HSPD-8: National preparedness. *Office of the Press Secretary, White House, December 17, 2003*. http://www.whitehouse.gov/news/releases/2003/12/20031217-6.html (accessed January 27, 2013).
18. Campbell, B. 2008. *Disasters, accidents, and crises in American history, a reference guide to the nation's most catastrophic events*, 145. New York, NY: Facts on File.
19. Center for Defense Information. 1982. President Reagan's Civil Defense Program. *The Defense Monitor*, 11(5), 1–8.
20. Civil Defense: The Johnson Administration. 2003. University of Richmond. https://facultystaff.richmond.edu/~wgreen/Ecdjohnson.htm (accessed December 20, 2012).
21. Civil Defense: The Kennedy Administration. 2003. University of Richmond. https://facultystaff.richmond.edu/~wgreen/Ecdkennedy.htm (accessed December 20, 2012).
22. Civil Defense: The Nixon Administration. 2003. University of Richmond. https://facultystaff.richmond.edu/~wgreen/Ecdnixon.htm (accessed December 20, 2012).
23. Civil Defense: The Reagan Administration. 2003. University of Richmond. https://facultystaff.richmond.edu/~wgreen/Ecdreagan.htm (accessed December 20, 2012).
24. Clark, G. 1985. *Judges and the cities: Interpreting local autonomy*, 172. Chicago, IL: The University of Chicago Press.
25. Coulter, M. 1975. *The South during reconstruction: 1865–1877*. Vol. 8. Baton Rouge, LA: Louisiana State University Press.
26. Department of Homeland Security. Critical Infrastructure Sectors; Homeland Security Presidential Directive 7 (HSPD-7). http://www.dhs.gov/critical-infrastructure-sectors (accessed January 29, 2013).
27. Department of Homeland Security. 2009. National Infrastructure Protection Plan. http://www.dhs.gov/national-infrastructure-protection-plan (accessed January 29, 2013).

28. Dorwart, J. 2008. *Invasion and insurrection: Security, defense, and war in the Delaware Valley 1621–1815*, 188. Danvers, MA: Rosewood Publishing.
29. Dowrick, D. 2009. *Earthquake resistant design and risk reduction*. 2nd ed. Hoboken, NJ: Wiley and Sons Publishing.
30. Eldredge, N. 2002. *Life on Earth: An encyclopedia of biodiversity, ecology, and evolution*, 354. Santa Barbara, CA: ABC-CLIO, Inc.
31. Else, D. 2009. *Defense production act: Purpose and scope*, 5. Washington, DC: Congressional Research Service.
32. Emergency. 2012. *Merriam-Webster Dictionary*. http://www.merriam-webster.com/dictionary/emergency (accessed December 19, 2012).
33. Executive Order 10427. 1953. Harry S. Truman Library and Museum. http://trumanlibrary.org/executiveorders/index.php?pid=669 (accessed December 21, 2012).
34. Federal Emergency Management Agency. 2003. *A citizen's guide to disaster assistance*, 3–2. Washington, DC: Emergency Management Institute, U.S. Government.
35. Federal Emergency Management Agency. 2012. About the agency. http://www.fema.gov/about (accessed December 22, 2012).
36. Fradin, J. and D. Fradin. 2010. *Hurricane Katrina*. Tarrytown, NY: Marshall Cavendish Benchmark.
37. Framingham, J. and M. Teasley. 2012. *Behavioral health response to disasters*, 39. Boca Raton, FL: CRC Press.
38. Freudenberg, N., S. Klitzman, and S. Saegert. 2009. *Urban health and society: Interdisciplinary approaches to research and practice*, 225. San Francisco, CA: Jossey-Bass Publishing.
39. Greene, C. and S. Kelly. 2000. *Through a night of horror: Voices from the 1900 Galveston storm*. Galveston, TX: Rosenberg Library.
40. Hardenstein, E. 1879. *The epidemic of 1878 and its homeopathic treatment: A general history of the origin, progress, and end of the plague in the Mississippi Valley*, 53-54. New Orleans, LA: J.S. Rivers Publishing.
41. History + Evolution of Emergency Management. 2012. Providence Rhode Island Emergency Management Agency and Office of Homeland Security. http://www.providenceri.com/PEMA/about/history-evolution-of-emergency-management (accessed December 19, 2012).
42. Hobhouse, H. 2005. *Seeds of change: Six plants that transformed mankind*, 232. Washington, DC: Shoemaker and Hoard.
43. Homeland Security National Preparedness Task Force. 2006. *Security: A short history of national preparedness efforts*, 5. Washington, DC: U.S. Government.
44. Katz, R. 2013. *Essentials of public health preparedness*, 48. Burlington, MA: Jones and Bartlett Publishing.
45. Kennedy, D. and L. Cohen. 2013. *The American Pageant: Since 1865*. Vol. 2, 469. Boston, MA: Wadsworth.

46. Lauber, P. 1993. *Volcano: The eruption and healing of Mount St. Helens.* New York, NY: Aladdin.
47. Lawson, R. 2003. *Portsmouth: An old town by the sea*, 109. Charleston, SC: Arcadia Publishing.
48. Lincove, D. 2000. *Reconstruction in the United States: An annotated bibliography*, 113. Westport, CT: Greenwood Press.
49. Mays, D. 2004. *Women in early America: Struggle, survival, and freedom in a new world*, 106. Santa Barbara, CA: ABC-CLIO Publishing.
50. Menzel, D. and H. White. 2011. *The state of public administration: Issues, challenges, and opportunities*, 215. Armonk, NY: M.E. Sharpe.
51. Miller, D. and J. Rivera. 2011. *Comparative emergency management: Examining global and regional responses to disasters*, 6. Boca Raton, FL: CRC Press.
52. Mirsky, R. 2005. Trekking through that valley of death—The Defense Production Act. *Innovation: America's Journal of Technology Commercialization.* http://www.innovation-america.org/trekking-through-valley-death-defense-production-act (accessed December 21, 2012).
53. Muschert, G. and J. Sumiala. 2012. *School shootings: Mediatized violence in a global age.* Bingley, UK: Emerald Group Publishing.
54. National Academy of Public Administration. 1993. *Coping with catastrophe: Building an emergency management system to meet people's needs in natural and manmade disasters*, 11. Washington, DC: *National Academy of Public Administration*, U.S. Government.
55. National Climatic Data Center. 2005. Hurricane Katrina. National Oceanic and Atmospheric Administration. http://www.ncdc.noaa.gov/special-reports/katrina.html (accessed December 19, 2012).
56. National Fire Protection Association. 2010. *Fire officer: Principles and practice*, 8. Sudbury, MA: Jones and Bartlett Publishers.
57. Northrup, C. 2003. *The American economy: A historical encyclopedia*, 84. Santa Barbara, CA: ABC-CLIO Publishing.
58. Peacock, J. 2003. *Reconstruction: Rebuilding after the Civil War*, 36. Mankato, MN: Capstone Press.
59. Pender, G. and H. Faulkner. 2011. *Flood risk science and management*, 500. West Sussex, UK: Blackwell Publishing.
60. Portsmouth New Hampshire Fire Department. 2012. Portsmouth fire: A chronological history of a New England seaport fire department. http://www.cityofportsmouth.com/fires/history.htm (accessed December 19, 2012).
61. Post Katrina Emergency Management Reform Act of 2006 Law & Legal Definition. 2012. U.S. Legal. http://definitions.uslegal.com/p/post-katrina-emergency-management-reform-act-of-2006/ (accessed December 19, 2012).
62. Purpura, P. 2011. *Security: An introduction*, 75. Boca Raton, FL: CRC Press.
63. Robertson, R. 2001. *Rotting face: Smallpox and the American Indian.* Caldwell, ID: Caxton Press.

64. Rogers, K. 2011. *Debunking Glenn Beck: How to save America from media pundits and propagandists*, 38–39. Santa Barbara, CA: ABC-CLIO Publishing.
65. Rosenberg, C. 1987. *The cholera years: The United States in 1832, 1849, and 1866.* Chicago, IL: The University of Chicago Press.
66. Saarinen, T., V. Baker, R. Durrenberger, and T. Maddock. 1984. *The Tucson, Arizona Flood of 1983*. Washington, DC: National Academy Press.
67. Schweikart, L. and M. Allen. 2004. *A patriot's history of the United States: From Columbus's great discovery to the war on terror*. New York, NY: Penguin Group.
68. Simpson, D. 2006. *9/11: The culture of commemoration*. Chicago, IL: The University of Chicago Press.
69. Smith, T. n.d. Emergency quotes. *Brainyquote.com*. http://www.brainyquote.com/quotes/keywords/emergency_3.html (accessed December 18, 2012).
70. Spingola, D. 2011. *The ruling elite: A study in imperialism, genocide, and emancipation*, 273. Bloomington, IN: Trafford Publishing.
71. Texas State Historical Association. 2012. Galveston's response to the hurricane of 1900. http://www.texasalmanac.com/topics/history/galvestons-response-hurricane-1900 (accessed December 14, 2012).
72. The Great Fever. 2012. Mississippi Public Broadcasting – Public Broadcasting Service. http://www.pbs.org/wgbh/amex/fever/peopleevents/e_1878.html (accessed December 17, 2012).
73. Trefousse, H. 1991. *Historical dictionary of reconstruction*, 49. Westport, CT: Greenwood Press.
74. U.S. Geological Survey. 2012. Massachusetts earthquake history. http://earthquake.usgs.gov/earthquakes/states/massachusetts/history.php (accessed November 19, 2012).
75. U.S. Geological Survey. 2012. Pennsylvania earthquake history. http://earthquake.usgs.gov/earthquakes/states/pennsylvania/history.php (accessed June 9, 2014).
76. U.S. Nuclear Regulatory Commission. 2013. History. http://www.nrc.gov/about-nrc/emerg-preparedness/history.html (accessed June 9, 2013).
77. USA Patriot Act (H.R. 3162) (October 24, 2001). EPIC—Electronic Privacy Information Center. http://epic.org/privacy/terrorism/hr3162.html (accessed January 29, 2013).
78. Veenema, T. 2013. *Disaster nursing and emergency preparedness for chemical, biological, and radiological terrorism and other hazards*, 4. New York, NY: Springer.
79. Webley, K. 2011. Mississippi River, 1927. *Time Magazine*. http://www.time.com/time/specials/packages/article/0,28804,2070796_2070798_2070780,00.html (accessed December 20, 2012).
80. White, R. 1991. *It's your misfortune and none of my own: A new history of the American West*, 199. Norman, OK: The University of Oklahoma Press.
81. White, S., R. Bingham, and E. Hill. 2003. *Financing economic development in the 21st century*, 34. Armonk, NY: M.E. Sharpe.

82. Wilson, J. 2001. The state of emergency management 2000: The process of emergency management professionalization in the United States and Florida. Doctoral diss., Florida International University, Miami, FL.
83. WWI Timeline. 2012. Mississippi Public Broadcasting—Public Broadcasting Service. http://www.pbs.org/greatwar/timeline/time_1917.html (accessed December 19, 2012).

PRINCIPLE HAZARDS FACING THE UNITED STATES

"The likelihood of a major natural disaster, flood, hurricane, or earthquake, affecting our communities was inevitable."

George Haddow and Jane Bullock,
Institute for Crisis, Disaster, and Risk Management

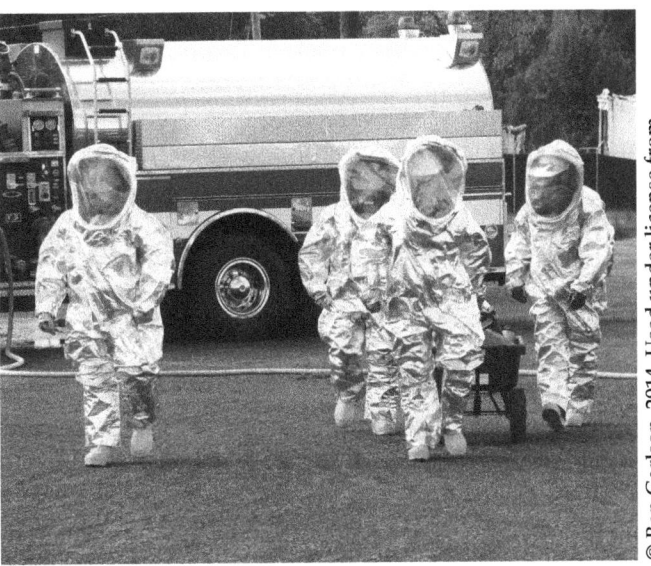

Figure 3.1 Both man-made and natural disasters necessitate a response.

Objectives

The objectives of this chapter are to:
- review historical incidents involving man-made and natural disasters;
- gain an overview of possible man-made and natural threats during modern times;
- define and identify man-made threats;
- define and identify natural threats; and
- understand the role of homeland security and emergency management regarding these threats.

Introduction

Nations around the globe, including the United States, face a wide range of national safety and security threats. Some of these threats result from naturally occurring events, such as severe weather, or earthquakes, that can quickly devastate a community or a region. Other threats are the result of the acts of man, some of which are intentional, such as the attacks of September 11, 2001, or accidental, such as the 2010 oil spill along the Mississippi Gulf Coast or the chemical disaster that occurred in Bhopal, India, in 1984. Figure 3.1 shows a hazardous materials response that might occur regarding such incidents.

The United States is one of the most open nations in the world. Sharing an international border that stretches 8,000 miles with Canada to the north and Mexico to the south, the United States also has over 13,000 miles of open coastal waters, power grids, ports, rail stations, airports, and river bridges, and each is a potential target.

Overview of Domestic Safety and Security

The lines between domestic and international security for the United States are unclear and indistinct. On one end of the spectrum, the Department of Defense operates within the executive branch of government under the command of the President of the United States. On the other end of the spectrum, local communities rely on the services of their state and local law enforcement, fire and emergency services, and corrections officials. In the middle are agencies including the Federal Bureau of Investigation, Drug Enforcement Administration, and the Central Intelligence Agency. Although recovery from disaster is unique to each community, depending on the amount and type of damage caused by the disaster and the resources that the community has ready or can receive, state governments have the legal authorities for emergency response and recovery and serve as the point of contact between local and federal governments.

With the end of the cold war, the United States found itself as the world's only remaining superpower, but challenged on multiple fronts: economically, diplomatically, scientifically, and militarily. The threats of the cold war were replaced with a wide variety of issues: illicit drug production, trafficking, and consumption; illegal immigration; and global terrorism and natural and accidental disasters. Various nations became battlefronts, centers of drug production, or havens for those deemed as terrorists. Natural and accidental events resulted in great damage and impacted the lives of citizens across multiple states.

As a response to these threats, National Guard personnel found themselves deployed to support border security initiatives. The Coast Guard and Navy responded to the oil spills, the Drug Enforcement Administration found its global mission expanded (as did most other federal law enforcement agencies), and states established emergency management and homeland security organizations. States challenged the national government on immigration. This was the world in which the United States faced new dangers and, to respond, new solutions had to be developed. One of the national responses to these evolving threats and challenges was the development of the Department of Homeland Security.

Modern Threats, Hazards, and Challenges

Emergency management planning today faces many critical obstacles, such as an imbalance of focus between homeland security and natural disaster management, the challenge of involving the public in preparedness planning, the lack of an effective partnership with the business community, cuts to emergency management (EM) funding, and questions surrounding the evolving organizational structure of the nation's emergency management system. Such obstacles need to be overcome if emergency management activities are to be successful in the years ahead.[1]

Natural Disasters

Natural disasters that have impacted the United States include, but are not limited to, drought and subsequent dust storms; earthquakes; extreme heat; fires, floods, landsides, and debris flow; severe weather and storms, to include hurricanes, and tornadoes, tsunamis, winter storms, and extreme cold; and diseases. Though some events have caused extensive property damage, rarely has the United States experienced extensive loss of life as a result of a natural disaster.

---

[1] George Haddow, Jane Bullock, and Damon Coppola, *The Future of Emergency Management* (Burlington, MA: Butterworth-Heinemann, 2011).

Incidents such as the New Madrid Earthquake of 1811–1812, the Memphis Yellow Fever Epidemic of the 1870s, the Johnstown Flood of 1889, the Galveston Hurricane of 1900, and the San Francisco Earthquake and Fire of 1906 are several major incidents that impacted the nation and overwhelmed local capabilities. Among each of these cases, with the exception of the New Madrid Earthquake, loss of life and extensive property damage occurred and exceeded the capabilities of local resources to respond.

Natural threats and hazards are continuous and may involve localized events as well as events that impact the entirety of the nation. Not all dangers result from human origins. Nature is full of deadly surprises and harmful calamities. One need to only review news articles to learn the details of natural disasters that impact the nation throughout most on any given year. Natural disasters are defined as follows:

> A natural disaster is a serious disruption to a community or region caused by the impact of a naturally occurring rapid onset event that threatens or causes death, injury or damage to property or the environment and which requires significant and coordinated multi-agency and community response. Such serious disruption can be caused by any one, or a combination, of the following natural hazards: bushfire; earthquake; flood; storm; cyclone; storm surge; landslide; tsunami; meteorite strike; or tornado."[2]

Certainly, many more natural events may be listed within the context of this definition. Regardless of the type of natural disaster, society is imperiled with loss of life occurring in the worst cases. Natural disasters have the potential of disrupting supply lines and logistics, impacting economic functioning and agricultural functions.

Drought

Drought is defined as "a period of dryness especially when prolonged" and a period that "causes extensive damage to crops or prevents their successful growth."[3] Drought has often affected the United States throughout its history locally, regionally, and nationally. During the last century, three of the worst droughts were the Dust Bowl of 1930, the 1950s drought, and the drought between the years 1987 and 1989.[4] During the 1930s, drought conditions were so severe that "soil, depleted of

---

[2]Australian Emergency Management, "Natural Disasters in Australia: Reforming Mitigation, Relief, and Recovery Arrangements," 2002, http://www.em.gov.au/Documents/Natural%20Disasters%20in%20Australia%20-%20Review.pdf (accessed February 11, 2013).

[3]"Drought," 2013, *Merriam-Webster Dictionary*, http://www.merriam-webster.com/dictionary/drought (accessed February 12, 2013).

[4]National Oceanic and Atmospheric Administration, "North American Drought: A Paleo Perspective," 2013, http://www.ncdc.noaa.gov/paleo/drought/drght_history.html (accessed February 12, 2013).

moisture, was lifted by the wind into great clouds of dust and sand which were so thick they concealed the sun for several days at a time."[5] The 1950s drought resulted in crop yields in some areas being reduced by approximately 50%.[6] The 1980s drought was severe and represented one of the most expensive natural disasters in the U.S. history. Cumulatively, the costs of the 1980s drought were approximately $39 billion.[7]

At the turn of the twenty-first century, China found itself suffering from a rapid deterioration of cropland resulting in part from decades of programs instituted to increase agricultural output and an expansion of urbanization and industrial development. The urbanization and industrial development contributed to the reduction in forestation and other vegetation that once provided moisture to the region. Additionally, three decades of satellite imagery from those areas of northern China have revealed the loss of thousands of lakes due to the reduction in annual rainfall and increased demands on the available water.[8]

During modern times, drought is a natural force that impacts communities across the nation. Figure 3.2 shows drought conditions for 2012 throughout the United States.

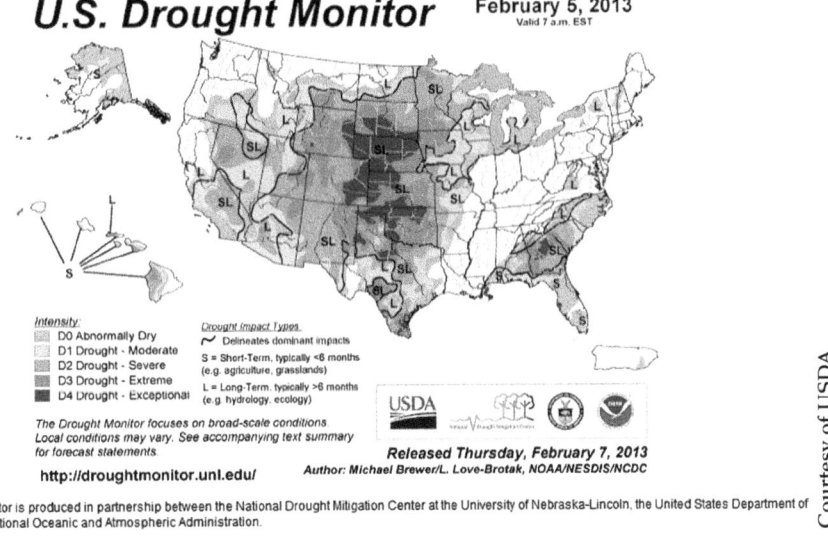

Figure 3.2 Drought monitor.

---

[5]Ibid.

[6]Ibid.

[7]Ibid.

[8]Reggie Royston, "China's Dust Storms Raise Fears of Impending Catastrophe," *National Geographic*, 2010, http://news.nationalgeographic.com/news/2001/06/0601_chinadust.html (accessed June 28, 2013).

Heat Wave

A heat wave represents a "prolonged period of excessive heat, often combined with excessive humidity."[9] During such conditions, temperatures may "hover 10 degrees or more above the average high temperature for the region and last for prolonged periods of time."[10] Heat waves are extremely dangerous and are often associated with loss of life. Within the United States, in average terms, events of "excessive heat" are responsible for claiming "more lives each year than floods, lightning, tornadoes and hurricanes combined."[11] Across the nation, during the 1980 heat wave, approximately 1,250 people perished with 700 victims alone in the city of Chicago.[12] During 2003, an August heat wave resulted in approximately 50,000 deaths.[13] Heat waves affect all segments of the nation. Figure 3.3 shows an example of the scopes of heat waves that often occur across the United States.

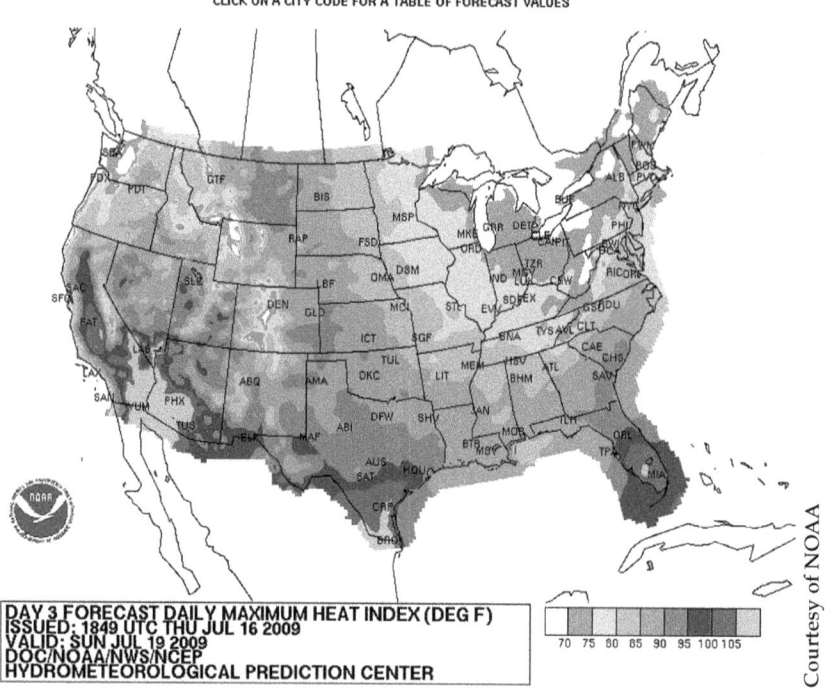

Figure 3.3 Heat index.

---

[9]Federal Emergency Management Agency, "Community Emergency Response Team Appendix 1-A: Hazard Lesson Plans," 2013, training.fema.gov/EMIWeb/downloads/04-Heat-PM-Rev2.doc (accessed February 12, 2013).

[10]Ibid.

[11]National Oceanic and Atmospheric Administration, "Heat: A Major Killer," 2013, http://www.nws.noaa.gov/os/heat/index.shtml (accessed February 12, 2013).

[12]Ibid.

[13]Ibid.

Tsunami

The word "tsunami" is derived from the Japanese words "tsu, meaning harbor, and nami, meaning wave."[14] A tsunami is generally caused by landslides or earthquakes beneath the ocean.[15] Although many individuals may think of Asian disasters involving large ocean waves, the United States is not immune to the dangers of tsunamis. The states of Alaska and Hawaii have felt the impacts of tsunamis as has the U.S. West Coast.[16] The deadliest tsunami to impact the United States occurred in 1964 off the coast of Alaska and resulted in many casualties.[17]

Earthquake

Earthquakes are quick displacements of "land/rock mass, which typically occurs along a fault line" that results in the forming of "seismic waves from the origin outward" and the releasing of energy.[18] Earthquakes are often expressed according to the characteristics of magnitude and intensity.[19] Incidences of earthquakes represent approximately "20% of insured catastrophic losses" and approximately one-third of "economic losses" that result from natural hazards.[20] Figure 3.4 presents an example of earthquake risk for the western United States.

Several parts of the United States rest upon fault lines that are prone to earthquakes. These areas include California, Alaska, and the central part of the nation running north and south along the Mississippi River. It was along the Mississippi River, during the period of 1811–1812, that a series of earthquakes, now known as the New Madrid Earthquakes, occurred. This fault line between St Louis, Missouri, and Memphis, Tennessee, was sparsely populated. As a result, it inflicted little personal injury, but did cause property damage including the destruction of the town of New Madrid, Missouri on February 12, 1812. In a letter dated January 13, 1814, Missouri Territorial Governor William Clark asked for federal relief for the inhabitants of New Madrid County. This was one of the earliest, if not the first, examples of a request for disaster relief from the United States federal government.

---

[14]Ibid.

[15]Eric Geist, Paul Earle, and Jill McCarthy, "Could it Happen Here? Tsunamis That Have Struck U.S. Coastlines," 2005, http://soundwaves.usgs.gov/2005/01/fieldwork2.html (accessed February 12, 2013).

[16]Ibid.

[17]Ibid.

[18]Eric Banks, *Catastrophic Risk Management: Analysis and Management* (West Sussex, England: John Wiley & Sons Publishing, 2005), 19.

[19]U.S. Geological Survey, "The Severity of an Earthquake," 2013, http://pubs.usgs.gov/gip/earthq4/severitygip.html (accessed February 12, 2013).

[20]Carlos Oliveira, Antoni Roca, and Xavier Goula, *Assessing and Managing Earthquake Risk: Geo-Scientific and Engineering Knowledge for Earthquake Risk Mitigation: Developments, Tools, and Techniques* (Dordrecht, The Netherlands: Springer Publishing, 2008), 388.

76 Foundations of Emergency Management

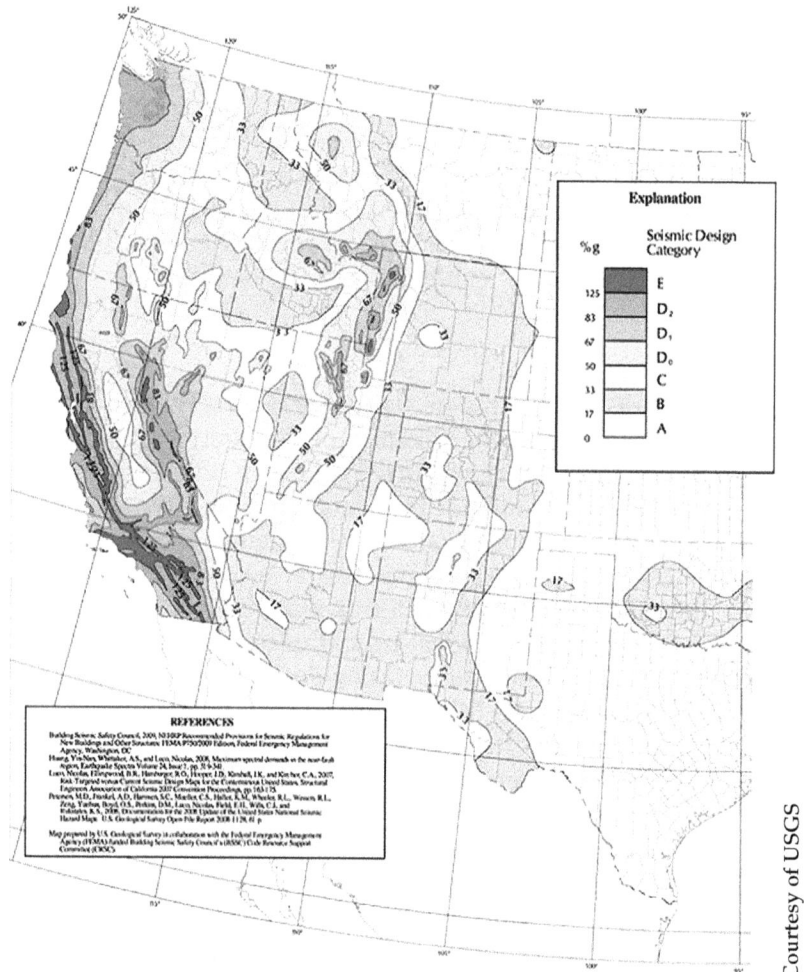

Figure 3.4 Example of earthquake regional threat.

This incident is not the largest recorded earthquake to strike the United States. Such an event was the 1964 Prince William Sound, Alaska earthquake and ensuing tsunami that caused $300 million in property damages and claimed around 128 lives. Earthquakes represent a silent danger to many areas within the nation. Figure 3.4 shows a graphical display of regional earthquake threats.

Volcano

Though most Americans do not consider the possibility of a volcano erupting to be a major threat, the reality is that volcanic activity is ongoing. A volcano is defined as "a vent in the crust of the earth or another planet or a moon from which usually

CHAPTER 3 Principle Hazards Facing the United States

Figure 3.5 Mount St. Helens.

molten or hot rock and steam issue" or "a hill or mountain composed wholly or in part of the ejected material."[21] Most volcanoes exist at the boundaries of he Earth's plates. It is estimated that about 550 volcanoes have erupted on the Earth's surface since recorded history.

History is full of examples of the impact of volcanic action upon communities. The famous eruption of Mount Vesuvius in 79 AD buried the towns of Pompeii and Herculaneum and resulted in an estimated loss of over 300 lives. More recently, the 1815 volcanic eruption in Tambora, Indonesia, resulted in the death of as many as 92,000. In 1883, a tsunami in Krakatau, Indonesia, killed an estimated 36,000 individuals while the 1902 eruption of Mount Pelee in Martinque may have resulted in the deaths of over 29,000.[22]

Notably, within the United States, a significant volcanic eruption occurred during 1980 when Mount St. Helens erupted in the state of Washington. This event was "the most studied volcanic eruption of the twentieth century."[23] Figure 3.5 shows the erupting of Mount St. Helens.

Tornado

Tornadoes are common throughout the United States. It is estimated that as many as 1,000 tornadoes annually occur in the nation.[24] A tornado is defined as "a violently

---

[21]"Volcano," 2013, *Merriam-Webster Dictionary*, http://www.merriam-webster.com/dictionary/volcano (accessed February 12, 2013).

[22]"Famous Volcanic Eruptions," 2013, University of Bristol, http://www.chm.bris.ac.uk/webprojects2003/silvester/Page6Famous.htm (accessed June 28, 2013).

[23]"Mt. St. Helens Eruption (1980)," 2013, San Diego State University, http://www.geology.sdsu.edu/how_volcanoes_work/Sthelens.html (accessed February 13, 2013).

[24]Ibid.

rotating column of air extending between, and in contact with, a cloud and the surface of the earth."[25] Tornadoes are usually associated with thunderstorms and may involve "speeds that exceed 200 mph" and that may "approach 300 mph" in "extreme cases."[26]

For homeland security and emergency response personnel, tornadoes always pose a great challenge. Often quickly forming and with the potential to both move rapidly and change course, large tornadoes can create a wide path of destruction over a vast area. As a result, search and rescue operations may be hampered not only by poor weather but also by downed power lines, trees, and other various debris, including wreckage of buildings and vehicles.

The deadliest tornado in world history was the Daultipur-Salturia Tornado in Bangladesh on April 26, 1989, which killed approximately 1,300 people.

Blizzard

According to the National Weather Service, a blizzard occurs when the following conditions are observed over a period of three or more hours: "sustained wind or frequent gusts to 35 miles an hour or greater; and considerable falling and/or blowing snow (i.e., reducing visibility frequently to less than ¼ mile)."[27] Blizzards occur throughout various geographic segments within the United States. During 2013, a blizzard occurred among the New England states that deposited well over 2 ft of snow and resulted in death and destruction of and calamity for many people and organizations.[28] This event caused the canceling of approximately "5,000 flights," and "knocked out power to more than 635,000 customers."[29] It also necessitated the evacuating of "coastal regions" and the suspending of postal service in "seven states."[30]

Fire

Throughout the history of the United States, the nation has faced the challenges and dangers of fire. In 2011, there were 1,389,500 fires reported in the United States. These fires caused 3,005 civilian deaths, 17,500 civilian injuries, and $11.7 billion in

---

[25]The Weather Channel, "Tornadoes: Definition & Prevalence," 2013, http://www.weather.com/outlook/wxready/articles/id-54 (accessed February 13, 2013).

[26]Ibid.

[27]National Weather Service, "Blizzard," 2013, http://w1.weather.gov/glossary/index.php?letter=b (accessed February 12, 2013).

[28]"PHOTOS: Blizzard 2013 Blasts Through Northeast," *Fox News Insider*, 2013, http://foxnewsinsider.com/2013/02/09/photos-blizzard-northeast-nemo-2013/ (accessed February 12, 2013).

[29]Mary Snow, Dana Ford, and David Ariosto, "Northeast Digs Out After Deadly Blizzard," *CNN News*, 2013 (accessed February 11, 2013).

[30]Ibid.

property damage. Of those, an estimated 484,500 were structure fires, causing 2,640 civilian deaths, 15,635 civilian injuries, and $9.7 billion in property damage.[31]

Any region of the United States is susceptible to incidents of fire. In 1871, an estimated 1,500–2,500 people were killed as a result of what became known as the Peshtigo Fire. Occurring the same day as the Great Chicago Fire, it burned over 1,875 square miles and destroyed 12 communities.[32] Regarding property damage and loss, the National Fire Protection Association identifies the Oakland Fire Storm of 1991 as the costliest wild land fire in the United States with damage estimated at over $1.5 billion.[33]

Even today, with the resources available to combat major fires, they still pose great challenges to emergency management professionals. During 2012, the U.S. wildfire season was so extreme that "the U.S. Department of Agriculture Forest Service ran out of money to pay for firefighters, fire trucks and aircraft that dump retardant on monstrous flames."[34] This funding situation necessitated the transferring of funds from fire prevention programs to satisfy outlying costs.[35] Eventually, Congressional intervention was necessary to generate approximately $400 million toward fire prevention.[36]

The United States is not the only nation that has experienced the impact of uncontrolled fires; in 1987, the Black Dragon fire in China burned a total of 18 million acres of forest along the Amur River.

Flood

Flash floods occur when excessively large quantities of rain occur "within a short period of time," generally less than 6 h, thereby causing water levels to increase and decrease quickly.[37] Flooding also occurs when mass amount of water, originating

---

[31]"Deadliest/Large-Loss Fire," 2012. National Fire Protection Association, http://www.nfpa.org/itemDetail.asp?categoryID=954&itemID=44745&URL=Research/Fire%20statistics/Deadliest/large-loss%20fires (accessed March 30, 2013)

[32]Stephanie Hemphill, "Peshtigo: A Tornado of Fire Revisited," *Minnesota Public Radio,* 2002, http://news.minnesota.publicradio.org/features/200211/27_hemphills_peshtigofire/ (accessed March 30, 2013).

[33]"Deadliest/large-loss Fire," 2012.

[34]Darryl Fears, "U.S. Runs Out of Funds to Battle Wildfires," 2012, http://articles.washingtonpost.com/2012-10-07/national/35500576_1_battle-wildfires-wildfire-suppression-worst-wildfire-season (accessed February 13, 2013).

[35]Ibid.

[36]Ibid.

[37]"Flash Flood," 2013, U.S. Geological Survey, http://ks.water.usgs.gov/waterwatch/flood/definition.html (accessed February 12, 2013).

from "a river or other body of water" either causes damage or represents a threat.[38] Flooding impacts many regions throughout the United States annually.

Some of the major flooding that has impacted the nation include the 1913 Ohio Flood that resulted in over 450 deaths and over $100 million in property damage and the 1927 Mississippi River Flood that impacted States from Missouri down to Louisiana had sparked a national level debate on finding solutions to combat flooding.[39]

Hail Storms

Hail is defined as "precipitation in the form of small balls or lumps usually consisting of concentric layers of clear ice and compact snow."[40] Therefore, a hailstorm represents "a storm accompanied by hail."[41] Damages from hail are not insignificant both economically and financially. Hail may fall to the earth with speeds up to 100 mph and may also "contain foreign matter, such as pebbles, leaves, twigs, nuts, and insects" within their icy content.[42] Annually, hail inflicts approximately $1 billion in damages to both crops and property.[43]

Dust Storms

A dust storm is an event that may arise surprisingly and without warning, and generally has the appearance of an "advancing wall of dust and debris which may be miles long and several thousand feet high."[44] Dust storms are quite dangerous even though they may last only for a short period of time. The conditions of dust storms often include blinded or extremely limited visibility, thereby increasing the dangers of travel. Figure 3.6 shows a dust storm encompassing Phoenix, Arizona.

---

[38] "Flood," 2013, U.S. Geological Survey, http://ks.water.usgs.gov/waterwatch/flood/definition.html (accessed February 12, 2013).

[39] Charles Perry, "Significant Floods in the United States During the 20th Century—USGS Measures a Century of Floods," *U.S. Geological Survey*, 2000, http://ks.water.usgs.gov/pubs/fact-sheets/fs.024-00.html#HDR1 (accessed March 30, 2013).

[40] "Hail," 2013, *Merriam-Webster Dictionary*, http://www.merriam-webster.com/dictionary/hail (accessed February 12, 2013).

[41] "Hail Storm," 2013, *Merriam-Webster Dictionary*, http://www.merriam-webster.com/dictionary/hail%20storm (accessed February 12, 2013).

[42] "Largest Hailstone in U.S. History Found," *National Geographic*, 2013, http://news.nationalgeographic.com/news/2003/08/0804_030804_largesthailstone.html (accessed February 13, 2013).

[43] Ibid.

[44] National Oceanic and Atmospheric Administration, "Motorist Beware!," 2013, http://www.nws.noaa.gov/om/brochures/duststrm.htm (accessed February 12, 2013).

Figure 3.6 Dust storm.

Landslides

A landslide represents the mass wasting of materials. Mass wasting is defined as the "down slope movement of soil and/or rock under the influence of gravity."[45] This movement occurs because of slope failure involving a moment in which "gravity exceeds the strength of the earth materials."[46] Notably, landslides have often occurred in the western United States where residents have endured much damage, including the loss of homes and property. During 2011, in the state of California, entire sections of roadways slid into the Pacific Ocean because of a landslide.[47]

Sinkholes

Sinkholes are "depressions or holes in the land surface" which result from "the dissolving of the underlying limestone."[48] Sinkholes may occur in open areas as well

---

[45]"What is a Landslide?" 2013, University of Wisconsin, http://www.geology.wisc.edu/courses/g115/projects03/emgoltz/definition.htm (accessed February 13, 2013).
[46]Ibid.
[47]"California Road Slides into Ocean After Landslide," *BBC News*, 2011, http://www.bbc.co.uk/news/world-us-canada-15819302 (accessed February 12, 2013).
[48]State of Florida, "Sinkholes," 2013, http://www.swfwmd.state.fl.us/hydrology/sinkholes/ (accessed February 12, 2013).

Figure 3.7 Sinkhole.

as underneath structures or roadways. The damages that result from sinkholes may be both excessive and costly.[49] Within the state of Florida, many lakes are actually "relic sinkholes."[50] Figure 3.7 shows a typical sinkhole.

Hurricanes and Cyclones

Cyclones are analogous with rotating storms (e.g., hurricanes, tornadoes). A cyclone is defined as "a storm or system of winds that rotates about a center of low atmospheric pressure, advances at a speed of 20 to 30 miles (about 30 to 50 kilometers) per hour, and often brings heavy rain."[51] During the first decade of the twenty-first century, the most notable example of a cyclone storm was Hurricane Katrina.

The deadliest natural disaster to impact the United States was the Galveston Hurricane of 1900. Striking on September 8, 1900, it caused between 6,000 and 12,000 deaths and devastated the city, destroying approximately 3,600 homes and buildings.[52] The 1926 Miami hurricane is considered the costliest hurricane to strike the United States.[53]

During 2012, Hurricane Sandy devastated the eastern seaboard of the United States and caused significant destruction among regions of New York and New

---

[49]Ibid.

[50]Ibid.

[51]"Cyclone," 2013, *Merriam-Webster Dictionary*, http://www.merriam-webster.com/dictionary/cyclone (accessed February 12, 2013).

[52]Amanda Ripley, "The 1900 Galveston Hurricane," *Time*, 2008, http://www.time.com/time/nation/article/0,8599,1841442,00.html (accessed April 1, 2013).

[53]Jay Barnes, *Florida's Hurricane History* (Chapel Hill, NC: The University of North Carolina Press, 2007).

Jersey that rarely experience major hurricane activity.[54] This storm "killed more than 100 people, destroyed whole communities in coastal New York and New Jersey, left tens of thousands homeless, crippled mass transit, triggered paralyzing gas shortages, inflicted billions of dollars in infrastructure damage and cut power to more than 8 million homes, some of which remained dark for weeks."[55] This event impaired the regional economy and impacted numerous organizations around the nation that had relationships with the impacted area.

Whirlpool

A whirlpool is a "swirling body of water usually produced by ocean tides," and more powerful whirlpools are known as maelstroms.[56] During the fifteenth century, experts speculate that a Malian fleet of ships was destroyed by a whirlpool while attempting to explore the Atlantic Ocean.[57] Out of the 200 vessels that were deployed, only one returned.[58]

Limnic Eruption

Limnic eruptions occur rarely. However, they represent situations in which carbon dioxide erupts suddenly from "water, posing the threat of suffocating wildlife, livestock, and humans."[59] The conditions that generate these events are rare.[60] During 1986, in the Cameroon region, a limnic eruption occurred that expelled approximately "1.6 million tons" of carbon dioxide from a lake at a speed of roughly 60 mph. The resulting cloud "displaced all the oxygen in several small villages, suffocating between 1,700 and 1,800 people, not counting all their livestock."

Major Accidents

The emergency management community must be prepared to respond to a wide range of accidents, both major and minor. Over the past several decades, major events such as the tragic 1984 Union Carbide industrial accident in Bhopal, India; the 1986 disaster at the Chernobyl Nuclear Power Plant in the Soviet Union; the 1989 Exxon Valdez oil spill in Prince William Sound, Alaska; the Deepwater Horizon oil

---

[54]"Hurricane Sandy: Covering the Storm," *The New York Times,* 2012, http://www.nytimes.com/interactive/2012/10/28/nyregion/hurricane-sandy.html (accessed February 11, 2013).

[55]Ibid.

[56]S.B. Bhagwat, *Foundations of Geology* (New Delhi, India: New Vision Publishing House, 2009), 532.

[57]Ed Wright, *Lost Explorers: Adventurers that Disappeared Off the Face of the Earth* (St Leonards NSW, Australia: Murdoch Books, 2008), 11.

[58]Ibid.

[59]Bhagwat, *Foundations of Geology,* 2009.

[60]Jamie Frater, *Ultimate Book of Bizarre Lists* (Berkeley, CA: Ulysses Press, 2010), 52.

spill in 2010; and the 2011 Fukushima Daiichi Nuclear Disaster have demonstrated the potential scope of an accidental disaster. Each of these events presented unique and very grave challenges to responders.

Triangle Shirtwaist Factory of 1911

One of the earliest major accidents in the U.S. history occurred on March 25, 1911, when a fire broke out on the 8th floor of the Triangle Shirtwaist Company located in New York City. Employing approximately 500 workers, the company was plagued by low pay and difficult working conditions.

When fire and rescue personnel arrived on scene, they found their ladders were too short to reach trapped workers, most of whom were women, on the 8th and 9th floors. Workers found that the main stairwell was too engulfed in smoke and fire to exit and the door to the only other exit door was locked, trapping the workers. About 20 escaped down the outside fire escape before it collapsed, killing several. Some tried to escape using the elevators, until they quit, trapping other workers. Sixty-two workers jumped from the 8th and 9th floors to their death on the street below. The fire resulted in the deaths of 146 young women. Prior to the terrorist attacks of 9/11, the Triangle Shirtwaist fire was the largest mass casualty event in New York City history. The owners were placed on trial, but no convictions resulted. Many lessons were learned from this disaster, which included regulations related to workplace safety.[61]

Texas City Explosion of 1947

The worst industrial disaster in the United States occurred on April 16, 1947, in Texas City, Texas, when the cargo of the *S.S. Grandcamp* and then the *S.S. High Flyer* exploded. The *S.S. Grandcamp* was loaded with 2,300 tons of explosive grade ammonium nitrate, the same type of explosive that would be used years later in the bombing of the Murrah Federal Building in Oklahoma City. The *S.S. High Flyer* held a cargo of over 2,000 tons of sulfur.

As the ship burned, firefighters were called to respond. As they battled the fire, onlookers gathered, including a group of schoolchildren. When the explosions occurred, all of the firefighters and practically all of the onlookers on their pier were killed as were many employees in the Monsanto Chemical Company and throughout the dock area.

There were 362 freight cars in the Texas City Terminal Co. yards, all of which were damaged and many of which were totally destroyed. Approximately 600 automobiles in the Monsanto parking lot and the dock area were practically a total

---

[61]Jennifer Rosenberg, "Triangle Shirtwaist Factory Fire," *20th Century History*, 2013, http://history1900s.about.com/od/1910s/p/trianglefire.htm (accessed March 30, 2013).

loss. It is estimated that over 500 were killed and over 2,000 injured, making the Texas City Explosion the worst industrial disaster in the United States history.[62]

1984 Union Carbide Industrial Accident

The 1984 Union Carbide industrial gas leak in Bhopal, India, is considered the world's worst industrial disaster. Bhopal is a city of approximately one million. In 1984, the plant was manufacturing Sevin at one quarter of its production capacity due to decreased demand for pesticides. At 11:00 pm on December 2, 1984, while most of the one million residents of Bhopal slept, an operator at the plant noticed a small leak of methyl isocyanate (MIC) gas and increasing pressure inside a storage tank. The vent-gas scrubber, a safety device designed to neutralize toxic discharge from the MIC system, had been turned off 3 weeks before.[63]

Over 500,000 people were exposed to methyl isocyanate gas and other chemicals. The toxic substance made its way in and around the communities located near the plant. Even though Bhopal had four major hospitals but there was a shortage of physicians and hospital beds and no plan for mass casualty emergency response. As casualties filled the hospitals, health-care professional desperately searched to identify the cause and soon linked the illness back to the Union Carbide plant. More than 20,000 people required hospital treatment for symptoms including swollen eyes, frothing at the mouth, and breathing difficulties. Similar to many accidents, the Union Carbide industrial gas leak resulted from a series of events and failures of safety measures that were expected to neutralize any potential industrial incident.

Following the disaster, environmental awareness and activism in India increased significantly. In 1986, the Indian Government enacted the Environment Protection Act, creating the Ministry of Environment and Forests and strengthening India's commitment to the environment. Under the new act, the Ministry of Environment and Forests was given overall responsibility for administering and enforcing environmental laws and policies. It established the importance of integrating environmental strategies into all industrial development plans for the country.

1986 Chernobyl Nuclear Power Plant Explosion

The Chernobyl Nuclear Power Plant disaster was a catastrophic nuclear accident that occurred on April 26, 1986, in Ukraine. The Chernobyl disaster is widely considered to have been the worst nuclear power plant accident in history and is one of the only two classified as a level 7 event on the International Nuclear Event Scale.

---

[62]Fire Prevention and Engineering Bureau of Texas, "Texas City, Texas, Disaster," 2013, http://www.local1259iaff.org/report.htm (accessed March 31, 2013).

[63]Shrivastava Paul, *Bhopal: Anatomy of a Crisis* (Cambridge, MA: Ballinger Publishing, 1987), 184.

Figure 3.8 Abandoned city of Chernobyl.

The Chernobyl accident was the result of a flawed reactor design that was operated with inadequately trained personnel. The resulting steam explosion and fires released at least 5 percent of the radioactive reactor core into the atmosphere and downwind. This resulted in the largest uncontrolled radioactive release into the environment ever recorded for any civilian operation, and large quantities of radioactive substances were released into the air for about 10 days.

The response to the fire and explosion reflected great confusion, which is often typical during major disasters. First responders had no idea of the degree of danger they faced. Two Chernobyl plant workers died on the night of the accident, and a further 28 people died within a few weeks as a result of acute radiation poisoning.

The scope of the disaster combined by the secrecy at the time in the Soviet Union further impacted operations to first stabilize the disaster and then implement steps toward recovery. Within the first few days of the disaster, the decision was made to evacuate the approximate 45,000 residents of the town of Pripyat. Later, other evacuations occurred with as many as 200,000 people being permanently displaced from their homes.[64] The legacy of Chernobyl remains as more nations look to the use of nuclear power as an alternative to other sources of traditional fuels. Figure 3.8 shows the abandoned city of Chernobyl.

1989 Exxon Valdez oil spill

On March 24, 1989, the *Exxon Valdez* spilled between 260,000 and 750,000 barrels of oil into Prince William Sound, Alaska. In response to the spill, the U.S. Congress

---

[64]"Chernobyl Accident 1986,"2013. World Nuclear Association, http://www.world-nuclear.org/info/Safety-and-Security/Safety-of-Plants/Chernobyl-Accident/#.UU8whVfyCSp (accessed March 24, 2013).

passed the Oil Pollution Act of 1990 (OPA). The legislation included a clause that prohibits any vessel that, after March 22, 1989, has caused an oil spill of more than 1,000,000 U.S. gallons in any marine area, from operating in Prince William Sound.[65] This was considered the most significant oil spill in the U.S. history until the 2010 Deepwater Horizon oil spill, which caused extensive damage to marine and wildlife habitats along the Gulf Coast of Mississippi and Louisiana.

Fukushima Daiichi Earthquake, Tsunami, and Nuclear Disaster

Though, technically a result of an act of nature, the Fukushima Daiichi earthquake, tsunami, and nuclear disaster revealed how a natural disaster could be compounded. The Fukushima Daiichi earthquake, tsunami, and nuclear disaster is the largest nuclear disaster since the Chernobyl disaster of 1986, and only the second disaster, along with Chernobyl, to measure level 7 on the International Nuclear Event Scale. There were no casualties caused by radiation exposure, though approximately 25,000 died due to the earthquake and tsunami. An analysis of the quantity of radiation released suggests that Japan may expect an increase in cancer-related deaths in the coming decades.

Terrorism

Foundations of Terrorism

Terrorism has always existed in some form or fashion. Terrorizing the civilian population is expected as a component of warfare. However, the word "terror" was not introduced into the English language until the French Revolution of the 1790s. The etymology of terror indicates a Latin origin. Terror, in English, is derived from the Latin verb *ters* from which the more modern words *terrere* and *deterre* were formed. Terrere means "to cause to tremble." Deterre means to "frighten."

These words are quite adequate to describe the crime that extends fear and foreboding far beyond the primary victims of each terror event. Fear, panic, and chaos are the intended consequences of the "urban revolutionary." Terrorism is a strong form of "coercive diplomacy."[66] Although the most obvious effects may manifest a body count, property loss, and damage, terrorism also produces political, economic, and psychological costs. In a war of nerves, the psychological costs may be the most useful to the terrorist.

---

[65]Ruth Musgrave and Judy Flynn-O'Brien, *Federal Wildlife Laws Handbook with Related Laws* (Lanham, MD: Government Institutes Publishing, 1998).
[66]Ward and Smith, 1987, p. 10–11.

Although terrorism is neither unique to our time nor a recent development in civilization, never before have so many terrorist groups operated over so much territory for so long a time with such impunity. Terrorist groups are independent because they adhere to their individual convictions, but are united in the intentional destruction of established order. Their actions create victims beyond the range of their intended targets. However, terrorists are also human and fallible, thereby affording society an opportunity to protect itself.[67]

Purpose of Terrorism

Terrorism is a form of intimidation that is intended to influence political responses and governmental behavior. Therefore, the purpose of terrorism exceeds the fulfillment of an immediate act. The purpose of a robbery, carjacking, or rape, even one that involves terror and intimidation, is immediate. The fears of injury, death, or rape are present during the assault. However, such immediate fears are normally the extent of the terror associated with the event (with the exception of harsh memories). Once the offender departs the scene of the event, the criminal usually has no further intent of victimization. However, acts of terrorism differ with respect to the extent of perpetrating fear and intimidation through time. Terrorism is quite unique because the terrorist wants both fear and intimidation to remain over extended periods of time. The terrorist also wants both enigma and publicity to be associated with any perpetrated events.

Often one hears or reads news articles about the "mindless violence" of terrorism. Therefore, it is believable that terror acts are spontaneous or even unplanned. However, such observations are categorically untrue. The events of terrorism are generally well-planned, well-rehearsed, and well-executed. An example of such meticulous, systematic, and methodical crafting and implementation was demonstrated during the terrorist attacks of September 11, 2001. The planning and training, required to successfully perform this massive act of terrorism, necessitated significant preparation and rehearsing.

Terrorism is often an awesome display of power and tactical skill. Although it is impossible to analyze pre-historical accounts of terrorism, the succeeding sections of this chapter review event records throughout ancient history. Each of these events of terror is indicative of the basic premises of intimidation and fear regarding the influencing of human behaviors through time.

Although many Americans identify international terrorism as the major security threat confronting the nation today, that may be debatable. Terrorism itself is a tactic that is most frequently used by a weaker adversary to instill fear and destroy public confidence in the governmental leadership that is attacked. From the perspective of

---
[67]Boltz et al., 1990, p. xiii.

international terrorism, history contains examples of powerful nations unsuccessfully engaging weaker adversaries. Insurgencies, terrorism, and small wars have proven the downfall of many countries. The threat from international terrorism, now challenging the country, is not sponsored by another nation. Instead, it originates from a state-less source and will be extremely difficult, if not impossible, to defeat as one would defeat a conventional enemy.

Currently, the international terrorism threat that the United States considers most dangerous is primarily driven by al Qaeda, a radical Islamist entity. Although efforts in the Global War on Terror have disrupted al Qaeda in Afghanistan, and subsequent operations have captured or killed many of its top leadership, its leaders have been replaced and its operational base has been relocated to Pakistan's Federally Administered Tribal Areas. While the United States must seek ways to defeat international terrorism, public support will prove the key factor influencing the ultimate success or failure of any American military campaign.

Most U.S. experts divide terrorism into two forms: international and domestic. International terrorist organizations are formed and usually plan operations outside the United States; domestic groups are those that are formed and operate within the United States. Both types of organizations can carry out acts of terror within the borders of America. For example, the al Qaeda attacks of September 11, 2001, are considered acts of international terrorism. On the other hand, the bombing of the 1996 Atlanta Olympics by U.S. citizen Eric Rudolph is considered an act of domestic terrorism.

International Terrorism

The events of September 11, 2001, were instigated and perpetrated by foreign terrorists. Their heinous actions unexpectedly and quickly inflicted mass damages and fatalities, and dramatically altered American lifestyles and perceptions of security. Although the nation has not endured any further events of mass terror on the scale of the 9/11 attacks, it must remain vigilant to ensure that national security is facilitated to protect the national infrastructure and citizenry. Numerous threats of international terrorism continue to relentlessly endanger the United States. During the time of the authorship of this text, package shipments, originating in the nation of Yemen, were posted and mailed to two Jewish synagogues within the United States. These packages were intended to be mail bombs. These bombs were "rigged inside computer printers shipped from Yemen and discovered in Britain and Dubai."[68] The nation of Yemen is recognized as an "incubator" for terrorist organizations.

---

[68]Jerome, 2010.

This attempted terror event exploited weaknesses within the international distribution and logistics system. Very few packages are thoroughly screened during transit. However, with respect to this scenario, the assistance of Saudi Arabia was a critical influence regarding the discovery of the mail bombs because of its network of informants and surveillance abilities. The assistance of Saudi Arabia was also beneficial in identifying terror threats that targeted France and instrumental in identifying terrorists in the nations of Pakistan, Iraq, Somalia, and Kuwait. These events are reminders that terrorism does not cease, and a variety of entities exist that desire to strike the domestic United States anonymously. Foreign terror organizations present a myriad of threats to U.S. security, and the assistance of other nations is often crucial to nullifying potential threats.

This incident is not indicative of a solitary organization that wishes to harm the United States and its interests. A variety of terrorist organizations exist whose goals involve the destruction of America and its interests. According to the U.S. Department of State, there are currently 46 foreign organizations that it has identified as terrorist organizations. Each of these organizations represents some form of terrorist threat against the United States and its interests. The designation as a terrorist organization occurs in "accordance with section 219 of the Immigration and Nationality Act (INA), as amended," and is an "effective means of curtailing support for terrorist activities and pressuring groups to get out of the terrorism business." Although the groups within the classification change periodically, the State Department provides its listing on its website.

Domestic Terrorism

Foreign terrorists are not the only ones who threaten U.S. security; indeed, the threats associated with domestic terrorism are vivid and dangerous. Domestic terrorism is conceptualized as those terror threats that originate within the United States. The contexts of domestic terrorism are varied and range from hate crimes to crimes of eco-terrorism. Domestic terrorism in the United States takes on many flavors, everything from groups that operate on the left of the political spectrum to those who appear on the right.

Special issues groups that have adopted violence as a method of advancing their agendas are not uncommon. As an example, eco-terrorism involves ideologies regarding the sanctity of animal life or the sanctity of the planet. Eco-terrorist organizations generally exist as independent cells that organizationally manifest the concept and structuring of leaderless resistance. The forms and tactics of eco-terrorism vary and are not limited to any solitary domain of environmentalism. Within the United States, a myriad of research laboratories, production facilities, resorts, retail outlets, and suburbs have been the targets of groups such as the Animal Liberation Front (ALF) and the Earth Liberation Front (ELF). Their tactics involve bombings, arson, harassment, and threats.

Eco-terrorism is "the use of or threat to use violence in protest of harm inflicted on animals and the world's biosphere."[69] Another definition indicates that eco-terrorism is the "use or threatened use of violence of a criminal nature against innocent victims or property by an environmentally oriented, sub-national group for environmental-political reasons, or aimed at an audience beyond the target, often of a symbolic nature."[70]

Although the ALF and the ELF have some international presences, the United States hosts an unknown quantity of terror cells that espouse their philosophies of eco-terrorism. Their dangerousness and ability to damage both persons and property must neither be ignored nor understated. During "2002, FBI domestic-terrorism section chief James Jarboe said the FBI had ranked the ALF and ELF as the top domestic-terrorism threat, and told Congress they had committed more than 600 criminal acts in the United States since 1996, resulting in damages topping $43 million."[71] Additionally, "since 1976, 1,100 criminal acts have been committed in the United States by radical environmentalist groups, resulting in more than $110 million in property damage alone, a figure that does not include the significant additional costs associated with lost research, increased security, and dampened productivity."[72] Both the ALF and the ELF "have attacked business interests as they view Corporate America as harming the environment and animals," and caused approximately "over $42 million in damages" during "1996–2002."[73]

Organizational aspects of eco-terrorism may be considered with respect to a strategic group perspective. The organizational characteristics of eco-terrorism exhibits the attributes of "leaderless resistance."[74] This strategy eliminates "all ideology extraneous to the very specific cause," and is advantageous for eco-terrorist organizations because it "creates an overlapping consensus among those with vastly different ideological orientations."[75] The organizational benefit is manifested because it incites "mobilizing a mass of adherents who would have never been able to work together in an organization," thereby facilitating a stronger mobilization potential to instigate actions for a "specific cause."[76]

Both the ALF and the ELF share the structuring and strategy of "leaderless resistance," and have "close ties" regarding agendas and ideological tenets.[77] The eco-terror events of both the ALF and the ELF "have been perpetrated in virtually

---

[69] Chalk, Hoffman, Reville, and Kasupski (2005, p. 47).
[70] Bennett (2007, p. 45).
[71] Hall (2009, p. 690).
[72] Jackson et. al. (2005, p. 142).
[73] Alexander (2004, p. 24).
[74] Joosse (2007, p. 351).
[75] Ibid.
[76] Ibid.
[77] Walsh and Ellis (2007, p. 357).

every region of the US against a wide variety of targets."[78] During 1993, the ELF and ALF declared "solidarity," and that increasing amounts of "convergence of leadership, membership, agendas and funding" have occurred between these organizations.[79]

Response to Terrorism

Regardless of whether a terrorist threat is domestic or foreign, the United States must craft effective, efficient, and robust counter-terrorism strategies to facilitate the security of the nation. The National Counterterrorism Center (NCTC) exists to enhance such U.S. security functions. The mission of the NCTC is to "Lead our nation's effort to combat terrorism at home and abroad by analyzing the threat, sharing that information with our partners, and integrating all instruments of national power to ensure unity of effort."[80] Within this mission are contained three salient functions of counterterrorism strategy: (1) analysis, (2) information sharing, and (3) integration of resources.

Analysis involves the integration of both domestic and foreign analysis "from across the Intelligence Community (IC) and produces a wide-range of detailed assessments designed to support senior policymakers and other members of the policy, intelligence, law enforcement, defense, homeland security, and foreign affairs communities."[81] The NCTC is also a component of the Homeland Threat Task Force, which is responsible for the facilitation of inter-agency collaboration and provides periodic updates regarding terror threats that challenge the nation. Examining threat aspects of WMDs is also within the domain of the NCTC.

The NCTC participates in the Terrorist Identities Datamart Environment (TIDE) system. Through the use of such technology, the NCTC "maintains a consolidated repository of information on international terrorist identities and provides the authoritative database supporting the Terrorist Screening Center and the USG's watch-listing system."[82] The NCTC also disseminates materials and resources among "approximately 75 USG agencies, departments, military services and major commands."[83] The NCTC further cooperates with other government and private entities to disseminate relevant counterterrorism information and intelligence resources. Such clients include the intelligence community, and "State, Local, and Private partners."[84] Other factions of government include the Department

---

[78]Leader and Probst (2003, p. 38).
[79]Ibid.
[80]NCTC, 2010.
[81]Ibid.
[82]Ibid.
[83]Ibid.
[84]Ibid.

of Homeland Security, the Federal Bureau of Investigation, and the Interagency Threat Analysis and Coordination Group.

The NCTC facilitates resource integration among a variety of entities. This integration occurs among numerous entities, including "diplomatic, financial, military, intelligence, homeland security, and law enforcement," and provides a method of ensuring the "unity of effort."[85] Through such integration, the NCTC contributes to the strategic planning initiatives regarding the War on Terror as well as contributes toward various and sundry functions of action planning among intelligence functions. Further, the NCTC is responsible for the assigning of "roles and responsibilities to departments and agencies as part of its strategic planning duties, but NCTC does not direct the execution of any resulting operations."[86]

Narco-Terror and Transnational Organized Crime

During 2003, 39 percent of the State Department's list of designated foreign terrorist organizations had some degree of connection with drug activities.[87] For example, the DEA obtained evidence that suggests that the Colombian-based Marxist-Leninist FARC organization trades cocaine for weapons and funds its activities with cash derived from cocaine sales. In addition, another Colombian terrorist organization the United Self-Defense Groups of Colombia (AUC) conducted 804 assassinations and 203 kidnappings during the first 10 months of 2000. Its leader claimed that 70 percent of AUC's funding came from drugs.[88]

However, of even greater concern to the United States, opium production in Afghanistan is at record levels. Many experts believe that drug money is helping to finance the Taliban's war against NATO in Afghanistan. Further links between opium cultivation and sales and al Qaeda are also suspected.[89] In a statement before Congress, the DEA Administrator provided evidence of links between narco-traffickers and terrorists:[90]

> In October 2001, a joint DEA/FBI investigation targeting two heroin traffickers in Peshawar, Pakistan led to the seizure of 1.4 kilograms of heroin

---

[85]Ibid.

[86]Ibid.

[87]Drug Enforcement Agency, "Drugs and Terrorism a Dangerous Mixture, DEA Official Tells Senate Judiciary Committee," 2003, http://www.usdoj.gov/dea/ongoing/narcoterrorism_story052003.html (accessed April 16, 2014).

[88]Ibid.

[89]Drug Enforcement Administration, "Statement of Karen P. Tandy, Administrator Drug Enforcement Administration, Before the Committee on International Relations, U.S. House of Representatives," February 12, 2004, http://www.usdoj.gov/dea/pubs/cngrtest/ct021204.htm (accessed April 16, 2014).

[90]Ibid.

in Maryland and identification of two suspected money launderers, one with suspected ties to al Qaida. Similarly, Operation Marble Palace in 2001 determined that several members of a targeted heroin trafficking organization had possible ties to the Taliban and that a connected bank account had been used to launder proceeds to alleged Taliban supporters in Pakistan.

Also, the DEA and other law enforcement agencies have multiple reasons to shut off the supply of illicit drugs: they are in and of themselves dangerous and their profits help to fund terrorist activities. Historically, discussions of organized crime (OC), referred to the Mafia or a small number of other criminal organizations. However, such considerations are no longer the case associated with modern criminal factions. According to the FBI:[91]

> International organized crime is defined as those self-perpetuating associations of individuals who operate internationally for the purpose of obtaining power, influence, monetary, and commercial gains, wholly or in part by illegal means, while protecting their activities through a pattern of corruption and violence. International organized criminals operate in hierarchies, clans, networks, and cells. The crimes they commit vary as widely as the organizational structures they employ.

Both globalization and technologies (e.g., the Internet) have increased the reach of OC. In addition to the traditional activities of gambling, prostitution, and drug sales, modern OC figures are also involved in the energy and financial sectors and cyberspace. They perpetrate sophisticated fraud schemes, traffic in humans for illegal purposes, and launder billions of dollars in illicit monies throughout the world. Officials fear that modern OC has the power to manipulate security exchanges to threaten the entire global economy.[92] In order to address this significant and growing problem, in 2007 the Attorney General unveiled *The Law Enforcement Strategy to Combat International Organized Crime*. This strategy addresses eight specific threats that include such twenty-first century realities as identity theft and cyber fraud schemes. It also calls for increasing cooperation between federal, state, local, and international law enforcement bodies.[93]

The U.S. military has already faced the frustrating task of participating in the protracted drug war against various transnational criminal organizations (TCOs)

---

[91]Federal Bureau of Investigation, "Department of Justice Launches New Law Enforcement Strategy to Combat Increasing Threat of International Organized Crime," 2008b, http://www.fbi.gov/pressrel/pressrel08/ioc042308.htm (accessed November 8, 2008).
[92]Ibid.
[93]Ibid.

operating out of Latin America, Asia, and the greater Middle East. The current focus of the U.S. Armed Forces is to provide wide-area surveillance, south of the U.S. border, in support of U.S. law enforcement efforts to interdict the flow of illegal drugs. Especially in Colombia and Mexico, U.S. law enforcement has been thwarted by the ability of the various drug cartels to mutate rapidly and suborn local governments. Several nations may become dominated politically and financially by powerful criminal enterprises. Certainly, a number of Latin American countries have had to struggle against this challenge.

Netwar, Cyberwarfare, Cybercrime, and Cyberterrorism

The term "netwar," developed by John Arquilla and David Ronfeldt, referring to the study of network-based conflict and crime, has become a reality of the twenty-first century. This type of threat exploits weaknesses within the virtual environment and may be quite difficult to detect and even more difficult to deter. Criminal, terrorist, and social activist organizations are most effective when they develop networking capabilities that are attuned to the information age. Many organizations have been successful in using the Internet to disseminate information that furthers their agendas. A variety of "net warriors" have used highly interconnected, organizational, and communication networks for the purposes of war, crime, or terror.

The concepts of cyberwar and netwar encompass a new spectrum of conflict that is emerging in the wake of the information revolution. Netwar includes conflicts that may be waged by two categories of entities. The first category includes terrorists, criminals, gangs, and ethnic extremists. The second category includes civil-society activists (i.e., cyber activists or World Trade Organization protestors). Netwar is distinguished by the networked organizational structure of its practitioners; in that way, it embodies the concept of leaderless resistance. Although seemingly leaderless, individual groups have the capacity to quickly come together and collaborate simultaneously through the execution of multiple, swarming attacks against U.S. interests.

A future enemy could conduct a slow-motion, strategic, and economic warfare campaign by using the virtual world against the private, economic interests of the United States. Likewise, technologically sophisticated criminal organization might be employed to conduct a campaign of crime against the United States and its major allies. New opportunities for high-performance crime could use rapid deployment of a wide range of cyber payments or electronic currency systems that facilitate the global transition to electronic commerce. To confront this new type of conflict, it is crucial for governments, military, and law enforcement to enhance their own networking capabilities.

Historically, crime has occurred globally across physical environments. Tangible materials and resources were used to originate, perpetrate, and manifest acts of crime. However, the virtual world demonstrates intangibility within its domain. Therefore, it represents new opportunities for criminals and terrorists to perpetrate their activities. Given the dangers of terrorism globally, U.S. national policy and security initiatives must be cognizant of the dangerousness of the virtual world. This notion is especially important regarding the susceptibility of the electronic infrastructure to acts of terrorism.

Cyberterrorism is defined as the "convergence of terrorism and cyberspace" involving "unlawful attacks and threats of attack against computers, networks, and the information stored therein when done to intimidate or coerce a government or its people in furtherance of political or social objectives" that results in "violence against persons or property, or at least cause enough harm to generate fear."[94] Examples include injurious or lethal attacks; contaminating of water supplies; economic losses; and aircraft crashes.[95] Depending upon the impact and severity of the incident, substantial attacks against critical national infrastructures may be categorized as acts of cyberterrorism.[96] Attacks that are deemed to be nuisances or disruptive to nonessential types of services may not necessarily be categorized as acts of cyberterror.[97] An examination of these notions yields an interesting observation: the basic concept of cyberterrorism is commensurate with the attributes of physical crime. Regardless of whether criminal activity occurs within the physical or virtual domains, both environments require an origin, method, resource, victim or target, and a motivating factor.

The events of September 11, 2001, involved massive, physical damages among multiple locations geographically. Criminal organizations are always seeking new methods of perpetuating their activities while escaping detection. The use of the virtual world provides a basis for inflicting significant damage within the infrastructures of America and its allies. As examples, mass power outages could be implemented through the interruption of service network infrastructures; criminal organizations could attempt to incite meltdowns at nuclear facilities; control systems of refineries and fuel networks could be attacked to cause massive amounts of throughput to occur, thereby instigating systems failure and potential explosions among the locations of the networked infrastructure; waterway control software could be attacked to cause failure and mass flooding among dam river sites; and critical components of the aviation and aerospace sectors could be attacked.

---

[94]Dorothy E. Denning, "Cyberterrorism: Testimony Before the Special Oversight Panel on Terrorism—Committee on Armed Services—U.S. House of Representatives," 2000, http://www.cs.georgetown.edu/~denning/infosec/cyberterror.html (accessed April 16, 2014).
[95]Ibid.
[96]Ibid.
[97]Ibid.

Theoretically, through the use of computer-based and digital technologies, such crimes may be committed remotely, from anywhere on the planet, against targets that are also located anywhere else on the planet. An offender need not be physically present at the target location to inflict significant damage through the use of computer systems and digital technologies. This notion is significant—never before has history demonstrated the potential of small groups or individuals to adversely affect mass populaces in such a significant manner.

The U.S. Government acknowledges the dangerousness of this situation. Because of its integration and dependence upon technology, the digital infrastructure of the United States must be protected against threats. During 2010, the U.S. military announced the formation of the U.S. Cyber Command (USCYBERCOM or CYBERCOM). The CYBERCOM "plans, coordinates, integrates, synchronizes, and conducts activities to: direct the operations and defense of specified Department of Defense information networks and; prepare to, and when directed, conduct full-spectrum military cyberspace operations in order to enable actions in all domains, ensure US/Allied freedom of action in cyberspace and deny the same to our adversaries."[98] Within CYBERCOM are various factions representing elements from the military services. These CYBERCOM components consist of the Army Forces Cyber Command (ARFORCYBER), the Twenty-Fourth Air Force, the Navy Fleet Cyber Command (FLTCYBERCOM), and the Marine Forces Cyber Command (MARFORCYBER).

The concept of leveraging computer technologies for an advantageous purpose is not new. Historically, the United States leveraged digital technologies to provide competitive and strategic advantages during the cold war. During the 1980s, the Central Intelligence Agency sabotaged computer control systems to instigate an explosion within the infrastructure of Soviet fuel systems involving the trans-Siberian pipeline.[99] During this period, the Soviets were covertly obtaining U.S. technologies through intermediaries in France and were implementing them within their national infrastructure, thereby facilitating a variety of technological improvements. Once American agencies learned of this situation, they responded by crafting logic bombs within computer components that were "designed to pass Soviet quality tests and then to fail in operation."[100]

The Soviets attempted to overtly obtain pipeline control system components, but were refused these items. They responded by planning to steal the pipeline software from a Canadian firm. However, the United States received a tip regarding the Soviet operation, and crafted a Trojan Horse, a logic bomb within the control

---

[98]USSC, "United States Strategic Command," 2010, http://www.stratcom.mil/factsheets/cc/ (accessed November 12, 2010).

[99]William Safire, "The Farewell Dossier," *The New York Times*, 2004, http://www.nytimes.com/2004/02/02/opinion/the-farewell-dossier.html (accessed November 12, 2010).

[100]Ibid.

software.[101] This logic bomb was designed to "reset pump speeds and valve settings," thereby generating fuel pressures that exceeded the capacity of the physical construction of the pipeline. Once operational, the flawed software instigated the "most monumental, non-nuclear explosion and fire ever seen from space."[102]

Indeed, technology was crucial to the successful U.S. outcome regarding the cold war. As the Soviets acquired a variety of technologies covertly, the United States placed "deliberately flawed designs for stealth technology and space defense."[103] Such disinformation caused Soviet scientists and researchers to expend many different efforts that wasted significant time and money.[104]

Modern aspects of cyberwar operate similarly. During the period of the authorship of this text, there is much debate regarding the Stuxnet worm that damaged Iranian nuclear facilities and impeded the progress of developing the Iranian nuclear program. These facilities employed German technologies (Siemens) as components of their control systems infrastructure. Stuxnet targeted Siemens equipment that controlled Iranian "oil pipelines, electric utilities, nuclear facilities and other large industrial sites."[105] Stuxnet was also discovered in the nations of Pakistan, India, Indonesia, and other countries, and is believed to be the product of a nation-state because of its sophistication.[106] Initial observations showed that approximately "60 percent of the 100,000 Stuxnet-infected computers worldwide were in Iran," that "under 1 percent of those infections were in the US—roughly 900 computers systems," and that "within that smaller group, about 5 percent of the infections (40–50 computers) were on Siemens industrial control systems."[107]

Stuxnet appeared also as an incident of "outright industrial sabotage or cyber warfare, created and unleashed not by rogue hackers but by a state."[108] No nation officially declared that it authored Stuxnet, but possible origins included the United States, Israel, Britain, France, Germany, China, and Russia.

After its discovery, Stuxnet was examined and tested by factions of the U.S. Government, including the Department of Energy and the Department of

---

[101]Ibid.

[102]Ibid.

[103]Ibid.

[104]Ibid.

[105]David E. Sanger, "Iran Fights Malware Attacking Computers," *The New York Times*, 2010, http://www.nytimes.com/2010/09/26/world/middleeast/26iran.html?_r=2&th&emc=th (accessed November 14, 2010).

[106]Ibid.

[107]Mark Clayton, "Stuxnet Worm: Private Security Experts Want US to Tell Them More," *The Christian Science Monitor*, 2010, http://www.csmonitor.com/USA/2010/1003/Stuxnet-worm-Private-security-experts-want-US-to-tell-them-more (accessed November 14, 2010).

[108]Bob Dreyfuss, "Cyberwar Against Iran: Is Obama Already at War with Tehran?," 2010, http://www.thenation.com/blog/155026/cyberwar-against-iran-obama-already-war-tehran (accessed November 11, 2014).

Homeland Security, and private enterprises (e.g., Symantec). An analysis of the Stuxnet software yielded an interesting observation: the malware was designed to fulfill specific purposes. German researcher Ralph Langner examined how Stuxnet identified its target, making it the first-known targeted cyber-missile. It was designed to home in on and destroy something in the reality of the physical world. Stuxnet was hailed as the first significant malware software designed to specifically target industrial systems.

During modern history, the Stuxnet incident is not the first use of rogue software to impede or alter the functioning of large-scale systems. The Greek cellular telephone network was compromised during 2005. In this incident, "a two-year investigation by the Greek government found an extremely sophisticated Trojan Horse program that had been hidden by someone who was able to modify and then insert 29 secret programs into each of four telephone switching computers," and that "the level of skill needed to pull off the operation and the targets strongly indicated that the culprit was a government."[109] It was further claimed that the 2007 Israeli attacks against an assumed Syrian nuclear reactor were enhanced through the use of software involving "sophisticated jamming technology."[110] Using this software nullified the radar systems, thereby allowing "Israeli aircraft" to be "unnoticed."[111]

The dangers of the virtual world are complex and dynamic. Historically, governments have sought advantages through the use of technology. The leveraging of these technological resources among governments and others is in its infancy. However, it will mature in due time in conjunction with technological advancements and new discoveries. American efforts to develop strategic and tactical advantages, through the use of virtual domains, are also in their formative years. Each will mature over time.

Weapons of Mass Destruction (WMD)/CBREN

Perhaps the greatest fear of anyone connected to homeland security today is that a weapon of mass destruction will make its way into the hands of a terrorist. It is known that terrorist groups, such as al Qaeda are desperate to obtain chemical, biological, and nuclear weapons. This realization is especially problematic because the technology to make such weapons becomes more easily obtained through time. Periodically, a new member is admitted to the "nuclear club." Currently, there is great concern that Iran, a nation that holds great animus toward both the United States and Israel, is refining and finalizing its ability to manufacture a nuclear

---

[109]Markoff, John. 2010. "A Silent Attack, but Not a Subtle One." *The New York Times*. http://www.nytimes.com/2010/09/27/technology/27virus.html?_r=0 (accessed June 9, 2014).

[110]Ibid.

[111]Ibid.

bomb. More troubling, Iran has close relationships with groups that the United States has designated as terrorist entities (e.g., Hezbollah). Given these concerns, a question must be posed: Should it manage to develop a nuclear weapon? Would Iran be willing to turn it over to a terrorist surrogate? Such a scenario worried the United States enough that it invaded Iraq in 2003 to depose Saddam Hussein and potentially prevent such an occurrence.

The acronym CBREN specifically defines various types of weapons of concern:

- C: Chemical
- B: Biological
- R: Radiological (nuclear radiation)
- E: Explosive
- N: Nuclear (capable of producing a nuclear explosion)

As should seem obvious, WMDs have the potential to devastate large portions of the American populace and national infrastructure. The dangers posed by this category are well known by the U.S. Government. During 2005, the National Counterproliferation Center (NCPC) was created to "counter the threats caused by the proliferation of chemical, biological, radiological, and nuclear weapons."[112] The mission of the NCPC is to "help lead the Intelligence Community in developing integrated strategies and actions both to counter current weapons of mass destruction threats and to anticipate and counter future WMD proliferation."[113]

The NCPC was created by the Intelligence Reform and Terrorism Prevention Act of the Bush Administration. The NCPC exists as the:

> Primary organization within the Intelligence Community for managing, coordinating, and integrating planning, collection, exploitation, analysis, interdiction and other activities relating to weapons of mass destruction, related delivery systems, materials and technologies, and intelligence support to United States Government efforts and policies to impede such proliferation.[114]

Because a variety of terrorist organizations seeks to obtain WMDs, the NCPC is of paramount importance for ensuring national security. The functions of the NCPC are not replicated among other government agencies or functions within the intelligence community. Instead, it exists as a unique organization. The personnel

---

[112]National Counterproliferation Center, "History of NCPC," 2014, http://www.counterwmd.gov/history.htm (accessed April 16, 2014).
[113]Ibid.
[114]Ibid.

of the NCPC represent a varied conglomeration of expertise and skills. These personnel are representatives from "Intelligence Community agencies, as well as the Department of State, the Department of Defense, and the Department of Energy's National Laboratories."[115] The functioning of the NCPC is integrative because it converges the initiatives of the intelligence community toward the satisfaction of the "counterproliferation priorities" of policy makers.[116] Six directorates comprise the basic organizational structuring of the NCPC: (1) Deputy Director for Intelligence and Action Integration; (2) Deputy Director for Requirements and Gaps Integration; (3) Deputy Director for Interdiction and Networks; (4) Deputy Director for Resource Management and Investment; (5) Deputy Director for Global Biological Threat; and (6) Deputy Director for Science and Technology.[117]

These directorates cooperate to deter and interdict potential WMD threats that endanger the nation. Five primary objectives are components of this strategic initiative. These objectives are as follows:

- Discourage interest by states, terrorists, or armed groups from acquiring, developing or mobilizing resources for WMD purposes.
- Prevent or obstruct state, terrorist, or other efforts to acquire WMD capabilities, or efforts by suppliers to provide such capabilities.
- Roll back or eliminate WMD programs of concern.
- Deter weapons use by those possessing nuclear, radiological, biological, and chemical weapons and their means of delivery.
- Mitigate the consequences of any use of WMD against the United States or its allies.

The strategic pursuit of these goals is a massive undertaking, requiring much effort, cooperation, information sharing, and the involvement of multiple agencies. Although the NCPC is the primary entity whose function concerns counter-proliferation, a variety of peer agencies assist with these efforts. Counter-proliferation partner agencies include the following federal entities:

- Central Intelligence Agency
- Federal Bureau of Investigation
- National Geospatial Intelligence Agency
- National Reconnaissance Office
- Department of Energy

---

[115]Ibid.
[116]Ibid.
[117]Ibid.

- Defense Intelligence Agency
- National Security Agency
- Department of State
- Department of the Treasury
- Department of Homeland Security

A significant benefit of such collaboration is the manifestation of multiple perspectives regarding threats and security issues. The use of multiple perspectives provides a greater array of approaches from which counter proliferation strategies may be crafted and influenced.

Because of globalization, increased risks pervade security issues regarding the threats associated with WMDs. A variety of terror organizations and rogue states desire to obtain and master WMD technologies. These concerns are of critical interest to U.S. policymakers. Threats and risks are presented through the mastering of nuclear technologies among emerging, national powers whose energy infrastructures may be converted to produce atomic weaponry. Biological threats and risks appear through the potential of rogue scientists to mutate relatively harmless strains of bacteria or viruses into deadly weapons capable of infecting and killing masses of peoples. Certainly, many other examples may be identified. Through the development of the NCPC, the United States hopes to gain the ability to diminish the risks of WMDS and provide additional security measures to protect the citizenry and the national infrastructure.

Chapter Comments and Summary

The role of the federal government regarding disaster prevention, mitigation, and response has evolved throughout the past 200 years. The Congressional Act of 1803 was the earliest effort to provide federal disaster relief after a fire devastated a New Hampshire town. From that point forward, assorted legislation provided disaster support. Between 1803 and 1950, the federal government intervened in approximately 100 incidents (earthquakes, fires, floods, and tornadoes).

Such events are severe reminders of the potency of natural hazards and disasters to affect negatively not only the impacted area but also the entire segments of the nation. These calamities necessitate emergency management and response and also involve various amounts of preparedness depending upon location and resources. Although they do not result from human activities, these natural events are just as deadly and as powerful as any threat that could be unleashed from human origins. This notion represents a dichotomy of threats: man-made events and natural disasters. Although these categories may differ with respect to their origins, their potentials to cause mass death, destruction, and suffering are comparable. Modern

American society must acknowledge the dangerousness of both categories and must craft methods to mitigate the effects of a range of incidents that may arise from many different origins.

Terrorism is a significant characteristic of the modern, multi-polar world that challenges U.S. security and influence globally. Regardless of whether terrorism threats are domestic or foreign, the United States must prepare itself for the potential of future incidents ranging from small, limited events to large-scale incidents that massively impact the citizenry and national infrastructure. Such planning must include the strategic aspects of possible incident responses through time. In order to facilitate such planning, government entities, such as the NCTC and the NCPC, are critical for safeguarding the nation, its infrastructure, its citizenry, and its interests both domestically and globally.

Although international terrorism remains a primary security concern of many American citizens, other international threats, including transnational crime, WMD/CBRNE trafficking, human trafficking, international cyber-security, and economic fraud pose dangers to our nation. The Global War on Terror remains a generational struggle, and it may continue for decades. Terrorists have declared their intention to acquire weapons of mass destruction (WMD/CBRNE) and use them against the United States and its allies. Weapons of mass destruction, in the hands of terrorists, pose a great threat to the entire world.

Terminology

All-Hazards
Domestic Terrorism
International Terrorism
Leaderless Resistance
Locality
Man-Made Disaster

Natural Disaster
Regional
Scope
Terrorism
Threat
Weapon of Mass Destruction

Thought and Discussion Questions

1. Perform some research regarding your locality. Determine the types of natural threats that endanger your community. Write a brief essay that highlights your findings.
2. Perform some research regarding your locality. Determine the types of man-made threats that endanger your community. Write a brief essay that highlights your findings.
3. Review some recent news articles regarding terrorism. Write a brief essay that critically analyzes the selected event. Based on your article, what measures do you believe can be taken to prevent similar acts of terror in the future?

4. This chapter introduces the concepts of domestic and international terrorism. Write a brief essay that compares and contrasts these two concepts. Within your response, incorporate some recent news items to substantiate your response.

References

1. AirDisaster. n.d. Feature: Hijack, part 1. http://www.airdisaster.com/features/hijack/hijack.shtml (accessed November 8, 2008).
2. Anti-Money. 2010. Anti-money laundering/combating the financing of terrorism. http://www.imf.org/external/np/leg/amlcft/eng/aml1.htm#typologies (accessed November 10, 2010).
3. Bacon, N. 2010. The cyberwar attack on Iran created by a foreign govt. not so subtle after all. http://www.newsnet14.com/2010/09/27/a-silent-attack-on-iran-not-so-subtle-after-all/ (accessed November 14, 2010).
4. Bakunin, M. 1870. Letters to a Frenchman on the present crisis. http://www.marxists.org/reference/archive/bakunin/works/1870/letter-frenchman.htm (accessed November 7, 2008).
5. Barnes, J. 1998. *Florida's hurricane history*. Chapel Hill: Univ. of North Carolina Press.
6. Benmelech, E. and C. Berrebi. 2007. Human capital and the productivity of suicide bombers. *Journal of Economic Perspectives* 21(3).
7. Boltz, F., Jr., K. Dudonis, and D. Schultz. 1990. *The counter-terrorism handbook: Tactics, procedures and techniques.* New York: Elsevier.
8. Bradley, M. 2005. *The secret societies handbook.* London: Cassell Illustrated.
9. Bruse, G. 1969. The stranglers: The cult of thugee and its overthrow in British India. New York: Harcourt, Brace and World.
10. Chernobyl Accident. 1986. World Nuclear Association. http://www.world-nuclear.org/info/Safety-and-Security/Safety-of-Plants/Chernobyl-Accident/#.UU8whVfyCSp (accessed March 24, 2013).
11. Clayton, M. 2010. Stuxnet worm: Private security experts want US to tell them more. *The Christian Science Monitor*. http://www.csmonitor.com/USA/2010/1003/Stuxnet-worm-Private-security-experts-want-US-to-tell-them-more (accessed November 14, 2010).
12. Counterterrorism Financing. 2010. Counterterrorism financing activities—criminal investigation. http://www.irs.gov/compliance/enforcement/article/0,,id=108256,00.html (accessed November 11, 2010).
13. Crenshaw, M. 1988. The subjective reality of the terrorist: Ideological and psychological factors in terrorism. In *Current perspectives on international terrorism*, ed. R. Slater and M. Stohl, 12–46. New York: St. Martins.
14. Crenshaw, M. 1992. Current research on terrorism: The academic perspective. *Studies in Conflict and Terrorism* 15: 1–11.

15. Deadliest/Large-Loss Fire. National Fire Protection Association (2/12). http://www.nfpa.org/itemDetail.asp?categoryID=954&itemID=44745&URL=Research/Fire%20statistics/Deadliest/large-loss%20fires (accessed March 30, 2013).
16. Deadliest Volcanic Eruptions. http://www.chm.bris.ac.uk/webprojects2003/silvester/Page6Famous.htm (accessed March 24, 2013).
17. Denning, D. 2000. Cyberterrorism: Testimony before the special oversight panel on terrorism—committee on armed services—U.S. house of representatives. http://www.cs.georgetown.edu/~denning/infosec/cyberterror.html (accessed November 11, 2010).
18. Dershowitz, A. 2002. *Why terrorism works: Understanding the threat, responding to the challenge.* New Haven, Connecticut: Yale University Press.
19. Dobson, C. 1974. *Black September: Its short violent history.* New York: McMillan Press.
20. Drama of the Desert: The week of the Hostages. Time, 0040781X, 9/21/1970, Vol. 96, Issue 12.
21. Dreyfuss, R. 2010. Cyberwar against Iran: Is Obama already at war with Tehran? http://www.thenation.com/blog/155026/cyberwar-against-iran-obama-already-war-tehran (accessed November 14, 2010).
22. Drug Enforcement Administration. 2003. Drugs and terrorism a dangerous mixture, DEA official tells Senate Judiciary Committee. http://www.usdoj.gov/dea/ongoing/narcoterrorism_story052003.html (accessed November 8, 2008).
23. Drug Enforcement Administration. 2004. Statement of Karen P. Tandy, Administrator Drug Enforcement Administration, before the Committee on International Relations, U.S. House of Representatives, February 12, 2004. http://www.usdoj.gov/dea/pubs/cngrtest/ct021204.htm (accessed November 8, 2008).
24. Eisenhower, J. S. D. 1993. *Intervention: The United States and the Mexican Revolution, 1913–1917.* New York: W. W. Norton.
25. Fendell, H. 2006. Israel commemorates 30th anniversary of Entebbe rescue. http://www.israelnationalnews.com/News/News.aspx/106568 (accessed November 8, 2008).
26. Federal Bureau of Investigation. 2008a. Putting Intel to work against ELF and ALF terrorists. http://www.fbi.gov/page2/june08/ecoterror_063008.html (accessed November 8, 2008).
27. Federal Bureau of Investigation. 2008b. Department of Justice launches new law enforcement strategy to combat increasing threat of international organized crime. http://www.fbi.gov/pressrel/pressrel08/ioc042308.htm (accessed November 8, 2008).
28. Fire Prevention and Engineering Bureau of Texas, 1947. Texas city disaster. http://www.local1259iaff.org/report.htm (accessed March 31, 2013).

29. Flood, S., ed. 1991. *International terrorism: Policy implications.* Illinois: Office of International Criminal Justice at the University of Illinois at Chicago.
30. Hemphill, S. 2002. Peshtigo: A tornado of fire revisited. Minnesota Public Radio (November 27). http://news.minnesota.publicradio.org/features/200211/27_hemphills_peshtigofire/ (accessed March 30, 2013).
31. Hersh, S. 2010. The online threat: Should we be worried about a cyber war? *The New Yorker.* http://www.newyorker.com/reporting/2010/11/01/101101fa_fact_hersh (accessed November 14, 2010).
32. Himmelsbach, R. P. and T. K. Worcester. 1986. *Norjak: The investigation of D. B. Cooper.* West Linn, OR: Norjack Project.
33. Howard, R. D., R. L. Sawyer, and N. E. Bajema. 2009. *Terrorism and counterterrorism: Understanding the new security environment.* 3rd ed. Boston, MA: McGraw Hill Publishing.
34. Hyde, M. and E. Forsyth. 1987. *Terrorism: A special kind of violence.* New York: Putnam's Sons Publishing.
35. IMF. 2010. The IMF and the fight against money laundering and the financing of terrorism. *International Monetary Fund.* http://www.imf.org/external/np/exr/facts/aml.htm (accessed November 10, 2010).
36. IslamOnline. n.d. Fatwa bank. http://www.islamonline.net/servlet/Satellite?pagename=IslamOnlineEnglishAsk_Scholar/FatwaE/FatwaE&cid=1119503545134 on 11/08/2008 (accessed November 8, 2008).
37. Jane's. n.d. Suicide terrorism: A global threat. *Jane's Terrorism and Security Monitor.* http://www.janes.com/security/international_security/news/usscole/jir001020_1_n.shtml (accessed November 9, 2008).
38. Jenkins, B. 1975. *High technology terrorism and surrogate warfare: The impact of new technology on low-level violence.* Santa Monica: RAND.
39. Jenkins, B. 1996. Terrorism trial begins in New York. http://www.cnn.com/US/9605/12/terror.plot/ (accessed November 8, 2008).
40. Jensen, C. and Y. Hsieh. 1999. Law enforcement and the millennialist vision: A behavioral approach. *Law Enforcement Bulletin* 68 (9).
41. Johnston, D. 1998. 17-year search, an emotional discovery and terror ends. http://query.nytimes.com/gst/fullpage.html?res=9E01E2DE1631F936A35756C0A96E958260&pagewanted=all (accessed November 8, 2008).
42. Kaplan, E. 2006. Tracking down terrorist financing. http://www.cfr.org/publication/10356/tracking_down_terrorist_financing.html#p7 (accessed November 11, 2010).
43. Kobetz, R. and H. H. A. Cooper. 1977. *Target terrorism: Providing protective service.* Gaithersburg, MD: The International Association of Chiefs of Police.
44. Lamb, Y. S. 2004. Birmingham bomber Bobby Frank Cherry dies in prison at 74. http://www.washingtonpost.com/wpdyn/articles/A614282004Nov18.html (accessed November 8, 2008).
45. Laqueuer, W. 2001. *A history of terrorism.* New Brunswick: Transaction Publishers.

46. Lewis, B. 1967. *The assassins: A radical sect in Islam.* New York: Oxford University Press.
47. Lockyer, H. 1958. *All the men of the Bible.* Michigan: Zondervan Publishing House.
48. Long, C. and W. Weissert. 2010. Risks grow for those whose lives straddle the border. http://news.yahoo.com/s/ap/20101110/ap_on_re_us/us_drug_war_mexico_americans_killed (accessed November 12, 2010).
49. Martin, G. 2006. *Understanding terrorism: Challenges, perspectives and issues.* 2nd ed. Thousand Oaks, CA: Sage Publications.
50. Michel, L. and D. Herbeck. 2001. *American Terrorist: Timothy McVeigh and the Oklahoma city bombing.* New York: Harper.
51. Mullins, W. 1988. *Terrorist organizations in the United States: An analysis of issues, organizations, tactics, and responses.* IL: Charles C. Thomas Publishing.
52. National Counterproliferation Center. 2014. History of NCPC. http://www.counterwmd.gov/history.htm (accessed April 16, 2014).
53. Nettleton, S. 2001. Kidnapped: pinned by the sword and the wall. *Colombia: War without end.* http://www.cnn.com/SPECIALS/2000/colombia.noframes/story/reports/kidnapped/index.html (accessed November 8, 2008).
54. O'Ferrell, P. 1972. *Ireland's English question: Anglo-Irish relations 1534–1970.* New York: Schoken Books.
55. Oil Pollution Act of 1990—Summary. *Federal Wildlife and Related Laws Handbook.* August 18, 1990. http://ipl.unm.edu/cwl/fedbook/statute_frame.htm. Retrieved March 10, 2008.
56. Perry, C. 2000. Significant floods in the United States during the 20th century—USGS measures a century of floods. USGS (12/10/10). http://ks.water.usgs.gov/pubs/fact-sheets/fs.024-00.html#HDR1 (accessed March 30, 2013).
57. Pillar, P. 2001. *Terrorism and U.S. foreign policy.* Washington D.C.: Brookings Institution.
58. PIstole, J. 2003. Before the House Committee on Financial Service Subcommittee on Oversight and Investigations, Washington D.C. http://www.fbi.gov/news/testimony/the-terrorist-financing-operations-section (accessed November 7, 2010).
59. Poland, J. 2005. *Understanding terrorism: Groups, strategies and responses.* 2nd ed. Upper Saddle River, NJ: Pearson/Prentice Hall Publishers.
60. Progress. 2010. Progress in the war on terrorist financing. http://www.ustreas.gov/press/releases/reports/js721.pdf (accessed November 11, 2010).
61. Ramsland, K. 2007. The kidnapping of Patty Hearst. *Crime Library.* http://www.crimelibrary.com/terrorists_spies/terrorists/hearst/1.html (accessed November 8, 2008).
62. Randall, W.P. 1965. *The Ku Klux Klan: A century of infamy.* Philadelphia: Chilton.
63. Rappaport, D. and Y. Alexander. 1982. *The morality of terrorism: Religious and secular justifications.* New York: Pergamon Press.

64. Ripley, A. 2008. The 1900 Galveston Hurricane. *Time* (September 15). http://www.time.com/time/nation/article/0,8599,1841442,00.html (accessed April 1, 2013).
65. Robinson, B. 2001. The gunpowder plot. http://www.bbc.co.uk/history/british/civil_war_revolution/gunpowder_robinson_01.shtml (accessed November 8, 2008).
66. Roig-Franzia, M. 2008. From Mexico, drug violence spills into U.S. *Washington Post.* http://www.washingtonpost.com/wp-dyn/content/article/2008/04/19/AR20080419019 (accessed November 9, 2008).
67. Rosenberg, J. Triangle Shirtwaist Factory Fire. *About.com,* 20th Century History. http://history1900s.about.com/od/1910s/p/trianglefire.htm (accessed March 30, 2013).
68. Royston, R. 2010. China's dust storms raise fears of impending catastrophe. *National Geographic News* (October 28). http://news.nationalgeographic.com/news/2001/06/0601_chinadust.html (accessed March 18, 2013).
69. Sachar, H. 2000. *Israel and Europe: An appraisal in history.* New York: Knopf Publishing Group.
70. Safire, W. 2004. The farewell dossier. *The New York Times*. http://www.nytimes.com/2004/02/02/opinion/the-farewell-dossier.html (accessed November 12, 2010).
71. Sanger, D. 2010. Iran fights malware attacking computers. *The New York Times.* http://www.nytimes.com/2010/09/26/world/middleeast/26iran.html?_r=2&th&emc=th (accessed November 14, 2010).
72. Shrivastava, P. 1987. *Bhopal: Anatomy of a crisis,* 184. Cambridge, MA: Ballinger Publishing.
73. Smith, T. 1976. Hostages freed as Israelis raid Uganda airport. *New York Times.* http://select.nytimes.com/gst/abstract.html?res=F60816FA38591B728DDDAD0894DF405B868BF1D3 (accessed November 8, 2008).
74. Taheri, A. 1987. *Holy terror: Inside the world of Islamic terrorism.* Chevy Chase, MD: Adler and Adler.
75. USSC. 2010. United States Strategic Command. http://www.stratcom.mil/factsheets/cc/ (accessed November 12, 2010).
76. Ward, R. and E. Smith. 1987. *International terrorism: Operational issues.* Chicago, IL: Office of International Criminal Justice, the University of Illinois at Chicago.
77. Washington Post. 2007. More attacks, mounting casualties. http://www.washingtonpost.com/wpdyn/content/graphic/2007/09/28/GR2007092802161.html (accessed November 8, 2008).
78. Weinberg, L. and P. Davis. 1989. *Introduction to political terrorism.* New York: McGraw Hill.
79. Weiss, M. 2005. Terrorist financing: U.S. agency efforts and inter-agency coordination. http://www.fas.org/sgp/crs/terror/RL33020.pdf (accessed November 11, 2010).

80. Whiston, W. 1991. *The works of Josephus: Complete and unabridged.* Peabody, MA: Hendrickson Publishers.
81. Wilkinson, P. and A. Stewart. 1987. *Contemporary research on terrorism.* Aberdeen, UK: The University Press.
82. Wright, R. 2008. Since 2001, a dramatic increase in suicide bombings. *washingtonpost.com.* http://www.washingtonpost.com/wpdyn/content/story/2008/04/18/ST2008041800913.html (accessed November 8, 2008).
83. Yonghe, Y. 2004. *Small sea travel diaries* (trans. Macabe Keliher). Taipei: SMC Publishing.

POWER AND POLICY OF STAKEHOLDERS AND FIRST RESPONDERS

"The likelihood of a major natural disaster, flood, hurricane, or earthquake, affecting our communities was inevitable."
— George Haddow and Jane Bullock,
Institute for Crisis, Disaster, and Risk Management

Figure 4.0 Emergency Response. Responding to emergencies necessitates a variety of responders, including firefighters.

OBJECTIVES

The objectives of this chapter are to:
- classify types of stakeholders,
- introduce the influences of stakeholder power,
- introduce stakeholder perspectives,
- discuss stakeholder influence regarding policy formulation, and
- introduce the basic notions of first responders.

Introduction

No community or locale is insusceptible to emergency scenarios. Across the nation, people are dependent upon their emergency services entities to provide efficient and effective actions before, during, and after calamities. This dependency reflects a relationship between the served community and the emergency management entities that are entrusted with protecting them. Further dependencies may be identified that exist among first-responder entities. For instance, fire departments may rely upon police departments to secure an area while they are fighting fire in a crowded downtown urban location, and an ambulance service may be necessary to retrieve casualties.

Such responses involve a consideration of human decisions and organizational policies that exist within the responding entities and also within society. All stakeholders are affected by such decisions and policies. Within the domain of emergency management, the saliency of a fiduciary relationship between service entities and those whom they serve is influenced by policy, power, and dependency types. Further, when considering the merits of such a relationship, it must be realized that it cannot be a one-way experience; instead, it must involve some aspects of collaboration and influence. Often, the complexities of relationships involving stakeholders and responder entities are influenced by policy and power within the emergency management domain, and involve numerous dependencies and partnerships among each of the factions. This notion reflects the "formal and informal relationships of emergency management."[1]

As the emergency response community works to prepare their jurisdictions for a wide range of potential contingencies, it is important to identify not only the critical elements of the infrastructure, but also the valuable assets that may be called upon when responding to the event. These assets may be considered within the contexts of stakeholders, first responders, and the served populace. Given the preceding notions, this chapter defines and discusses the concepts of stakeholders and first responders, their scopes of power, and their respective aspects of policy.

---

[1] Julie Todaro, *Emergency Preparedness for Libraries* (Lanham, MD: Rowman and Littlefield Publishing, 2009), 123.

People may influence the policies of their first-responder organizations. Emergency management is a necessary function of local governments that is supported by state and federal governments. Among many locales, sheriffs, police chiefs, medical examiners, fire chiefs, city commissioners, aldermen, and mayors are elected officials. Governors, lieutenant governors, and other state officials may also be elected by the populace. Federally, people vote for candidates who also impact national policies. Through the act of voting, stakeholders influence policy by electing those people who influence the crafting of policy. Some folks may not be elected, but work directly for a first-responder organization or may be appointed within a position whereby they may have influence regarding organizational policies.

Defining Stakeholders, First Responders, and Policy

Stakeholders may be defined according to numerous perspectives. A stakeholder is defined as "one who is involved in or affected by a course of action" that may be taken by an individual or an organization.[2] Given this notion, a stakeholder may be perceived as an entity that has a vested interest in the successfulness of the other party in whom he or she has a stake interest. Within the domain of emergency management, these notions are reflected within the context of a somewhat different stakeholder perspective. According to the Federal Emergency Management Agency (FEMA), a stakeholder is defined as "people who have, or think they have, a personal interest in the outcome of a policy."[3]

Policies may be considered from various perspectives. Within the context of public administration, a policy represents the "decisions and direction of a government as well as the ideas which lead to actions to achieve certain aims and objectives."[4] Policies are the foundations of organizational missions and mandates, and any obligation to policy makers supersedes any obligations to superiors within an organizational chain-of-command.[5] All citizens and residents are affected by the policies that influence their respective locales and are influenced by the policies of their first-responder organizations. Any first-responder entities are also

---

[2]"Stakeholder," *Merriam-Webster Dictionary*, 2013, http://www.merriam-webster.com/dictionary/stakeholder (accessed May 21, 2013).

[3]Federal Emergency Response Agency, "Chapter 2: Emergency Management Stakeholders," 2013, http://www.google.com/url?sa=t&rct=j&q=%22emergency%20management%22%20stakeholder&source=web&cd=1&cad=rja&ved=0CCsQFjAA&url=http%3A%2F%2Ftraining.fema.gov%2FEMIWeb%2Fedu%2Fdocs%2Ffem%2FChapter%25202%2520-%2520Emergency%2520Stakeholders.doc&ei=VP-bUZT_Ium0yAGa9oCABQ&usg=AFQjCNGQRhVrLenextMUT_MtBntj92gF3Q&bvm=bv.46751780,d.eWU (accessed May 21, 2013).

[4]Roland Axtmann, *Understanding Democratic Politics: An Introduction* (Thousand Oaks, CA: Sage Publishing, 2003), 114.

[5]Terry Cooper, *The Responsible Administrator: An approach to Ethics for the Administrative Role*, 5th ed. (Hoboken, NJ: Wiley Publishing (Jossey-Bass Imprint), 2011).

affected externally by such policies, and usually internally exhibit their own policies organizationally.

First-responder entities are those entities that arrive first at the scene of an incident. Conceptually, the definition of a first responder varies with the perspective of the sponsoring organization. From a medical perspective, a first responder is the "first medically trained person to arrive on the scene."[6] From an investigative perspective, a first responder is someone "who comes from the forensic laboratory or from a particular agency for initial investigation."[7] From a law enforcement perspective, a first responder is "the patrol officer who responds to the initial complaint."[8] These few definitions show that perspectives of first responders are domain-specific.

Within each of these domains are policies that are commensurate with the unique characteristics of each of the individual first-responder entities. During emergencies, law enforcement entities must adhere to their policies regarding "outlining security, transportation, and destination options" as first responders.[9] If fire is suspected within a structure, a fire department policy may indicate that "attempts will be made to access the building, contact a responsible party, and determine if a fire condition exists."[10]

These differences in policy show that different approaches exist among first responders when arriving upon the scene of an incident. Therefore, the affected citizenry and victims receive different treatment based upon the policy of the first-responder organization. Everyone is affected by the policies of their respective first-responder organizations. Given this notion, within the context of emergency management, a stakeholder may be perceived to be "an individual who is affected by the decisions made (or not made) by emergency managers and policymakers in his or her community."[11] Therefore, everyone is a stakeholder within the context of emergency management. Hence, the whole of society has a stake in the

---

[6]David Schottke, *First Responder: Your First Response in Emergency Care* (Sudbury, MA: Jones and Bartlett Publishing, 2007), 15.

[7]EC Council, *Computer Forensics: Evidence Collection and Preservation* (Clifton Park, NY: Cengage Course Technology, 2010), 4–2.

[8]Aric Dutelle, *An Introduction to Crime Scene Investigation* (Sudbury, MA: Jones and Bartlett Publishing, 2011), 5.

[9]Jack Digliani, *Reflections of a Police Psychologist* (Bloomington, IN: XLibris, 2010), 217.

[10]J. Curtis Varone, *Legal Considerations for Fire and Emergency Services* (Clifton Park, NY: Delmar Cengage Publishing, 2012), 328.

[11]Federal Emergency Response Agency, "Chapter 2," http://www.google.com/url?sa=t&rct=j&q=%22emergency%20management%22%20stakeholder&source=web&cd=1&cad=rja&ved=0CCsQFjAA&url=http%3A%2F%2Ftraining.fema.gov%2FEMIWeb%2Fedu%2Fdocs%2Ffem%2FChapter%25202%2520-%2520Emergency%2520Stakeholders.doc&ei=VP-bUZT_Ium0yAGa9oCABQ&usg=AFQjCNGQRhVrLenextMUT_MtBntj92gF3Q&bvm=bv.46751780,d.eWU (accessed May 21, 2013).

CHAPTER 4 Power and Policy of Stakeholders and First Responders 115

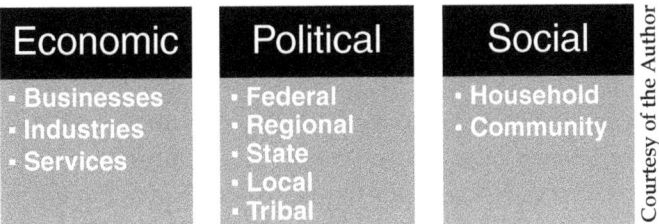

Figure 4.1 Societal Factions. Stakeholders exist throughout the entirety of society.

successfulness of first responders, the quality of their decisions, and the robustness of their policies within the domain of emergency management.

Identifying Stakeholders

Society is comprised of different factions of emergency management stakeholders. These factions may be identified economically, politically, and socially. These three perspectives are presented in the following figure.

The Economic Perspective

The economic composition of society involves a consideration of resource allocation and use regarding the goods and services that exist among locales. Business represents the entity through which most economic activity occurs. Businesses transactions provide medical services and pharmaceuticals; food and drink; automobiles, emergency vehicles, and transportation services; news broadcasts; utilities; clothing; and provide for the essential needs and wants of humans.

Maintaining the activity of businesses is essential for economic functioning. An emergency incident has the potential of disrupting business activity, thereby interrupting the flows of goods and services that satisfy the wants and needs of humans. Businesses often depend on each other for their existence and survival. For example, law enforcement organizations rely upon the successfulness of the automobile industry in order to have police cars. The capability of driving a police vehicle is dependent upon the successfulness of a fuel center to provide gasoline. The fuel center relies upon the successfulness of the oil and gas sectors of the economy. The successfulness of the oil and gas sectors of the economy relies upon the transportation industry to transport and to distribute fuel to refineries, gas stations, and fuel centers. The transportation industry relies upon the successful provision and maintenance of roadways, seaways, and airways.

Certainly, many more linkages and relationships exist within the context of business activities and economics. However, these considerations are indicative of the notion that business segments of the economy are interrelated, and manifest innumerable dependencies and relationships with each other. A failure in any one area may have the potential of affecting negatively and impeding other areas. This observation is referred to as a condition of cascading effects.

Disasters may interrupt the flows of goods and services that are necessary for the functioning of society and the provision of emergency goods and services. Without fuel, vehicles become immobile. Without electricity, electronic communications systems are immaterial. During calamities, businesses may have few or no resources to provide to society and emergency management entities. Therefore, when a disaster strikes, business leaders economically make decisions to allocate their scarce resources for the benefit of emergency operations.

Emergency management has a stake in the successfulness of business activity. From an emergency management perspective, businesses are critical resources when disaster strikes. Businesses often provide resources that are necessary to conduct emergency services operations. A recent example involves a 2013 emergency in Oklahoma. On May 20, 2013, an EF-5 category tornado devastated the town of Moore, Oklahoma.[12] This storm was the equivalent of a cateogy-5 hurricane, and involved winds of approximately 200 miles per hour.[13] It was approximately two miles wide, and carved a trail approximately "17 miles long."[14] The following image shows the devastation that was caused by this tornado.

The Anheuser-Busch Corporation, a large corporation that normally produces alcoholic beverages, converted some its product lines to produce water.[15] It provided 2,156 cases of water, equivalent to 51,744 cans, "for use by relief workers and residents" in the affected region in and around Moore.[16] Additionally, it provided financial assistance, in the amount of $25,000, through the American Red Cross, to assist with the "relief efforts."[17] Anheuser-Busch has a long history of such benevolence, beginning in 1906 when it provided donations to assist the American Red Cross during the aftermath of the San Francisco earthquake.[18] Since 1988, the

---

[12]"Search for Oklahoma Tornado Survivors Nearly Complete as Residents Return to Assess Damage," *Fox News*, 2013, http://www.foxnews.com/weather/2013/05/21/children-among-dead-after-twister-strikes-near-oklahoma-city/ (accessed May 21, 2013).

[13]Ibid.

[14]Chelsea Carter, Brian Todd, and Michael Pearson "Oklahoma Tornado Cuts 17-Mile Path: Deadly Twister had Winds of 200 mph or Greater," *KOCO News*, 2013, http://www.koco.com/news/national/oklahoma-tornado/-/9844074/20231722/-/145goa/-/index.html (accessed May 21, 2013).

[15]Anheuser-Busch, "Anheuser-Busch Delivers More Than 51,000 Cans of Water to Assist Tornado Relief Efforts in Oklahoma City," 2013, http://anheuser-busch.com/index.php/anheuser-busch-delivers-more-than-51000-cans/ (accessed May 21, 2013).

[16]Ibid.

[17]Ibid.

[18]Ibid.

Figure 4.2 Aftermath of the 2013 Moore, Oklahoma Tornado

corporation "has donated more than 72 million cans of emergency drinking water following natural and other disasters."[19]

Such relationships between business and emergency management are not uncommon. During the aftermath of Hurricane Sandy in 2012, the Wal-Mart Corporation and PepsiCo Corporation donated a variety of goods to the affected region.[20] These contributions are described as follows:[21]

> "Wal-Mart donated six trailer truck loads containing 60 pallets of dry food and beverages, 5,895 cases of cleaning supplies and 2,051 board games ... PepsiCo donated five trailer truck loads of beverages and three trailer truck loads of snacks, totaling more than 100,000 cases of products."[22]

---

[19]Ibid.
[20]Kim Souza, "Wal-Mart Answers Call to Help Sandy Victims," *City Wire*, 2012, http://www.thecitywire.com/node/24897#.UZw2pJy87eY (accessed May 21, 2013).
[21]Ibid.
[22]Ibid.

Large corporations are not the only players within the context of emergency management and stakeholders. During emergencies, communication is imperative to facilitate mitigation, response, and recovery operations. Therefore, local news media, such as television stations, radio stations, Internet providers, newspapers, and magazines are valuable resources for disseminating information and for providing emergency communication. Electricity is also a significant aspect of emergency management. Utilities, both public and private, are essential sources of power that facilitate emergency operations during the aftermaths of calamities. Small businesses also are deemed as valuable sources of goods and services. For example, a family hardware store may choose to donate a variety of supplies (e.g., gloves, lumber, etc.) to support emergency operations.

Given these examples, emergency management has a vested interested in the successfulness of businesses. Conversely, business also has a vested interest in the successfulness of emergency management. Emergency management response and recovery operations are necessary for preserving life and rebuilding infrastructure. They are also essential for sparking economic functioning that facilitates business activity thereby generating the flow of goods and services within a locale. For example, following the 2011 tornado that devastated Joplin, Missouri, the recovery phase of emergency management facilitated the reopening of 10 businesses within a year of the incident.[23]

The Political Perspective

The political perspective of identifying stakeholders involves considerations of individuals, groups, agencies, and organizations whose expertise and qualifications have both the potential of influencing policy and contributing meaningfully to emergency management endeavors. Such entities may or may not have government affiliation(s). Individuals represent subject matter experts whose experiences, credentials, and educations are commensurate with policy-crafting and -managing emergencies. These individuals may also have influence politically at various levels of government or may be in positions professionally that lend themselves to the crafting of emergency management policy. Examples range widely from vulcanologists to medical doctors.

Organizational stakeholders may or may not represent government entities. Such organizations may often represent professional occupations or may represent a group of like-minded individuals that have some influence regarding emergency management and policy crafting. For instance, after the 2010 oil spill in the Gulf of Mexico, the United States Conference of Mayors petitioned the federal government

---

[23]David Bieto and Daniel Smith, "Tornado Recovery: How Joplin is Beating Tuscaloosa," *The Wall Street Journal,* 2012, http://online.wsj.com/article/SB10001424052702303404704577309220933715082.html (accessed May 21, 2013).

as follows: "to provide timely initiatives, policies and actions to mitigate the devastation wrought by this release of pollution and advance the immediate rebuilding of impacted ecosystems and economies."[24]

The political perspective of identifying stakeholders also involves a consideration of government entities that contribute toward emergency management activities and the crafting of emergency management policy. These entities involve considerations of federal, regional, state, local, and tribal organizations that have a vested interest in the success of emergency management. They also contribute toward the performing of emergency management activities either directly or indirectly. Because emergency incidents are all considered to be local events, they are managed primarily at the state and local levels. Federal resources are generally supportive entities through which assistance is rendered.

Numerous federal stakeholders exist within the domain of emergency management. This text does not consider an exhaustive listing of such agencies because of the immensity of the array of government agencies. However, the examples herein are representative of the federal organizations that may be employed within the emergency management domain. In no certain order, some notable examples are given as follows:

- *U.S. Geological Survey (USGS)*—This agency may be used to generate topographical maps of affected regions.
- *National Oceanic and Atmospheric Administration (NOAA)*—Weather information may be obtained from this agency.
- *Environmental Protection Agency (EPA)*—This agency is beneficial regarding environmental issues (e.g., contamination, radiation, etc.).
- *National Homeland Security Research Center (NHSRC)*—This agency is a subset of the EPA. It contributes toward quickly detecting and containing contamination, mitigating impacts of such events, and recovering from these events.[25]
- *Federal Aviation Administration (FAA)*—This agency is concerned with all matters related to aviation.
- *Centers for Disease Control and Prevention (CDC)*—This agency is concerned with a variety of issues ranging from pandemics to bioterrorism.
- *U.S. Forest Service*—This agency is within the U.S. Department of Agriculture. It is concerned with a myriad of functions ranging from fighting forest fires to managing public lands among national forests.

---

[24]U.S. Conference of Mayors, "2010 Gulf Oil Spill," 2013, http://www.usmayors.org/features/oil-spill/documents/resolution-2010GulfOilSpill.pdf (accessed May 22, 2013), 2.

[25]National Homeland Security Research Center, "About the National Homeland Security Research Center," 2013, http://www2.epa.gov/aboutepa/about-national-homeland-security-research-center-nhsrc (accessed May 21, 2013).

- *U.S. Fire Administration (USFA)*—This agency is within the FEMA subset. Its mission is to "provide national leadership to foster a solid foundation for our fire and emergency services stakeholders in prevention, preparedness, and response."[26]
- *National Hurricane Center (NHC)*—This center is associated with various facets of researching and exploring hurricanes and cyclones in the Atlantic Ocean and the Pacific Ocean.

Various regional stakeholders exist that contribute to emergency management and the crafting of policy. Such entities may or may not be federal agencies. For example, beginning in 2014, the Nuclear Energy Institute plans to sponsor regional centers that are associated with nuclear emergencies.[27] Located in Memphis, Tennessee and Phoenix, Arizona, these regional entities will exhibit the capabilities of providing auxiliary emergency equipment to nuclear sites inside a 24-hour period thereby managing losses of electricity- and cooling-relevant water supplies.[28] Specifically, these centers are expected to deliver a complete array of "portable safety equipment, radiation protection equipment, electrical generators, pumps and other emergency response equipment to an affected site within the first 24 hours after an extreme event."[29]

Regional centers also may have a medical training emphasis within the context of emergency management. An example of this type of center is the Regional Center for Disaster Preparedness Education (RCDPE) sponsored by Arkansas State University. The RCDPE is primarily an educational entity through which courses of study may be completed in homeland security, emergency management, and disaster life support.[30] This center enjoys national recognition by the American Medical Association and the National Disaster Life Support Foundation.[31] Such regional medical training centers are a source of personnel that are prepared to support all stages of preparedness, response, mitigation, and recovery. Depending upon their experience, credentials, and professional positions, they also may be capable of influencing the crafting of emergency management policy.

An example of a regional center that is concerned with earthquakes is the Center for Earthquake Research and Information (CERI) sponsored by the University

---

[26]U.S. Fire Administration, "About the U.S. Fire Administration (USFA)," 2013, http://www.usfa.fema.gov/about/ (accessed May 21, 2013).

[27]Nuclear Energy Institute, "Industry Developing Regional Response Centers to Deliver Emergency Equipment," 2013, http://www.nei.org/newsandevents/newsreleases/industry-developing-regional-response-centers-to-deliver-emergency-equipment (accessed May 22, 2013).

[28]Ibid.

[29]Ibid.

[30]Arkansas State University, "Regional Center for Disaster Preparedness Education," 2013, http://www.astate.edu/college/conhp/departments/disaster-preparedness/ (accessed May 22, 2013).

[31]Ibid.

of Memphis. The CERI exists as a "clearinghouse for the acquisition and dissemination of earthquake information in the central U.S."[32] Specifically, it serves a purpose of providing "both scientific and non-scientific information about earthquakes and earthquake preparedness in a variety of formats to all segments of the population in the central and eastern U.S. who might ultimately be affected by a significant earthquake."[33] Certainly, the subject matter experts associated with such a regional center may have the capacity of influencing emergency management policy.

Throughout the states and among the U.S. territories, numerous stakeholder organizations exist through which emergency management is employed and through which policies are influenced and crafted. Common examples include departments of public safety, departments of health, departments of transportation, universities, emergency management agencies, and so forth. Others may consist of "metropolitan planning organizations/councils of government, flood control districts, and coastal zone agencies, geological services agencies, and soil conservation agencies."[34] Certainly, many more organizations and agencies may be listed.

Tribal entities are similar to those discussed within the preceding materials. Tribal organizations represent Native American entities through which policies are crafted and emergency management functions are performed. Examples of tribal entities consist of the National Tribal Emergency Management Council, Tribal Risk and Emergency Management Advisory, and Northwest Tribal Emergency Management Council. Such organizations have either a national or regional emphasis concerning their activities and influences. Tribal considerations may also have a smaller scope and magnitude. For example, the Seminole Tribe of Florida facilitates emergency management as follows: "a coordinated response to an unusual emergency situation, which includes activating our Emergency Operations Center (EOC) and Tactical Operations Centers (TOCs), which is how we coordinate the preparation and recovery efforts at all our Tribal Reservations and Communities."[35]

The political perspective represents a variety of entities that all have a stake in the successfulness of emergency management. Such organizations and individuals must cooperate as necessary to facilitate preparedness, mitigation, response, and recovery activities. They must also cooperate when fashioning policy that

---

[32]"Seismic Resource Center," Center for Earthquake Research and Information, University of Memphis, 2013, http://www.ceri.memphis.edu/awareness/ (accessed May 22, 2013).
[33]Ibid.
[34]Federal Emergency Response Agency, "Chapter 2," http://www.google.com/url?sa=t&rct=j&q=%22emergency%20management%22%20stakeholder&source=web&cd=1&cad=rja&ved=0CCsQFjAA&url=http%3A%2F%2Ftraining.fema.gov%2FEMIWeb%2Fedu%2Fdocs%2Ffem%2FChapter%25202%2520-%2520Emergency%2520Stakeholders.doc&ei=VP-bUZT_Ium0yAGa9oCABQ&usg=AFQjCNGQRhVrLenextMUT_MtBntj92gF3Q&bvm=bv.46751780,d.eWU (accessed May 21, 2013).
[35]Seminole Tribe of Florida, "Seminole Tribe of Florida Emergency Management," 2013, http://www.stofemd.com/ (accessed May 22, 2013).

externally affects emergency management, and must also craft policies internally that affect uniquely each respective organization.

The Social Perspective

The social perspective consists of community stakeholder factions that have a vested interest in the successfulness of emergency management and policy. Households are comprised of individuals that suffer during emergencies. Heads of households are responsible for rendering decisions that affect themselves and others within the residence. Households are responsible for obtaining materials (e.g., food supplies, medical supplies, etc.) through which they may endure emergencies.

Households are responsible for crafting their unique policies that influence their activities and behaviors during emergencies. Households are responsible for creating, maintaining, and executing checklists and family emergency plans that reflect their relevant actions during emergencies. Essentially, households must make every attempt to ensure their self-sufficiency during emergencies. The following figures show typical templates that may be used to craft household plans and policies.

Figure 4.3 Family Plan Page #1.

CHAPTER 4 Power and Policy of Stakeholders and First Responders

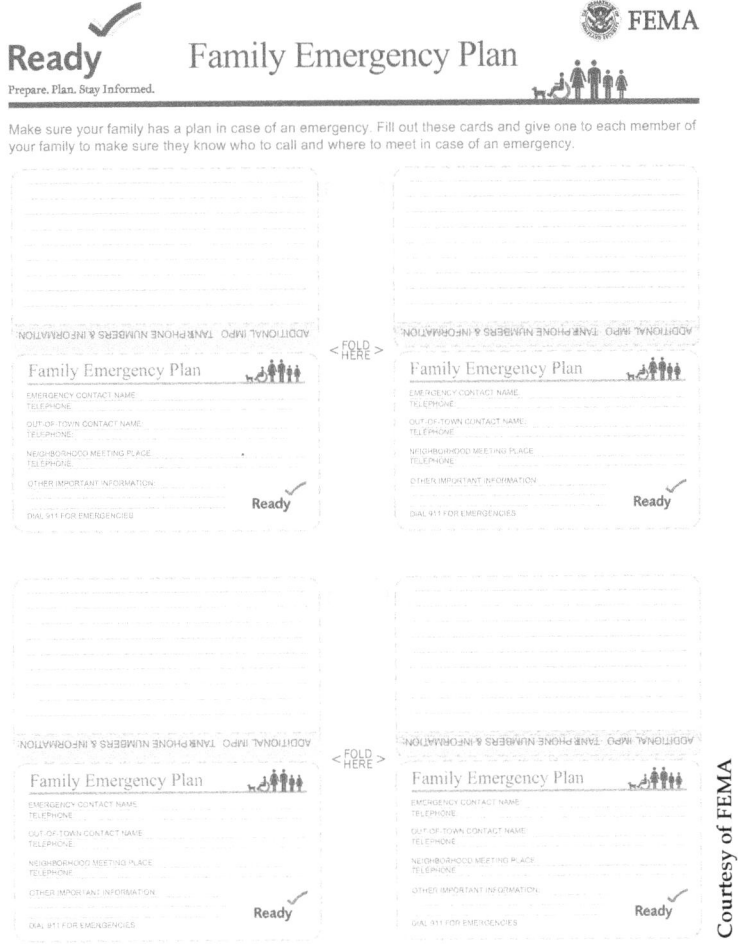

Figure 4.4 Family Plan Page #2.

Neighborhoods are comprised of households. Neighborhoods must cooperate among their respective households to exercise emergency management practices, and to craft policies that impact their respective communities. Within such plans and policies, neighborhoods should also accommodate considerations for elderly residents, disabled individuals, pets, supplies, and equipment. For example, within Los Angeles, California, the Montecito Heights neighborhood maintains emergency plans and policies.[36] Three primary areas are contained within these items: (1) basic plan identifying "threats and risks to the neighborhood, neighborhood assets, and a community disaster response strategy," (2) functional annexes detailing tasks

---

[36]City of Los Angeles Emergency Management Department, *Mosher/Homer Neighborhood Disaster Plan* (Los Angeles, CA: Arroyo Seco Neighborhood Council, 2013).

such as search and rescue and communication, and (3) hazard-specific annexes that detail "responses to specific hazards such as floods or earthquakes."[37]

Often, such plans and policies are crafted in conjunction with community emergency response teams (CERTs). Through CERTs, first responders may receive training and organization assistance that benefits neighborhoods and communities during periods of emergency. CERT training encompasses a variety of topics, including the assessing neighborhood damage and casualties, search and rescue, first aid, and triage.

The societal perspective also incorporates factions of nongovernmental and religious organizations; nonprofit organizations; and various other organizations that may exist within a community that may contribute to emergency management functions and the crafting of various policies. For example, during the aftermath of the 2011 tornadoes that devastate Tuscaloosa, Alabama, fraternities and sororities at the University of Alabama contributed to the emergency management effort through the UA Greek Relief organization. This organization distributed approximately "52,000 meals to victims and volunteers after the devastating tornadoes."[38] During the following 2011 football season, the organization solicited financial donations to benefit the efforts of recovery.[39]

Nonprofit organizations may be formed specifically to satisfy the needs of the ongoing crisis or during its aftermath. Following the devastation Tuscaloosa, Toomer's for Tuscaloosa was formed to satisfy the specific needs of victims. This scenario is described as follows:[40]

> "Within 24 hours of the tornado, the Facebook group Toomer's for Tuscaloosa had been created, and the first truckload of donations was on its way to the tornado victims in Tuscaloosa. . . . within hours, those needs were almost always met — whether it was oxygen tanks or insulin, diapers or meals."[41]

Considerations of nonprofit entities also accommodate the traditional contributions of familiar organizations that render meaningful service within the context of emergency management. Certainly, the Salvation Army, the American Red Cross, and the United Way are among the most notable of stakeholders.

---

[37]Ibid, 6.

[38]Jake Gray, "UA Greeks Raise Funds for Tornado Relief," *Tuscaloosa News*, 2011, http://www.tuscaloosanews.com/article/20111024/NEWS/111029886 (accessed May 22, 2013).

[39]Ibid.

[40]Lydia Avant, "Nonprofits sprung up to meet needs after tornado." *Tuscaloosa News*, 2012, http://www.tuscaloosanews.com/article/20120428/news/120429736 (accessed May 22, 2013).

[41]Ibid.

The social perspective involves considerations of stakeholders ranging from grass roots efforts to well-established, historical organizations that facilitate emergency management endeavors. It also ranges from the smallest of organizational units, representing individuals and households, to some of the largest of well-known relief and assistance organizations (e.g., the Salvation Army). Regardless of the size or resources of a benevolent organization, effectively implementing emergency management activities is dependent upon stakeholder successfulness. In turn, these stakeholders all have a vested interest in the outcome of the emergency management effort.

Power Influences, Stakeholders, and Policy

Management theory has long investigated the characteristics of power within and among organizations. Power is defined as "the ability to marshal the human, informational, and material resources to get something done."[42] This notion is relevant within the context of emergency management regarding preparedness, mitigation associated with the experiencing of calamities, and during the periods of response and recovery. Although the definition of power indicates that it is the ability to accomplish something, the American domain of emergency management exhibits various restraints and permissions regarding the exercising of power among the people, the states, and the federal government.

Emergencies are viewed as local incidents regardless of where they may occur within society. This perspective of locality is derived from the dissemination of power that exists throughout the federal, state, and local constructs of American government. Ultimately, the source of power rests with the American people, and its attributes are delineated within the U.S. Constitution. The Constitution is both a prescriptive and descriptive document that expresses constraints and permissions regarding the source of power and the mechanisms through which it is wielded and exercised. Within the Constitution, enumerated powers are given to the federal government, state powers are restrained to an extent, power not permitted to the federal government is delegated to the states or the people, and conflicts between federal and state powers are favored by the federal government.

These allocations of power are salient regarding the functioning of emergency management. During calamities, law enforcement organizations are often first responders. However, policing power is not granted to the federal government within the Constitution. No national police force exists within American society. Instead, the power necessary for exercising the police function is vested to the

---

[42]Morgan McCall, "Power, Influence, and Authority: The Hazards of Carrying a Sword," *Technical Report*, 10 (Greensboro, NC: Center for Creative Leadership, 1978), 5.

states through the Tenth Amendment as follows:[43] "The powers not delegated to the United States by the Constitution, nor prohibited by it to the states, are reserved to the states respectively, or to the people."[44]

This stating of the Tenth Amendment facilitates the power mechanism through which the individual states become the catalysts for emergency management among localities. Although the power of the federal government may be diminished regarding emergency management, the individual states may wield great power when dealing with emergencies. The individual states may empower their leaders to perform a variety of tasks that are associated with emergency management functions ranging from policing and security to facilitating mass public evacuations.

Because of the delineation of powers expressed within the U.S. Constitution, emergencies are perceived as local events, and they generate first responses initially from the affected incident areas. Coordination also occurs among the states and affected localities. Therefore, the responding entities and their necessary resources are the products of emergency management stakeholders among localities and state agencies.

These observations of government power impact the crafting policy by federal, state, and local stakeholders in the emergency management domain. Among the states, laws may exist requiring the maintaining of emergency management policy. The federal National Incident Management System (NIMS) requires that such policies be crafted and maintained among localities. For instance, in Washington County, Maine, an emergency management policy is maintained that fosters "a process for declaring and terminating a declared emergency and establishes the responsibilities of various municipal officials in oversight of emergency operations."[45]

Another consideration of power involves the declaring of a state of emergency during calamities. The Stafford Act authorizes the President of the United States to "issue major disaster or emergency declarations in response to catastrophes in the United States that overwhelm state and local governments" thereby resulting in the provision of a wide range of possible federal assistance resources.[46] However, a federal emergency declaration or federal assistance may not occur unless a request is first made by a state governor.[47] This scenario is indicative of the dichotomy of

---

[43] Thomas Gardner and Terry Anderson, *Criminal Law*, 11th ed. (Belmont, CA: Wadsworth Cengage Publishing, 2012), 11.

[44] Legal Information Institute, "Amendment X," Cornell University, 2013, http://www.law.cornell.edu/constitution/tenth_amendment (accessed May 22, 2013).

[45] Washington County Council of Governments "Local Emergency Planning," 2013, http://www.wccog.net/local-emergency-management.htm (accessed May 22, 2013).

[46] Francis McCarthy, *Federal Stafford Act Disaster Assistance: Presidential Declarations, Eligible Activities, and Funding* (Washington, DC: Congressional Research Service, 2011), 1.

[47] Ibid.

powers that exists among American government constructs and that is delineated by the U.S. Constitution. It is commensurate with the notion that emergencies are local events, and state and local agents are empowered to act as coordinating entities and first responders.

These notions of power permeate the emergency management domain, and are not limited to solely the considerations of federal, state, and local governments. Instead, the concept of power is applicable to all stakeholders, their relationships, and their interactions within the emergency management domain. Adapted from the Federal Emergency Management Agency, the following figure shows the power relationships that exist within the emergency management domain.[48]

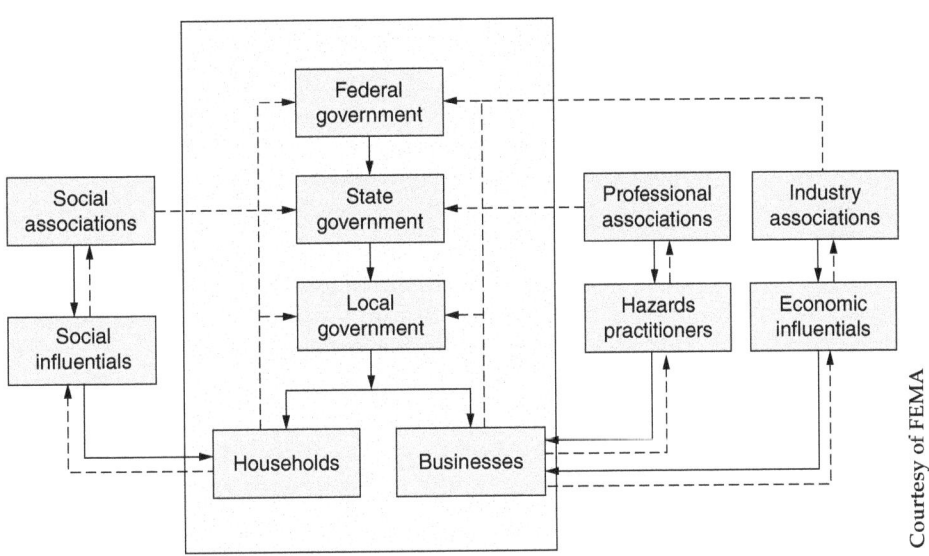

Figure 4.5 Power Relationships in the Emergency Management Domain.

---

[48]Federal Emergency Response Agency, "Chapter 2," http://www.google.com/url?sa=t&rct=j&q=%22emergency%20management%22%20stakeholder&source=web&cd=1&cad=rja&ved=0CCsQFjAA&url=http%3A%2F%2Ftraining.fema.gov%2FEMIWeb%2Fedu%2Fdocs%2Ffem%2FChapter%25202%2520-%2520Emergency%2520Stakeholders.doc&ei=VP-bUZT_Ium0yAGa9oCABQ&usg=AFQjCNGQRhVrLenextMUT_MtBntj92gF3Q&bvm=bv.46751780,d.eWU (accessed May 21, 2013), 41.

Examining this figure shows the complexities of relationships within the emergency management domain. Stakeholder influence regarding influencing policy or other organizations may be exercised through a variety of methods, and involves numerous perspectives of peer stakeholders. For example, because FEMA is such a large federal organization and serves a variety of stakeholders, it may be perceived differently by firefighters, police, emergency workers, businesses, legislators, state and local government officials, insurance companies, and so forth. Below are common perceptions that are associated with such diverse perspectives:

- "Governors, mayors, city managers, and county executives typically champion federal emergency management, but at time tend to be critical of FEMA as well."[49]
- "Major construction firms, the building trades, and economic development interests benefit from the post-disaster, federally subsidized reconstruction, but between disasters, these groups often perceive FEMA as a de facto regulator that sometimes impedes economic development in the interest of public safety."[50]
- "Property insurance corporate officials are generally enthusiastic about federal efforts in disaster loss reduction, but they tend to be suspicious that the federal government, perhaps with FEMA's encouragement, will nationalize certain lines of the insurance they sell and maintain. Insurers and FEMA generally agree on the need to promote disaster mitigation, but each sometimes disagrees on what and how this should be done."[51]

Gaining policy influence is not necessarily an easy task for stakeholders. Federally, it may involve the use of professional organizations, Congressional hearings, lobbying groups, and so forth. Within the states and locally, it may occur through similar means. During recent years, lobbying has become a method of attempting to influence policy by systematically "pressuring governmental officials to make decisions that comport with the interest of the group pursuing the desired action."[52] However, when considering lobbying, some stakeholders are prevented directly from Congressional lobbying because of "conflict-of-interest laws, tax rules that put nonprofit contributions in jeopardy of losing deductibility if the organization lobbies, and laws that restrict lobbying by public employees."[53] Additionally, the

---

[49]Claire Rubin, *Emergency Management: The American Experience 1900–2010*, 2nd ed. (Boca Raton, FL: Taylor and Francis, 2012), 118.

[50]Ibid.

[51]Ibid.

[52]Todd Donovan, Christopher Mooney, and Daniel Smith, *State and Local Politics: Institutions and Reform*, 2nd ed. (Boston, MA: Cengage, 2011), 200.

[53]Rubin, *Emergency Management*, 118.

Table 4.1 Effectiveness Combinations of Authority and Ability

Effectiveness	Authority	Ability
None	No	No
Effective	No	Yes
Marginal	Yes	No
Highly Effective	Yes	Yes

power and influence of stakeholders also tends to be unequal. For example, the clientele of the U.S. Department of Health and Human Services often has greater power and influence politically than do "FEMA clientele."[54]

Perspectives of power relationships and policy influence within the emergency management domain may be considered from the notions of authority and ability. Authority is defined as a "power to influence or command thought, opinion, or behavior" and as "freedom granted by one in authority."[55] Ability is the "quality or state of being able," the "physical, mental, or legal power to perform," or a "natural aptitude or acquired proficiency."[56] Clearly, authority and ability are two distinct concepts. Per its definition, authority represents an empowerment toward achieving some goal or objective, but it does not mean that one is capable of actually achieving results through time. Ability represents the capacity to accomplish a goal or objective. Per the definition of "ability," it is unassociated conceptually with power—it merely represents the capacity of getting results.

These notions of authority and ability may be considered from their combinations that may contribute toward an effectiveness of influencing other entities. Table 4.1 highlights this notion.

If a stakeholder has no authority and no power, then practically no effect may be felt when attempting to influence policy or when influencing peer organizations. If a stakeholder has no authority, but has the ability to accomplish results, then some amount of effectiveness may be generated. Conversely, if a stakeholder has authority, but has no ability, then little, if any, effectiveness may be generated. The final combination is the most powerful combination because authority coupled with ability is capable of producing effectiveness.

These considerations of effectiveness invoke examinations of the types of power that are exercised to generate results through time. Power may be categorized as

---

[54] Ibid.
[55] "Authority," *Merriam-Webster Dictionary*, 2013, http://www.merriam-webster.com/dictionary/authority (accessed May 22, 2013).
[56] "Ability," *Merriam-Webster Dictionary*, 2013, http://www.merriam-webster.com/dictionary/ability?show=0&t=1369250343 (accessed May 22, 2013).

follows: (1) reward power; (2) coercive power; (3) legitimate power; (4) referent power; and (5) expert power.[57] These forms of power are very briefly itemized as follows:

- *Reward Power*—This involves the ability and the willingness to provide either intrinsic or extrinsic rewards to motivate a party to perform.
- *Coercive Power*—This type of power is punitive and coercive in order to motivate another party to perform.
- *Legitimate Power*—This is based on the right and authority of someone within the organizational chain-of-command to issue orders and directives, and others recognize the legitimacy of the formal authority.
- *Referent Power*—This is derived from a desire of some entity to please another party "with whom they identify with because" of perceptions associated with "reputation, personal characteristics, style, or values."[58] This power is also derived from the attributes of the person possessing it (e.g., charisma).
- *Expert Power*—This involves a consideration of credibility associated with an entity, and may involve perceptions of competency and experience.

Law enforcement organizations are often first responders within the context of emergency management. These categories of power may be considered briefly from the perspective of the law enforcement domain. The following itemizations are introductory. Readers desiring a greater knowledge and deeper discussion of power are encouraged to consult texts in management theory, human resources management, or organizational behavior.

- *Reward Power*—This power category involves providing some benefit or compensation that is associated with motivating performance. Examples include commendations, medals, excellence awards, paid overtime opportunity, positive performance evaluations, and so forth.[59]
- *Coercive Power*—This power category involves the punitive aspects of policing. If an officer disobeys an order, then a reprimand may be issued. Other cases may involve the constraining or withholding of overtime hours, refusing transfer requests, or "assigning an individual to perform an unpleasant or less desirable task or duty."[60]
- *Legitimate Power*—This power category is associated with positions within the departmental chain-of-command. For instance, when within the same

---

[57]Trevor Amos, Adrian Ristow, Liezel Ristow, and Noel Pearse, *Human Resource Management*, 3rd ed. (Cape Town, South Africa: Juta Publishing, 2008), 197.
[58]Ibid.
[59]Vincent Henry, *The Compstat Paradigm: Management and Accountability in Policing, Business, and the Public Sector* (Flushing, NY: Looseleaf Law Publications, 2003), 52.
[60]Ibid., 50.

command structure, captains may direct lieutenants, these lieutenants may direct sergeants, these sergeants may direct patrol officers.[61] Officers residing within the upper echelon of the command structure wield the greatest amount of legitimate power whereas patrol officers situated at the bottom of the chain-of-command have little power within the organization.

- *Referent Power*—This power is derived from personal characteristics, such as charisma, congeniality, approachability, likability, and so forth. Officer wielding this power may "achieve compliance without alienating others and without 'pulling rank' simply because people who like them will gladly do what they ask."[62]
- *Expert Power*—This power is associated with education, skills, training, experience, and particular knowledge of a specific subject. Although someone may be merely a low-ranking patrol officer, having the knowledge of a "scarce or particularly valuable skill" generates a greater amount of power "relative to others of the same rank."[63] Examples include speech writing ability, possessing a law degree, computer programming skills, and so forth.[64]

Within the emergency management domain, the dynamics of power are not only applicable within organizations, but are also applicable externally between and among different stakeholder organizations. The dynamics of the emergency management domain contain both abrasive and collegial relationships and influential attributes that impact the crafting of policy through various ebbs, flows, and shifts of power.

For example, government entities may have influence regarding households, but these households also influence government. Legislators within government may develop laws, regulations, and policies that affect society and emergency management. However, households are comprised of individuals that vote during elections, and may either vote someone out of office if they are dissatisfied or may elect or reelect someone with whose performance and policy positions they are satisfied. Such interactions represent the directionality of power.

The relationships exhibited within the emergency management domain are complex and dynamic. They affect every facet of emergency management and its various stakeholders. The interactions between and among stakeholders often involve various amounts of power. Such power involves a consideration of its different forms that are used to achieve any desired outcomes that are associated with a myriad of varying goals and objectives. Policy is influenced by the power plays that occur among stakeholder factions within the emergency management domain.

---

[61]Ibid.
[62]Ibid., 54.
[63]Ibid., 57.
[64]Ibid.

Although the exercising of stakeholder power often involves a vicious cycle generating both predictable and unpredictable behaviors and outcomes among a practically innumerable array of entities, its relevancy within the emergency management domain cannot be understated or ignored.

Crafting Policy

Crafting policy is both an art and a science. Qualitatively, it necessitates considerations of stakeholders and involves the use of power throughout its phases. Quantitatively, it may incorporate various research studies and involve specific periods for completing its process phases. Crafting policy is not a random endeavor; instead, it involves a systematic methodology through which it is produced through time.

A typical process methodology for crafting policy is presented within the following figure.[65]

Policy Terminology	Stage 1: Agenda Setting	Stage 2: Policy Formulatio	Stage 3: Policy Adoption	Stage 4: Policy Implementation	Stage 5: Policy Evaluation
Definition of Policy Stage	Establishing which problems will be considered by public officials	Developing pertinent and acceptable proposed courses of action for dealing with a public problem	Developing support for a specific proposal so that a policy can be legitimized or authorized	Applying the policy by using government's administrative machinery	Determining whether the policy was effective and what adjustments are needed to achieve desired outcomes
Typical objective	Getting the government to consider action on a problem	Generating alternative solutions to the problem	Getting the government to accept a particular solution to the problem	Applying the government's policy to the problem	Evaluating effectiveness and identifying improvements

Figure 4.6 Policy Process. Policy is crafted methodically via an established process.

[65]Federal Emergency Response Agency, "Chapter 2," http://www.google.com/url?sa=t&rct=j&q=%22emergency%20management%22%20stakeholder&source=web&cd=1&cad=rja&ved=0CCsQFjAA&url=http%3A%2F%2Ftraining.fema.gov%2FEMIWeb%2Fedu%2Fdocs%2Ffem%2FChapter%25202%2520-%2520Emergency%2520Stakeholders.doc&ei=VP-bUZT_Ium0yAGa9oCABQ&usg=AFQjCNGQRhVrLenextMUT_MtBntj92gF3Q&bvm=bv.46751780,d.eWU (accessed May 21, 2013), 42.

Crafting policy may necessitate several iterations of process methodology, and some phases of the methodology may occur simultaneously. Despite this consideration, the crafting of policy may be segregated into five distinct phases: (1) agenda setting, (2) policy formulation, (3) policy adoption, (4) policy implementation, and (5) policy evaluation.

Agenda Setting

The initial stage of the policy process involves the setting of agendas. Focusing the attention of politicians and of the general public toward acknowledging the dangers of potential hazards is often problematic. Some politicians and members of the public may find the domain of emergency management to be immaterial and uninteresting. However, when a disaster strikes, such opinions may change quickly. The damage inflicted by disasters may range from nominal to severe. Regardless, localities must plan for disaster with respect to the all-hazards concept regarding potential endangerments that may impact them. Therefore, emergency managers must strive to create an awareness of hazards among their jurisdictions. When creating this awareness, emergency managers must make every effort to ensure that hazards are identified and contained within the political agenda of their locality.

Getting the government to consider action on a problem is often a daunting task. Numerous issues may be competing with emergency management for limited discussion time. Some administrators may not perceive emergency management as a top agenda priority, and may give it only a cursory consideration. Local school issues may require much attention. Municipal maintenance, such as sewers and utilities, may also require time within an agenda. Such issues are common, daily considerations of municipalities whereas disaster may be highly infrequent events. Therefore, emergency management may be perceived as less important.

Despite these observations, there are tactics emergency managers may employ to interject items into an agenda. If a nearby community, municipality within the region, or neighboring state has experienced a disaster, then this experience may be leveraged as a focusing event through which attention may be drawn to emergency management concerns. Similarly, using a publicized disaster as a window of opportunity to generate municipal discussion is another tactic. Both methods may result in having emergency management items included within an agenda.

Policy Formulation

After emergency management items are included within an agenda, emergency management personnel must be prepared to discuss these items. Unpreparedness and ignorance of the listed items shows a lack of professionalism and poor knowledge of the issues that are to be considered. Emergency management personnel must be well-prepared and knowledgeable, and must succinctly explain their issues and positions. When doing so, a variety of different views and perspectives may be considered when discussing agenda items. Numerous solutions and recommendations may be voiced regarding emergency management issues.

When discussing policy agenda items, emergency management personnel should be prepared to consider the following:

- Threats and vulnerabilities that may impact the locality;
- Risk assessment and evaluation;
- Threat zones;
- Emergency services capabilities;
- Emergency management capabilities;
- Specific definitions of policy goals and objectives;
- Regulations (e.g., land-use, building construction, etc.);
- Hazard awareness programs;
- Funding and financial matters;
- Potential emergency management scenarios and solutions;

Certainly, many more considerations may be listed. Each locality has its own specific, unique characteristics that will affect the discussions regarding policy formulation. No two municipalities are identical; many differences may exist regarding the adopting of policies within their respective locales. However, these considerations are a few attributes that must be acknowledged within the context of emergency management when crafting policy.

Any crafted policy is useless if it is not adopted and implemented by a municipality. Therefore, emergency management entities may consider cooperation with other stakeholders to craft policies that may be acceptable to the municipality. Practically every segment and component of the locality, both weak and strong, may contribute meaningfully and positively to the crafting of policy. By forging such collaborations and having a good amount of stakeholder inclusiveness, advocates for the proposed policy may be gained. By doing so, more stakeholders have a vested interest in the strength and quality of the proposed policy, and whether it is adopted and implemented within the locality.

The goals of this process phase are to get the government to accept a particular solution to the problem and to develop strong levels of support for a specific proposal so that a policy can be legitimized or authorized. Cooperation among stakeholders contributes toward these goals through generating numerous solutions and alternative solutions to emergency management problems. Because of the diversity of numerous stakeholders, a variety of perspectives are contained within the proposed policy. Therefore, stakeholders may be much more willing to cooperate toward having the policy accepted and authorized by the municipality.

Policy Adoption

This phase of the policy process necessitates a consideration of persuading the municipality to adopt the proposed policy. Stakeholders and their supporters may exert pressure toward convincing municipal leaders to accept and to adopt the proposed policy. Arguments must be presented that are supportive of adopting the proposed policy. These arguments may be supplemented by various analytical resources that strengthen the case for policy adoption. For example, a cost–benefit analysis may be incorporated to show the financial and economic attributes associated with policy adoption.

During the discussion of policy adoption, various factions of the community should voice supportive opinions and reasons that advocate adopting the proposed policy. Input from churches, government officials, neighborhood associations, businesses within the locality, and other relevant parties are useful resources for presenting cases toward the adopting of policy. The legal authority of a policy depends on whether it is accepted and recorded with the municipality. It must be legitimized and authorized. Therefore, all efforts should be made by these stakeholder groups to effectively convince municipal leaders to formally accept, adopt, and record the proposed policy appropriately.

Administrative concerns of municipal meetings are relevant toward the adoption of policy. Although this notion may seem insignificant, it is quite important. Meetings usually have a limited amount of time allocated to discuss each of the pertinent issues that affect a municipality. Therefore, emergency management personnel that are responsible for presenting the proposed policy must ensure that they arrive in a timely fashion, commence their presentation on time, professionally and properly present the proposed policy, and finish their presentation within the allocated time constraint. Adhering to these tenets reduces the chance of policy rejection.

Policy Implementation

The implementing of policy represents a significant milestone in the policy process because it signifies that a policy has been legitimized, authorized, and recorded thereby giving it legal justification. Although a policy may be operational, no one can be expected to comply with its tenets if they are ignorant of them. Therefore, the policy must be disseminated thoroughly among the pertinent stakeholders. Each of the stakeholders must also disseminate the policy throughout their respective organizational components. By doing so, an awareness and familiarity with the policy and its tenets is achieved.

Although a policy may be adopted successfully, it is ineffective if it is not implemented. Therefore, emergency management agents must ensure that the policy is actually exercised throughout their scope of operations. Various issues may arise when implementing policy. For instance, any stakeholders that were policy adversaries during the preceding stages of the policy process may be slow to implement the policy. In some instances, they may also seek to undermine the implementing of the policy. Emergency management advocates of the policy must be mindful of such potential dissention and resistance when implementing policy.

Therefore, emergency management personnel must be vigilant to manage the implementing of policy and must be mindful of potential problems that may arise with its implementation. If any problems are identified, then solutions must be accommodated quickly to ensure that the progression of implementing policy is unimpeded. If as many stakeholders as possible are positively included toward the crafting of policy during the preceding policy phases, then it is much more likely that the implementation of the policy will occur in accordance with legislative mandates.

Within the United States, government structuring has the potential of affecting policy. The American system of government is primarily a dichotomy of federal and state structures. From the federal perspective, state governments may either hamper or support the implementation of federal policies. For example, depending upon the situation, states may simply choose to comply with the policy. In contrast, from the state perspective, the federal government may support or hamper the implementation of state policies. For example, the federal government may retain funding or resources that are necessary within the state policy.

If a stakeholder is strongly committed to the goals expressed by the policy, then the chances of it devoting significant time and resources toward a successful implementation of the policy are high. Despite the best efforts, stakeholders must have available the resources that are necessary for implementing the policy successfully. Such commitment and the availability of resources are crucial for implementing a policy effectively and efficiently. If legislators are aware of the seriousness of the problems that are addressed by the policy, then the chances of their providing sufficient authority, capacity, and resources toward its successful implementation are bettered. This notion is very true when considering targeted stakeholders that may have the ability to resist the implementing of the policy.

Policy Evaluation

Evaluation is the final phase of the policy-crafting process. Evaluation is defined as analyzing "policies and programs through the conduct of systematic inquiry that describes and explains the policies' and programs' operations, effects, justifications, and social implications."[66] Within the policy process described herein, these observations are collected through a feedback mechanism during this process phase. The collected data may be analyzed, conclusions may be drawn, and recommendations may be offered regarding the policy. Such recommendations may include continuing the policy without modification, modifying the policy, or terminating the policy.

Policies must be evaluated periodically. Each locale may establish its own timing requirements for evaluating its policies. Some periods of evaluation may be quarterly, some may occur every six months, some may be annually, and others may happen every other year. Regardless of the frequency of policy evaluation, it must be leveraged as a tool through which the policy is improved through time.

Stakeholder Decisions

A decision is defined as "the act or process of deciding."[67] Emergency management is full of decisions. Stakeholders make decisions regarding their respective organizations; they make decisions regarding their relationships and interactions with peer entities; they make decisions during preparedness, mitigation, response, and recovery; and they make decisions regarding the crafting of policy. Essentially, they make decisions unceasingly.

The dynamics of these decisions may be quite complex or inanely simple, and the amount of risk associated with decisions may be quite high or may be very little, if any. Decisions are generally rendered "within an organizational context," and also with a greater scope regarding "the organization's environmental context."[68] This notion is relevant for all stakeholders. They must render decisions that affect their internal operations as well as decisions that affect their decisions with external entities.

Human decisions may be viewed from three perspectives: strategic, tactical, and operational. Strategic decisions generally affect stakeholders and policies for a period that averages approximately five years or more. This type of decision involves a long-term consideration of guiding decisions among stakeholder organizations. An example is the third volume of the *New York State Comprehensive Emergency Management Plan* maintained by the New York Disaster Preparedness Commission. An excerpt from this item is given as follows:

---

[66]Melvin Mark, Gary Henry, and George Julnes, *Evaluation: An Integrated Framework for Understanding, Guiding, and Improving Policies and Programs* (San Francisco, CA: Jossey-Bass Publishing, 2000), 3.

[67]"Decision," *Merriam-Webster Dictionary*, 2013, http://www.merriam-webster.com/dictionary/decision (accessed May 22, 2013).

[68]Frada Burstein and Clyde Holsapple, *Handbook on Decision Support Systems 1: Basic Themes* (Berlin, Germany: Springer-Verlag, 2008), 29.

"long-term recovery activities will be based on a newly prepared or updated post-disaster recovery action plan and will be integrated with existing community master plans, capital development, and hazard mitigation plans."[69]

This excerpt shows a long-term consideration of activities versus the maintaining of plans that affect those activities. It is unspecific regarding how these recovery activities are to be crafted and implemented. Instead, it only provides a generic consideration of crafting and implementing activities that are based upon plans that will exist at the point in time when a disaster occurs. Regardless, its allusion to a long-term condition represents a strategic decision.

The next consideration of decisions is tactical. These types of decisions require time, but the amount of time is generally less than five years, and is usually not immediate within the context of the decision domain. Decisions affiliated with annual inspections of materials and resources are a primary consideration of this category. For example, the Kansas Insurance Department requires the annual auditing of financial reports.[70] Because this period is neither immediate nor strategic, it is classified as tactical because it involves only one calendar year.

The final category of decisions is operational. This type of decision generally has the most familiarity because it represents the day-to-day, immediate decisions that are rendered within a stakeholder environment. A typical example of such a decision would be decisions regarding what material is covered during a personnel shift meeting. Although such decisions may seem insignificant (in some cases), they are the basis of conducting operations activities within the organization.

Decisions may also have varying amounts of complexity. They also may occur during situations in which very little information is available to support the decision. Basically, an emergency services manager must make a decision based on very little (if any) available information. This type of decision is referred to as an unstructured or weakly structured decision.[71] Conversely, there are times when a full range of information assets are available to support the rendering of a decision. This type of decision is called a structured decision. A third consideration is a decision state between having available little (or no) information and having available a good amount of information to support a decision. When this state occurs, the available information necessary for framing a decision may have a range of incompleteness. This type of decision is known as a partially structured decision.

---

[69]State of New York, *New York State Comprehensive Emergency Management Plan* (Albany, NY: New York Division of Homeland Security and Emergency Services, 2012), 10.

[70]Kansas Insurance Department, *Policy and Procedure Requiring Annual Audited Financial Reports*, 2012, http://www.ksinsurance.org/legal/regulations/Article%201/40-1-37_Attachment.pdf (accessed May 23, 2013).

[71]Doss, D., Sumrall, W., and Jones, D. *Strategic Finance for Criminal Justice Organizations* (Boca Raton, FL: CRC Press, 2012), p. 9–11.

These notions of varying amounts of information through which to frame a decision may also be considered from the perspectives of risk that may exist within the decision domain. Every decision involves some amount of risk. A structured decision has less risk because a good amount of information may be known about the decision domain. For example, if an emergency services manager is choosing from among a group of personnel to promote someone, and only one person may be promoted to a higher position, then a good amount of information is available from training records, performance reviews, and so forth in order to frame the decision.

Conversely, a weakly structured decision has greater risk because little, if any, information may be available to frame the decision. In this instance, emergency managers may have to make a decision using their best guess regarding what is known. The partially structured case has varying amounts of risk depending on how much is known about the decision domain. For example, when considering a public awareness campaign, emergency managers may decide which radio and television stations would generate the highest and best use of their advertising monies.

Within the emergency management domain, all stakeholders must make decisions. Whether one is considering which employee is best suited for a certain task or which evacuation route is the most expedient, all decisions have some amount of uncertainty. Regardless, all stakeholders within the emergency management domain must realize that at some point, they will be responsible for making decisions involving varying amounts of information and risk.

Chapter Comments and Summary

This chapter provides an overview of policies and power that affect stakeholders and first responders within the emergency management domain. It identifies stakeholders and introduces the types of stakeholder decisions that affect the emergency management domain. Numerous stakeholders exist within the emergency management domain ranging from each individual member of the general public to vast government organizations, such as FEMA.

Within the set of stakeholders is the category of first responders. These entities are the first to arrive at the scene of an incident. They represent a variety of organizations, groups, or individuals. Common examples include police departments, paramedics, and emergency medical technicians. Each first-responder entity perceives differently the emergency management domain. Further, each first-responder entity has different perspectives of emergency management policy.

Policy within the emergency management domain represents both an art and a science. It is not developed sporadically; instead, it is crafted through the exercising of a methodical, systematic process that consists of five phases. These phases consist of the following: (1) agenda setting, (2) policy formulation, (3) policy adoption,

(4) policy implementation, and (5) policy evaluation. When crafting policy, many stakeholders should be included within the process. Including a large quantity of stakeholders contributes to a greater level of policy robustness and potential acceptability.

Society is comprised of different factions of emergency management stakeholders. These factions may be identified economically, politically, and socially. An example of an economic faction represents businesses. An example of the political category is the relationships and power influences that exist between and among stakeholders. Socially, examples range from CERTs to neighborhood emergency plans.

Authority and power are considerations of relationships that exist within the emergency management domain. Power simply represents the ability of accomplishing results successfully through time with respect to some desired goals and objectives. Although a stakeholder may have authority, it may be powerless to achieve results. Therefore, power and authority are neither identical nor synonymous.

Decisions permeate the emergency management domain. Stakeholders render decisions that have strategic, tactical, and operational perspectives. Further, risk also is affiliated with decisions. Every decision has some amount of risk. When rendering decisions, emergency managers may be confronted with varying amounts of information from which to frame a problem. This situation involves considerations of structured, partially structured, and unstructured decisions.

Terminology

- Business
- Cascading effects
- Constitution
- Decision
- Disaster
- Economic perspective
- Evaluation
- Fiduciary obligation
- First responder
- Household
- Influence
- Interrelated
- Lobbying
- Mitigation
- Neighborhood
- Nuclear energy institute
- Operational
- Policy
- Policy process
- Political perspective
- Power
- Regional centers
- Relationship
- Risk
- Social perspective
- Stafford act
- Stakeholder
- Strategic
- Tactical
- Tenth amendment

Thought and Discussion Questions

1. This chapter introduces that concept of stakeholders. Perform some research within your locality, and determine what stakeholders of emergency management exist. Write a brief essay that describes each stakeholder, and that explains why each identified entity is a stakeholder.
2. This chapter introduces the concept of power. Spend some time reviewing news articles and interviewing the leaders of your local emergency service entities. Determine what types of power are exercised within your locality. Write a brief essay that highlights your findings.
3. This chapter introduces a process methodology for crafting policy within the emergency management domain. Do some research regarding alternative process models that may exist for crafting policy. Compare and contrast them with the model presented herein. Write a brief essay that highlights your findings.
4. Review the individual phases of the policy process described herein. Based on your assessments of each stage, which do you believe is the most important? Also, are they all equally important? Write a brief essay that expresses and justifies your opinion.

References

1. Ability. 2013. *Merriam-Webster Dictionary*. http://www.merriam-webster.com/dictionary/ability?show=0&t=1369250343 (accessed May 22, 2013).
2. Amos, T., A. Ristow, L. Ristow, and N. Pearse. 2008. *Human resource management*. 3rd ed. Cape Town, South Africa: Juta Publishing.
3. Anheuser-Busch. 2013. Anheuser-Busch delivers more than 51,000 cans of water to assist tornado relief efforts in Oklahoma City. http://anheuser-busch.com/index.php/anheuser-busch-delivers-more-than-51000-cans/ (accessed May 21, 2013).
4. Arkansas State University. 2013. Regional Center for Disaster Preparedness Education. http://www.astate.edu/college/conhp/departments/disaster-preparedness/ (accessed May 22, 2013).
5. Authority. 2013. *Merriam-Webster Dictionary*. http://www.merriam-webster.com/dictionary/authority (accessed May 22, 2013).
6. Avant, L. 2012. Nonprofits sprung up to meet needs after tornado. *Tuscaloosa News*. http://www.tuscaloosanews.com/article/20120428/news/120429736 (accessed May 22, 2013).
7. Axtmann, R. 2003. *Understanding democratic politics: An introduction*, 114. Thousand Oaks, CA: Sage Publishing.
8. Bieto, D. and D. Smith. 2012. Tornado recovery: How Joplin is beating Tuscaloosa. *The Wall Street Journal*. http://online.wsj.com/article/SB10001424052702303404704577309220933715082.html (accessed May 21, 2013).

9. Burnstein, F. and C. Holsapple. 2008. *Handbook on decision support systems 1: Basic themes*, 29. Berlin, Germany: Springer-Verlag.
10. Carter, C., B. Todd, and M. Pearson. 2013. Oklahoma tornado cuts 17-mile path: Deadly twister had winds of 200 mph or greater. *KOCO News*. http://www.koco.com/news/national/oklahoma-tornado/-/9844074/20231722/-/145goa/-/index.html (accessed May 21, 2013).
11. City of Los Angeles Emergency Management Department. 2013. *Mosher/Homer neighborhood disaster plan*. Los Angeles, CA: Arroyo Seco Neighborhood Council.
12. Cooper, T. 2011. *The responsible administrator: An approach to ethics for the administrative role*. 5th ed. Hoboken, NJ: Wiley Publishing (Jossey-Bass Imprint).
13. Decision. 2013. *Merriam-Webster Dictionary*. http://www.merriam-webster.com/dictionary/decision (accessed May 22, 2103).
14. Digliani, J. 2010. *Reflections of a police psychologist*, 217. Bloomington, IN: XLibris.
15. Donovan, T., C. Mooney, and D. Smith. 2011. *State and Local Politics: Institutions and Reform*. 2nd ed., 200. Boston, MA: Cengage.
16. Doss, D., Sumrall, W., and Jones, D. 2012. *Strategic finance for criminal justice organizations*. Boca Raton, FL: CRC Press.
17. Dutelle, A. 2011. *An introduction to crime scene investigation*, 5. Sudbury, MA: Jones and Bartlett Publishing.
18. EC Council. 2010. *Computer forensics: Evidence collection and preservation*, 4-2. Clifton Park, NY: Cengage Course Technology.
19. Federal Emergency Response Agency. 2013. Emergency management stakeholders. http://www.google.com/url?sa=t&rct=j&q=%22emergency%20management%22%20stakeholder&source=web&cd=1&cad=rja&ved=0CCsQFjAA&url=http%3A%2F%2Ftraining.fema.gov%2FEMIWeb%2Fedu%2Fdocs%2Ffem%2FChapter%25202%2520-%2520Emergency%2520Stakeholders.doc&ei=VP-bUZT_Ium0yAGa9oCABQ&usg=AFQjCNGQRhVrLenextMUT_MtBntj92gF3Q&bvm=bv.46751780,d.eWU (accessed May 21, 2013).
20. Gardner, T. and T. Anderson. 2012. *Criminal law*. 11th ed. Belmont, CA: Wadsworth Cengage Publishing.
21. Gray, J. 2011. UA Greeks raise funds for tornado relief. *Tuscaloosa News*. http://www.tuscaloosanews.com/article/20111024/NEWS/111029886 (accessed May 22, 2013).
22. Henry, V. 2003. *The compstat paradigm: Management and accountability in policing, business, and the public sector*. Flushing, NY: Looseleaf Law Publications.
23. Kansas Insurance Department. 2012. Policy and procedure requiring annual audited financial reports. http://www.ksinsurance.org/legal/regulations/Article%201/40-1-37_Attachment.pdf (accessed May 23, 2013).

24. Legal Information Institute. 2013. Amendment X, Cornell University. http://www.law.cornell.edu/constitution/tenth_amendment (accessed May 22, 2013).
25. Mark, M., G. Henry, and G. Julnes. 2000. *Evaluation: An integrated framework for understanding, guiding, and improving policies and programs*, 3. San Francisco, CA: Jossey-Bass Publishing.
26. McCall, M. 1978. *Power, influence, and authority: The hazards of carrying a sword*, Technical Report. Issue 10, p. 5 Greensboro, NC: Center for Creative Leadership.
27. McCarthy, F. 2011. *Federal Stafford Act disaster assistance: Presidential declarations, eligible activities, and funding*, 1. Washington, DC: Congressional Research Service.
28. National Homeland Security Research Center. 2013. About the National Homeland Security Research Center. http://www2.epa.gov/aboutepa/about-national-homeland-security-research-center-nhsrc (accessed May 21, 2013).
29. Nuclear Energy Institute. 2013. Industry developing regional response centers to deliver emergency equipment. http://www.nei.org/newsandevents/newsreleases/industry-developing-regional-response-centers-to-deliver-emergency-equipment (accessed May 22, 2013).
30. Rubin, C. 2012. *Emergency management: The American experience 1900-2010*. 2nd ed., 118. Boca Raton, FL: Taylor and Francis.
31. Search for Oklahoma tornado survivors nearly complete as residents return to assess damage. 2013. *Fox News*. http://www.foxnews.com/weather/2013/05/21/children-among-dead-after-twister-strikes-near-oklahoma-city/ (accessed May 21, 2013).
32. Schottke, D. 2007. *First Responder: Your First Response in Emergency Care*, 15. Sudbury, MA: Jones and Bartlett Publishing.
33. Seismic Resource Center. 2013. Center for Earthquake Research and Information, University of Memphis. http://www.ceri.memphis.edu/awareness/ (accessed May 22, 2013).
34. Seminole Tribe of Florida. 2013. Seminole Tribe of Florida Emergency Management. http://www.stofemd.com/ (accessed May 22, 2013).
35. Souza, K. 2012. Wal-Mart answers call to help Sandy victims. *City Wire*. http://www.thecitywire.com/node/24897#.UZw2pJy87eY (accessed May 21, 2013).
36. Stakeholder. 2013. *Merriam-Webster Dictionary*. http://www.merriam-webster.com/dictionary/stakeholder (accessed May 21, 2013).
37. State of New York. 2012. *New York State Comprehensive Emergency Management Plan*, 10. Albany, NY: New York Division of Homeland Security and Emergency Services.

38. Todaro, J. 2009. *Emergency preparedness for libraries*, 123. Lanham, MD: Rowman and Littlefield Publishing.
39. U.S. Conference of Mayors. 2010. 2010 Gulf oil spill. http://www.usmayors.org/features/oilspill/documents/resolution-2010GulfOilSpill.pdf (accessed May 22, 2013).
40. U.S. Fire Administration. 2013. About the U.S. Fire Administration (USFA). http://www.usfa.fema.gov/about/ (accessed May 21, 2013).
41. Varone, J. C. 2012. *Legal considerations for fire and emergency services*, 328. Clifton Park, NY: Delmar Cengage Publishing.
42. Washington County Council of Governments. 2013. Local emergency planning. http://www.wccog.net/local-emergency-management.htm (accessed May 22, 2013).

ROLES IN EMERGENCY MANAGEMENT

"There cannot be a crisis next week. My schedule is already full."[1]

Henry Kissinger

Figure 5.1 Train derailment of an Amtrak Passenger Train, Home Valley, Washington, April 3, 2005. "There were 106 passengers and 9 Amtrak employees on board. Thirty people (22 passengers and 8 employees) sustained minor injuries; 14 of those people were taken to local hospitals. Two of the injured passengers were kept overnight for further observation; the rest were released. Track and equipment damages, in addition to clearing costs associated with the accident, totaled about $854,000."[2]

---

[1]Henry Kissinger, "Crisis Quotes," *Brainyquote,* http://www.brainyquote.com/quotes/keywords/crisis.html (accessed June 12, 2013).

[2]National Transportation Safety Board, "Railroad Accident Brief," 2013, http://www.ntsb.gov/investigations/fulltext/RAB0603.html (accessed June 11, 2013).

> **OBJECTIVES**
>
> The objectives of this chapter are to
> - discuss federal, state, and local emergency management;
> - discuss leadership among multiple levels of emergency management;
> - discuss incident complexity;
> - discuss organizational alignment in emergency management; and
> - discuss directives, strategy, and guidance.

Introduction

Emergency management has emerged from the obscurity of identity crisis maturing into a formidable unity in decision-making mechanism during significant events and incidents. Emergency management operates in a dynamic environment. For many years, local emergency management suffered from lack of local support and funding as a priority as compared to local police and fire departments. Emergency management further struggled as part-time, poorly trained, or as an additional responsibility to an existing agency. In most local jurisdictions, emergency management was not historically accepted as an integral component of public safety and certainly not perceived to be a strategic community partner. The professional standing of emergency management is currently, and since the events of September 11, 2001 and Hurricane Katrina, becoming a matured partner in the preparedness community. Emergency management is applicable from the smallest incidents to cataclysms of international scope and magnitude. For instance, Figure 5.1 shows a typical, small incident may occur anywhere within American society – the wrecking of a train.

The Tenth Amendment at the least implies that the responsibility for emergency preparedness rests with the states. Local government units are political subdivisions of the state. The conventional preparedness initiatives have been established in partnership with the states. As early as 1979, the National Governor's Association studied emergency response and adopted standards that are currently in use although in revised language to accommodate changes particularly in the aftermath of the attacks of September 11, 2001 and Hurricane Katrina in 2005.

Those two events, one being an act of terrorism and the other a major natural disaster set the stage for Presidential Directives that began to position the federal government in more of an intervention role with state and local levels of government. The inception of the U.S. Department of Homeland Security has continued a trend of more federal input into the national preparedness efforts. Preparedness is a vital aspect of any potential emergency. Having some levels of preparedness is far better than having no preparedness whatsoever. Figure 5.2 alludes to this notion.

Figure 5.2 Preparedness.

Regarding disasters, one must consider the periods of risk, periods of threats, and political influences that may impede "a wide range of social, political, and organizational processes."[3] From "pre-impact preparedness measures"[4] to the "appreciative gap"[5] that often occurs in the aftermath of the disaster or crisis that results from greater separation between those who formulate and craft policy and those who implement the policy—a gap that leadership must bridge after an event occurs. Emergency management acts as a facilitator and point of coordination to bridge and close gaps within the disparity of multi-disciplinary, multi-jurisdictional dynamics in a cycle of preparedness inclusive of planning, response, and recovery initiatives. Thus, Hoetmer's definition of emergency management represents the "discipline and profession of applying science, technology, planning and management to deal with extreme events that can injure or kill large numbers of people, do extensive damage to property, and disrupt community life."[6]

The notion of comprehensive emergency management (CEM) is widely referenced within the literature of the domain. This framework outlines four basic components: preparedness, response, recovery, and mitigation. The preparedness phase of the framework includes emergency planning, training, warning systems, and public information. The response phase encompasses emergency plan activation, resource deployment, evacuation and sheltering, and search and rescue. The third phase, recovery, deals with damage assessments, debris removal, reconstruction, and disaster assistance.

---

[3] Bain and Hart, 2003:545.
[4] Cutter, 2003:439.
[5] Bain and Hart, 2003:551.
[6] Longley, 1999:845.

Finally, the fourth phase, mitigation, employs risk mapping, hazard management, research and development, and building codes and ordinances. In the practitioner's environment, it becomes critical to delineate these phases—primarily so that planning, training, and awareness do not become secondary to response or so that preparedness is ongoing after the urgency of the disaster and initial response. In fact, today in the practitioner's environment, and in consideration of this particular framework as stated in the early 1990s, the greater overlap appears between the preparedness phase and the mitigation phase where mitigation is described in this framework as the "overarching goal" occurring "between disasters."[7]

It is useful at this point to consider an accepted definition of disaster. A disaster is "a hazard occurrence resulting in significant injury or damage."[8] For instance, "a flood is a natural hazard; flood risk is defined in terms of the hundred-year flood; the people and buildings located within the hundred-year flood zone are vulnerable, and a flood disaster is a flood that injures a number of people, or causes significant damage."[9]

In the aftermath of the attacks on the United States on September 11, 2001, government agencies and researchers began to present and test alterations to the emergency response cycle. One example involves schematic outlining response and mitigation as the major components with response consisting of rescue, relief, recovery, and mitigation consisting of reconstruction and cycling back to preparedness.[10] This cycle is closely aligned with the cycle of crisis relative to crisis managed developed by Fink (1986) in that the preparedness phase is aligned with Fink's prodromal phase which precedes prevention and positioned nearest to the reconstruction phase or in Fink's cycle, the resolution phase. Currently, the U.S. Department of Homeland Security through its Federal Emergency Management Agency (FEMA) describes the emergency management components as prevention, mitigation, response, and recovery.

Theoretical Framework

Emergency operations centers (EOCs) and EOC directors are by design tasked to deal with multiple complex issues simultaneously. More traditional leadership often fail in the environments as fundamental assumptions of "organizational theory and practice" assuming "that a certain level of predictability and order exists in the world" are usually not applicable in these circumstances.[11] The Cynefin framework, employed worldwide, "sort's issues facing leaders into five contexts defined

---

[7]Ibid, p. 848.
[8]Ibid.
[9]Ibid.
[10]Cutter, 2003:440.
[11]Snowden and Boone, 2007:1.

by the nature of the relationship between cause and effect."[12] Leaders and decision makers "operate in ordered domains (simple and complicated)" and "unordered contexts (complex and chaotic)" based on the following characteristics:[13]

1. Simple—within the realm of "known knowns" where often decisions or right answers are "self-evident" and "undisputed." Problems occurring in this context are usually due to "oversimplification" . . . entrained thinking . . . or complacency" and "most frequent collapses into chaos occur because success has bred complacency."[14]
2. Complicated—within the realm of "known unknowns" where decisions are no longer simple and "right answers" may be many. This context often requires "expertise" to "diagnose" the specifics of the problem or issue.[15] Problems or obstacles to effective leadership occurring in this domain are usually due to "entrained thinking" by the experts; failure to acknowledge innovation from nonexperts resulting in "lost opportunities;" and "analysis paralysis" often the result of "entrained thinking or ego."[16]
3. Complex—within the realm of "unknown unknowns" where "right answers" may not exist or will not be readily apparent. Obstacles to effective leadership in the complex domain often center around the "temptation to fall back into traditional command and control management styles . . . and, leaders who try to impose order into a complex context will fail, but those who set the stage, step back a bit, allow patterns to emerge, and determine which ones are desirable will succeed. They will discern many opportunities for innovation, creativity, and new business models."[17]
4. Chaotic—within the realm of "unknowables" where "searching for right answers would be pointless" as "only turbulence" exists. The terrorist attacks of 9/11 are categorized within this domain. Problems or pitfalls for leadership often become apparent following incidents as many successful leaders in the chaotic domain are unable to adjust to "context shifts" and "shift styles to match," for example, the downgrading or reconstitution of order from chaos to complexity.[18] However, this domain does provide the "best place for leaders to impel innovation."[19]
5. Disorder —where it is "particularly difficult to recognize when one is in it" and the "way out of this realm is to break down the situation into constituents parts

---

[12]Ibid, p. 2.
[13]Ibid., p. 5.
[14]Ibid., p. 2.
[15]Ibid., p. 3.
[16]Ibid., p. 3.
[17]Ibid., p. 4.
[18]Ibid.
[19]Ibid., p. 5.

and assign each to one of the other four realms. Leaders can then make decisions and intervene in contextually appropriate ways."[20]

The most effective leaders know that each domain defined in the Cynefin framework along with each disaster, crisis, or emergency requires constant shifting of styles associated with rendering decisions to align with the emergency. One must consider that in addition to the uniqueness of an incident and its relationship to a particular domain, the incident and/or the response can exist in any of the domains and "can evolve" into any of the other domains.[21]

Decision-making support models often are developed for and operate primarily only in the simple and complicated domains, failing to transition effectively into the complex and chaotic domains while leadership should ultimately retain public trust for long-term handling of crisis or emergency incidents. Insightfully, decision makers need the benefit of an integrated view—that is consideration of the technical/nontechnical as well as the expert/nonexpert—to accomplish the optimal "multi-disciplinary" discussion required in the most effective emergency operations and decision making.[22]

Effective emergency management is clearly something that requires not only the sharing of data and information but also the ability to manage information of different types so that it can be accessed at the point of need. Knowledge management involves both people (ways of working and organizational cultures) and technology as an enabler to support people and organizational requirements for information.

The key considerations are those of awareness (knowing what is available, what the quality and potential applications are, and how and where to access it), capacity (the skills base to source, analyze, and disseminate data and information), communication (the technical and human channels to ensure that awareness is maintained, standards are observed, and data and information can move freely as required), and interoperability (the ability of technical as well as human systems to work seamlessly together to provide information as when and where required).[23]

Leadership in Emergency Management

Leaders in the emergency response environment are charged to make decisions "of the more challenging type i.e. they are complex, contingent, relate to multiple objectives that are defined by a range of groups, they are commonly unstructured at the outset of an incident and a range of levels of command and control are involved."[24]

---

[20]Ibid., p. 6.
[21]French and Niculae, 2004:10.
[22]Ibid., p. 12.
[23]MacFarlane, 2005:23.
[24]Ibid., p. 24.

Also, a decision support system must exhibit transparency regarding understandable output for nontechnical managers. This output must not be filtered through vast bureaucratic layers of "technical support" personnel.[25] Characteristics of emergency decisions are often unexpected, unscheduled, unplanned, unprecedented, and unpleasant. Such characteristics are unique from other public–private sector decision contexts. These additional characteristics include uncertainty, complexity, time, pressure, a dynamic event that is innately unpredictable, information and communication problems (overload, paucity, or ambiguity), and heightened levels of stress for participants, coupled with potential personal danger.[26]

It is suggested herein that the emergency response environment often requires innovation and that includes innovative leadership. Figure 5.3 shows Army Lieutenant General Russel Honore whose leadership influenced the response endeavors of Hurricane Katrina. Meta-leadership is defined as "overarching leadership," going beyond leadership and beyond the job description. Meta-leadership requires

Figure 5.3 Lieutenant General (LTG) Russel Honore, U.S. Army (Ret.).

Courtesy of the U.S. Army

---

[25]Cutter, 2003:442.

[26]MacFarlane, 2005:25.

moving from "silo-thinking" to out-of-the-box thinking.[27] Meta-leadership is summarized as the "capacity to envision a sum that is larger than its parts and then to find a way to communicate, inspire, and persuade broader participation."[28]

Homeland security (and emergency management) leaders, as meta-leaders must know courage to confront resistance; curiosity combined with courage to allow imagination; organizational sensibilities enabling navigation of the system; persuasion to inspire and stimulate other; conflict management to accomplish consensus building or on the spot dispute resolution; crisis management for quick thinking and inventive impression; emotional intelligence for steadiness based on intrinsic strength; persistence for enduring a rigorous, often nonrewarding environment; and they must know the value of people to people networking (Marcus, Dorn and Henderson, 2006:48–53). Essentially, one must convene people and let them discover.

During the aftermath of Hurricane Katrina, LTG Honore led "federal troops to help rescue thousands still stranded in New Orleans days after the storm."[29]

This idea of discovery is well aligned with tenants of meta-leadership and the domains of the "unknown" as defined in complexity theory.[30] Discovery, as a tool of leadership, may be elected or optimal in certain situations; however, with complex problems often associated with homeland security and disaster management, discovery may be required to effectively identify a solution. Convening people in these environments includes diverse personalities, and leading some personalities toward a common goal can often be described as planned manipulation.

Trust, or the lack thereof, in the EOC process and leadership can define success or may become a main culprit accounting for command problems in incident response. Often, mistrust at the executive level and nearly an "absolute trust" in agency command levels are evidenced, especially related to delivery of services in an integrated partnership during routine and exigent circumstances. An insecure agency head can easily bastardize the platform established for an EOC. Failure in understanding of the EOC function—a conduit function opposed to a command function—also attributes to ineffectiveness.

Command and control problems may be directly associated with varying styles of leadership and, as highlighted in *Deconstructing Army Leadership*, by Paparone, an "assumption that only a hierarchy can produce effective leadership perpetuates the myth that the most senior person knows more than others."[31] Such a notion

---

[27]Marcus, Dorn and Henderson, 2006:44.
[28]Ibid, p. 47.
[29]Greg Bluestein, "Army General Recalls Katrina Aftermath," *The Washington Post*, 2006, http://www.washingtonpost.com/wp-dyn/content/article/2006/09/07/AR2006090700163.html (accessed June 11, 2013).
[30]Snowden and Boone, 2007:3.
[31]Paparone, 2004:4.

indicates that "we do not know if those who occupy such positions are leaders until followers demonstrate they choose to follow," citing the old adage, "If you think you're leading but no one is following, you're only taking a hike."[32]

A major component of success in managing any crisis or disaster is realization related to the importance of an environment of planning rather than "a" plan. The planner recognizes that: (1) any pre-disaster data collection is critical; (2) updated, current data are critical and are an ongoing process; (3) no two incidents are the same nor do similar incidents require similar response; and (4) technology is constantly changing and improving and having the best available tools on call is an obvious advantage. As for learning—a process of growth and change—it is not the idea of change or crisis that is the challenge, it is, however, the willingness to accept the responsibility of managing the change or crisis, recognizing that change and crisis may be positive rather than negative. Further, politics is a challenge in any environment of change or challenge, but that should be an expectation.

Leadership in disaster response must exemplify courage to confront resistance often revealed through diverse personalities; competencies enabling systems navigation; valuing of people in encouraging discovery; and persistence and intrinsic strength as an agent of change in emerging environment. The journey continues; meta-leaders in response environments are on a continual journey through an emerging and complex environment.

Organizational Alignment

Not since the National Security Act of 1947 has there been such a dramatic re-organizing of U.S. Government until the creation of the U.S. Department of Homeland Security (DHS) in 2002. Creating the DHS was not the beginning of the U.S. involvement in homeland security initiatives. Prior to this time, many governmental agencies were engaged in matters of homeland security. President George Bush and the U.S. Congress, following the attacks of 9/11, acted to create an agency of government that would coalesce national efforts of homeland security into a seamless approach to national preparedness. The U.S. Commission on National Security/21st Century (the Hart-Rudman Commission), chartered by the Secretary of Defense in 1998, found that "the assets and organizations that now exist for homeland security are scattered across more than two dozen departments and agencies and all fifty states."[33]

FEMA was created as an independent function of government during President Jimmy Carter's administration in 1979. Shortly after DHS was created, FEMA was moved into the DHS. The transition of FEMA into DHS has been a challenge. In some measure, the question has become "FEMA in or FEMA out"—should FEMA

---
[32]Ibid.
[33]Wise and Nadar, 2002:10.

remain within DHS, or should it be an independent agency that reports directly to the president as it was before the DHS was created? Is it even possible to restore FEMA's capabilities to deal with natural and technological disasters? The collaborative approach that guided FEMA's programs in the 1990s may be lost. Thus far, each administration has retained FEMA under the umbrella of the DHS. FEMA is one of 22 agencies absorbed into the DHS as the new department emerged.

National Framework: Local, State, and Federal

All incidents are considered to be local events within the context of emergency management. Although this perspective is advocated by many, one must also consider the hierarchies that exist when events of different magnitudes and scopes occur. Events may affect only the citizenry of a small locale or they may affect the entirety of the nation depending upon their severity. Therefore, different perspectives exist regarding the hierarchies that may be involved with emergency management. These viewpoints consist of local, state, and federal perspectives.

Local Emergency Management

Most events and incidents originate in local communities. Local communities not only represent the population base of the United States they also represent the locations for critical infrastructure. In addition, most of the critical infrastructure is owned by private industry. Local first responders—police, fire, emergency management, emergency medical, public health personnel, and health-care providers are trained in responding to the people and infrastructure unique to their communities. Local responders in partnership with citizens are the "first boots" on the ground in planning, training, response, and recovery. Local emergency management officials are responsible for coordinating resources and assets necessary in responding to a disaster and in recovery efforts following a disaster. As the DHS has evolved and adapted to change, the local community remains the foundation for preparedness efforts in the broad spectrum of the "all-hazards" environment. All-hazards encompass the natural disasters and the man-caused incidents.

Emergency management builds from the bottom upwards. The incident or event erupts in the local community and first responders, along with citizens, respond. If the event causes the resources of the local community to be exhausted, the local executive leaders request assistance from the state. If the event or incident overwhelms the capability of the state, the state then requests assistance from the federal government. The reality is that once a local jurisdiction is engaged in an incident/event it is forced to take ownership of the consequence from eruption to resolution. The state and federal response is not in place to assume ownership—they are in place to assist in meeting the needs of the local jurisdiction in managing the event.

After the attacks against the United States on 9/11, and the reaction of government—specifically, the creation of the U.S. Department of Homeland Security—the idea of a national preparedness initiative emerged. National preparedness is intended to network federal, state, tribal, and local response entities into a cohesive organism focused on securing the nation from acts of terrorism and recovering communities when natural disasters strike. The idea or concept of national preparedness is different from a federal, top-down program.

The national preparedness initiatives begin at the local level of government where the first tier of responsibilities for mitigation, prevention, response, and recovery begin. Some posit the notion that "the national emergency management system is a complex network that includes federal, state, and local government agencies; special districts and quasi-governmental organizations, nonprofit organizations; volunteers (both organized and spontaneous); and private sector firms that provide services and products under contract. Thus, one of the primary responsibilities of the emergency manager involves anticipating how and where to obtain additional resources in the event that the locality is overwhelmed by the demands of a disaster."[34]

In 2005, Hurricane Katrina devastated the Gulf Coast states of Alabama, Mississippi, and Louisiana. In the hurricane's aftermath, the reactive public policy changes initiated significant re-organization of the DHS and FEMAs role in national preparedness. These changes were codified in the Post-Katrina Emergency Reform Act of 2006. This began a trend toward more state and federal involvement in local events.

State Emergency Management

State emergency management agencies are established through the legislative branch and under the supervision of the governor. In most states, the governor is given the authorities for emergency management and granted emergency powers. State emergency management agency's mission is to protect life and property within the boundaries of their respective states. Local governments are political subdivisions of the state. Although local governments have autonomy, they remain a subdivision of the state and are subject to the state constitution and laws governing the respective state. State emergency management agencies are recognized through FEMA as representing the state and local jurisdictions in matters involving support and assistance from the federal government.

During an event or incident, and when local resources are overwhelmed, local governments request assistance from the state through the state emergency management agency. In the case of a major disaster, a request from the local government official will be proffered for disaster assistance. The process begins with the chief elected official from the local government declaring a state of emergency. That declaration

---

[34]Rubin, 2007:36.

is then followed with a written request for assistance to the governor—through the state emergency management agency. If warranted, the governor then forwards the request to the president through the FEMA system. When this process is completed, it usually results in the president issuing a Presidential Declaration of Disaster. Once this process is accomplished, the local jurisdiction becomes eligible for disaster assistance funds. The disaster assistance funds are grant funds administered by FEMA.

One direct link that state emergency agencies sustain with local emergency management is through the administration of the Emergency Management Performance Grant (EMPG). This grant provides FEMA the authority to administer grants to states and their political subdivisions (local emergency management) for promoting national preparedness in an all-hazards environment. Conceptually, this grant supports efforts in maintaining a comprehensive system for preparedness. The EMPG has become a vital support for local emergency management in partnership with the state.

Most of the state emergency management agencies originated from the days of civil defense. Many of them were titled Office of Civil Defense (OCD) until FEMA was created in 1979. From 1979 through the 1980s, many states began changing the names to emergency management. Figure 5.4 shows the typical concepts that are associated with a state emergency management agency.

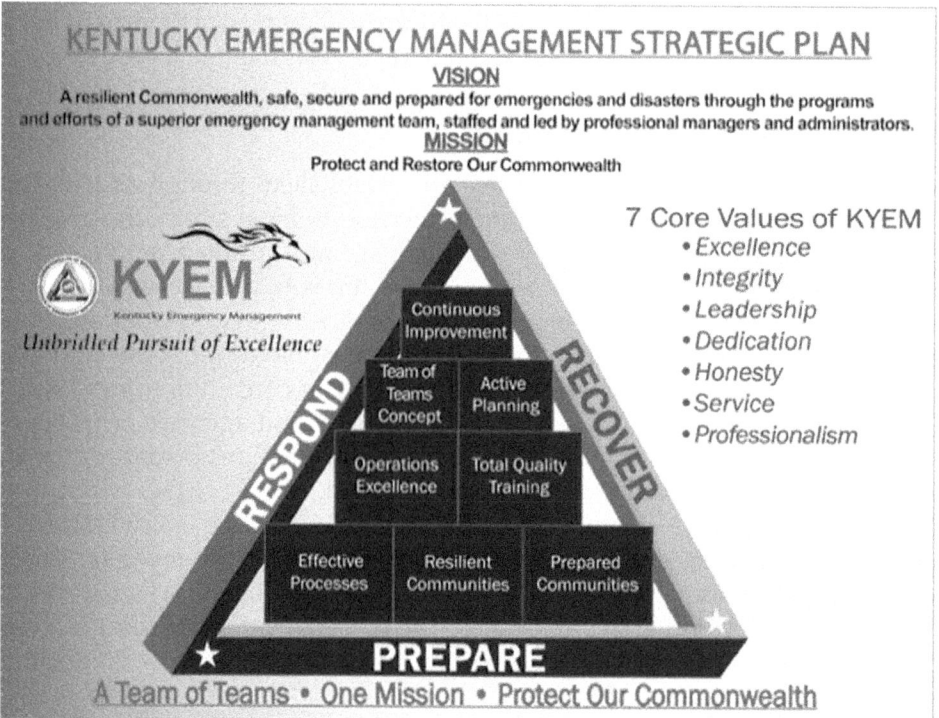

Figure 5.4 State emergency management attributes.

Federal Emergency Management

Before 1979, the U.S. Department of Housing and Urban Development (HUD) was the lead agency for providing disaster relief, preparedness, and civil defense. Executive Orders 12127 and 12148 in 1979 created the FEMA. The transfer of duties is given thusly:[35]

1. from the Department of Commerce the U.S. Fire Administration and National Fire Academy for Fire Prevention Control;
2. from the Department of Housing and Urban Development's Federal Insurance Administration federal flood, riot, and crime insurance programs;
3. from the Office of Science and Technology Policy oversight of the Federal Emergency Broadcast system;
4. from the Department of Defense the functions of the old Defense Civil Preparedness Agency;
5. from the General Services Administration the Federal Preparedness Agency; and
6. from the Department of Housing and Urban Development the Federal Disaster Insurance Administration.

Additionally, "the reorganization transferred national functions related to earthquake hazards, dam safety, natural and nuclear disaster warning systems, and terrorist incidents to FEMA. The reorganization attempted to centralize the responsibility for planning and coordinating these functions within one separate, independent lead agency."[36]

This model of reorganization operated until the advent of the U.S. Department of Homeland Security in 2002. By 2003, FEMA had been subsumed into the DHS no longer acting as an independent agency. FEMA becoming a component of DHS has been a challenge. It is difficult to have discourse focusing solely on emergency management resulting from the integration of emergency management into the homeland security model.

At the time of the creation of DHS and the subsuming of FEMA, the nation was recovering from and coping with the shock of the attacks of 9/11. There was a focus on the conceptual War on Terror and the antiterrorist activities absorbing the attention of the nation's political and social environments. Terrorism was the preoccupation. The mission of DHS was articulated in these three initiatives: prevent terrorist attacks, reduce vulnerability to terrorism, and maximize the recovery from acts of terrorism.

In the first *National Strategy for Homeland Security*, published in 2002, the strategy described homeland security as "a concerted national effort to prevent terrorist attacks within the United States, reduce America's vulnerability to terrorism, and

---
[35]Mushkatel and Weschler, 1985:49.
[36]Ibid., p. 50.

minimize the damage and recover from attacks that do occur."[37] The problem is evident: the mission did not encompass natural disasters yet FEMA is operating under the auspices of DHS. This problem in policy and mission is not insignificant. This uneasy environment between FEMA and DHS was prevalent when on August 30, 2005 Hurricane Katrina—at that time the most destructive natural disaster in the United States—crushed the Gulf Coast states.

Because the federal response to Hurricane Katrina identified critical gaps in capabilities, the U.S. Congress passed the Hurricane Katrina Reform Act of 2006. FEMA was re-positioned in DHS and tasked to administer the preparedness grants program replacing the Office for Domestic Preparedness. According to the Government Accounting Office, "Hurricane Katrina severely tested disaster management at the federal, state, and local levels and revealed weaknesses in the basic elements of preparing for, responding to, and recovering from a catastrophic disaster."[38] The GAO report further states:

> The Post-Katrina Act was enacted to address various shortcomings identified in the preparation for and response to Hurricane Katrina. The act enhances FEMA's responsibilities and its autonomy within DHS. FEMA is to lead and support the nation in a risk-based, comprehensive emergency management system of preparedness, protection, response, recovery, and mitigation. Under the act, the FEMA Administrator reports directly to the Secretary of Homeland Security; FEMA is now a distinct entity within DHS; and the Secretary of Homeland Security can no longer substantially or significantly reduce the authorities, responsibilities, or functions of FEMA or the capability to perform them unless authorized by subsequent legislation. The act further directs the transfer to FEMA of many functions of DHS's former Preparedness Directorate. The statute also codified FEMA's existing regional structure, which includes 10 regional offices, and specified their responsibilities."[39]

This legislative action had far-reaching impact on the emergency management and homeland security—preparedness programs and initiatives on a national scale. This action also ushered in creation of and revision to a myriad of policy shifts including mission and practices.

---

[37] White House, 2002.
[38] GAO, 2009:2.
[39] Ibid., p. 3.

Tribal Governance and Preparedness

Many tribal governments exist along with local jurisdictions. The frameworks that guide national preparedness efforts network local, state, tribal, and federal agencies and jurisdictions. Tribal public safety personnel and officials work in concert with local jurisdictions in accomplishing assessments and plans according to identified threats and risk.

Directives, Strategy, and Guidance

In 2003, the president issued Homeland Security Presidential Directive 5 (HSPD-5) which set forth a requirement for a National Incident Management System (NIMS) and a National Response Plan. A foundational Presidential Directive was issued in 2003, HSPD-8. This directive called for a national strategy to include all-hazards in the national preparedness initiatives. Further, a pivotal element in HSPD-8 was to create a National Preparedness Goal that would evolve into a model for measuring preparedness. These HSPDs continue to guide national preparedness initiatives although they have been revised and at times re-named. For example, the National Response Plan is now the National Response Framework (NRF) and HSPD-8 has been revised by President Obama and is now identified as Presidential Policy Directive 8. Its basic concepts remain.

The NIMS is a systematic model for managing disasters. In simple terms, NIMS is the conductor of the orchestra. It is an enlarged incident command system. "The National Incident Management System (NIMS) identifies concepts and principles that answer how to manage emergencies from preparedness to recovery regardless of their cause, size, location or complexity. NIMS provides a consistent, nationwide approach and vocabulary for multiple agencies or jurisdictions to work together to build, sustain and deliver the core capabilities needed to achieve a secure and resilient nation."[40]

The NRF is a design for managing and coordinating response during events/incidents whether man-made or natural disasters—all-hazards. The NRF outlines the responsibilities of the emergency support functions (ESFs) and their relationship to the EOC. The ESFs represent the partners in response: law enforcement, fire, emergency medical, public health, public works, health care, transportation, and several other disciplines. These functions are staffed by emergency support coordinators who are designated personnel from their respective disciplines specially trained in the operations of the EOC. The NRF provides a standard protocol for the EOC that is mirrored at each level of government. For example, the local EOC

---

[40]FEMA, 2013.

has an ESF 1, the state EOC has an ESF 1, and the national operations center has an ESF 1. The NRF provides organizational tools to assist in managing disasters in a systematic model.

The historical challenge regarding the conflict in the mission of DHS is evolving in an attempt to capture the requirements of the comprehensive network of homeland security. In recent years, there has been a move away from the original focus of combating terrorism to include natural hazards. The *Quadrennial Homeland Security Report* (QHSR) outlines five homeland security missions:

Mission 1: Preventing terrorism and enhancing security

    Goal 1.1: Prevent terrorist attacks

    Goal 1.2: Prevent the unauthorized acquisition or use of chemical, biological, radiological, and nuclear materials and capabilities

    Goal 1.3: Manage risks to critical infrastructure, key leadership, and events

Mission 2: Securing and managing our borders

    Goal 2.1: Effectively control U.S. air, land, and sea borders

    Goal 2.2: Safeguard lawful trade and travel

    Goal 2.3: Disrupt and dismantle transnational criminal organizations

Mission 3: Enforcing and administering our immigration laws

    Goal 3.1: Strengthen and effectively administer the immigration system

    Goal 3.2: Prevent unlawful immigration

Mission 4: Safeguarding and securing cyberspace

    Goal 4.1: Create a safe, secure, and resilient cyber environment

    Goal 4.2: Promote cybersecurity knowledge and innovation

Mission 5: Ensuring resilience to disasters

    Goal 5.1: Mitigate hazards

    Goal 5.2: Enhance preparedness

    Goal 5.3: Ensure effective emergency response

    Goal 5.4: Rapidly recover

The Stafford Act

One of the most prominent programs in the homeland security and emergency management realm is the disaster assistance funds. The Robert T. Stafford Relief and Emergency Assistance Act of 1988 (42 U.S.C. 5195) states,

> The purpose of this Act is to provide a system of emergency management preparedness for the protection of life and property in the United States from hazards and to vest responsibility for emergency preparedness jointly in the federal government and the several States and their political subdivisions . . . The federal government shall provide necessary direction, coordination, and guidance and shall provide necessary assistance as authorized in the Act so that a comprehensive emergency preparedness system exists for all hazards.

Disaster grants, through the Stafford Act, assist local communities by providing public and individual assistance in the days immediately following a disaster. Each Presidential Disaster Declarations are different depending on the results of the damage assessments and identified needs. There are two broad categories in most of the declarations: public assistance and individual assistance. Public assistance will assist local governments in restoring damage related to public infrastructure: utilities, water treatment systems, debris removal, and essential public services.

Often, the public assistance will provide reimbursement for overtime to public employees who are deployed to work in the areas damaged during the event. When individual assistance is eligible, it will provide for short-term assistance to individuals and families. This assistance may come in the form of rental assistance, food replacement, and assistance in replacing personal property.

The Stafford Act is a critical guiding tool for local governments. In the aftermath of Hurricane Katrina, the Mississippi Gulf Coast, especially the three most southern counties, faced the challenge of having no revenue for an undetermined length of time. The revenue from the Casino industry was interrupted. The business community was unable to operate in the immediacy of the storm. Neighborhoods were destroyed along with the businesses that supported the local economy. For the short term, it was grants from the Stafford Act that provided those communities—government, business, and families—with the ability to recover and rebuild.

Unfortunately, many elected officials are unfamiliar with the Stafford Act and often fail to understand how it operates and sustains government when no other assistance is available. When public safety communications is nonexistent because of damage; when the County building is destroyed; when the courthouse is under water or destroyed by winds; when the police headquarters building is destroyed and the police cruisers have been destroyed; when the fire houses are not able to be occupied and the fire trucks are destroyed—and the realization becomes evident that there is not enough "rainy day funds" to replace those resources and assets—where is help? This is the place where the Stafford Act is initiated to assist government, business, and families in restoring their communities.

Presidential Policy Directive 8, which replaces the HSPD-8, defines how the United States will prepare for threats/hazards that pose the greatest risk. The National Preparedness Goal is articulated in PPD- 8: "A secure and resilient nation with the capabilities required across the whole community to prevent, protect against, mitigate, respond to, and recover from the threats and hazards that pose the greatest risk."[41] The goal outlines the five mission areas that the homeland security/emergency management community plans from and performs on as a part of their preparedness projects and programs. The five areas are prevent, protect, mitigate, response, and recovery. The network of responders from local, state, tribal, and federal agencies work toward performing on these common mission areas.

There is a plethora of frameworks, guidance documents, HSPDs, Presidential Policy Directives, laws, and regulations that constitute the parameters for emergency management. A few of the foundational documents have been highlighted in this chapter as a survey of basic principles that support the national preparedness efforts.

Chapter Comments and Summary

The landscape irrevocably changed for civilian first responders following the attacks of September 11, 2001. Emergency management continues to emerge as a vital and essential participant in forming the comprehensive planning initiatives relative to the threats from man-caused and natural disasters. The environment for emergency managers and homeland security coordinators is complex and requires the continued professionalization of those serving in local, state, tribal, and federal agencies. The networking of partnerships is nonnegotiable. The threats to the safety and welfare of communities are ever present.

The events of 9/11, Hurricane Katrina, Hurricane Sandy, and the bombings in Boston, Massachusetts, during the Boston Marathon in 2013 are constant reminders that threats exist. Complacency is not an option. Local communities remain the venue as targets. Maintaining and sustaining national preparedness is paramount in the role of government to provide for the safety, security, and welfare of its citizens. From the old paradigm of civil defense to the new paradigm of securing the homeland, emergency management continues its presence in preparing local communities for the inevitable.

---

[41]FEMA, 2013.

Terminology

Civil defense
Cynefin framework
Decision
Decision framework
Decision model
Emergency operations centers
Hierarchy
Leadership
Local incident
Management
National incident
Organizational alignment
Stafford Act
State Emergency Management Agency
State incident

Thought and Discussion Questions

1. This chapter introduced the importance of leadership within the context of emergency management. Perform some research concerning three leaders who were responsible for leading emergency management efforts during significant events. Examine their individual characteristics as leaders. Write a brief essay that compares and contrasts their leadership abilities.
2. Do some research concerning the basic concepts of both leadership and management. Determine 10 characteristics that highlight both concepts. Write a brief essay that compares and contrasts these two concepts within the context of emergency management.
3. Contact your local state emergency management agency. Determine what types of threats they believe have a good risk of affecting your locale, and how such emergencies may be managed if a disaster strikes. Write a brief essay that highlights your findings.

References

1. Bain, A. and P. Hart. 2003. Public leadership in times of crisis: Mission impossible? *Public Administration Review* 63 (5): 544–553.
2. Bluestein, G. 2006. Army General recalls Katrina aftermath. *The Washington Post.* http://www.washingtonpost.com/wp-dyn/content/article/2006/09/07/AR2006090700163.html (accessed June 11, 2013).
3. Cutter, S. L. 2003. GI science, disasters, and emergency management. *Transactions in GIS* 7 (4): 439–445.
4. DHS. 2010. *Quadrennial Homeland Security review report: A strategic framework for a secure homeland.* Washington, DC: Department of Homeland Security. http://www.dhs.gov/xlibrary/assets/qhsr_report.pdf (accessed June 9, 2014).

5. FEMA. 2013. U.S. Department of Homeland Security, National Incident Management System. http://www.fema.gov/national-incident-management-system. (accessed June 9, 2014).
6. FEMA. 2013. U.S. Department of Homeland Security, National Preparedness Goal. http://www.fema.gov/national-preparedness-goal.
7. Fink, S. 1986. *Crisis management: Planning for the inevitable.* New York, NY: American Management Association.
8. French, S. and C. Niculae. 2004. Believe in the model: Mishandle the emergency. http://www.iscram.org/dmdocuments/ISCRAM2004_InvitedSF.pdf. (accessed June 9, 2014).
9. Hammer, R. J., J. S. Edwards, and E. Tapinos. 2012. Examining the strategy development plan through the lens of complex adaptive systems theory. *Journal of the Operational Research Society* 63 (2012): 909–919.
10. Kissinger, H. Crisis quotes. *Brainyquote.* http://www.brainyquote.com/quotes/keywords/crisis.html (accessed June 12, 2013).
11. Cova, T. (1999) 'GIS in emergency management', in Longley, P. A., Goodchild, M. F., Maguire, D. J. and Rhind, D. W. (eds.). Geographical Information Systems: Volume 2-Management issues and Applications. John Wiley & Sons, New York. pp. 848–858
12. MacFarlane, R. 2005. A guide to GIS applications in integrated emergency management. Emergency Planning College, Cabinet Office. https://www.gov.uk/government/uploads/system/uploads/attachment_data/file/61203/gis_guide_acro6.pdf (accessed June 9, 2014).
13. Marcus, L. J., B. C. Dorn, and J. M. Henderson. 2006. Meta-leadership and national emergency preparedness: A model to build government connectivity. *Biosecurity and Bioterrorism* 4 (2): 44.
14. National Transportation Safety Board. 2013. Railroad accident brief. http://www.ntsb.gov/investigations/fulltext/RAB0603.html (accessed June 11, 2013).
15. Paparone, C. R. 2004. Deconstructing army leadership. *Military Review.* http://www.au.af.mil/au/awc/awcgate/milreview/paparone.htm (accessed June 9, 2014), p. 4.
16. Snowden, D. J. and M. E. Boone. 2007. A leader's framework for decision making. *Harvard Business Review.* pp. 1–9. http://opmexperts.com/A%20Leader's%20Framework%20for%20Decision%20Making%20-%20HBR.pdf (accessed June 9, 2014).
17. Wise, C.R. and R. Nader. 2002. Organizing the federal system for homeland security: Problems, issues, and dilemmas. 44 *Public Administration Review.* Vol. 62, Special issue. http://www.jstor.org/stable/3110169 (accessed June 9, 2014).

18. Waugh, W. L., Jr. and G. Streib. 2006. Collaboration and leadership for effective emergency management. *Public Administration Review*. Vol. 66, Special issue: Collaborative public management, 131–140. http://www.jstor.org/stable/4096577 (accessed June 9, 2014).
19. White House. 2002. National Strategy for Homeland Security, Office of Homeland Security, Washington DC. www.ncs.gov/library/policy_docs/nat_strat_hls.pdf (accessed June 9, 2014).

STRATEGY

It isn't that they can't see the solution. It is that they can't see the problem.

C.K. Chesterton

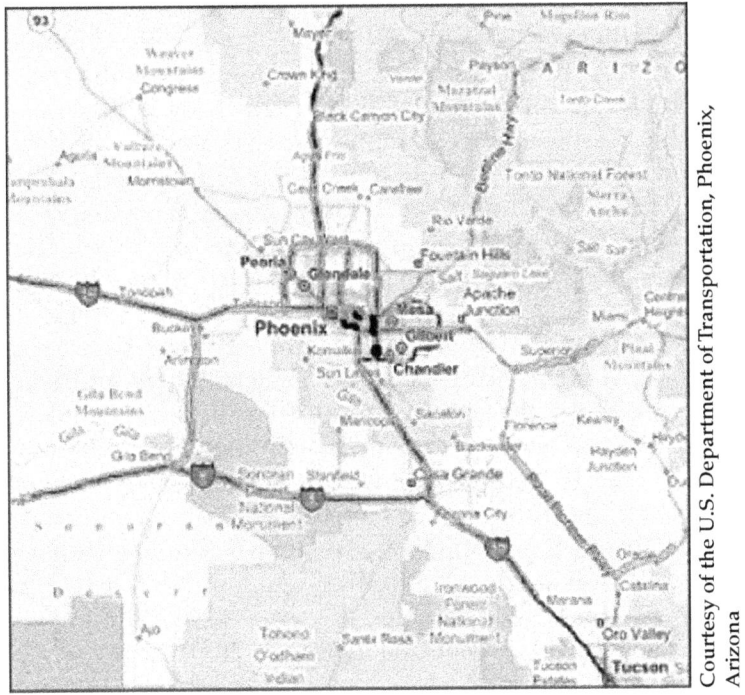

Figure 6.1 U.S. Department of Transportation, Phoenix, Arizona, Mass Evacuation Plan, "a statewide evacuation and reception strategy."[1]

---

[1] U.S. Department of Transportation. "Highway Evacuations in Selected Metropolitan Areas: Assessment of Impediments," 2013, http://ops.fhwa.dot.gov/eto_tim_pse/reports/2010_cong_evac_study/phoenix_az.htm (accessed June 13, 2013).

> **OBJECTIVES**
>
> The objectives of this chapter are to:
> - Discuss the concept of strategy;
> - Introduce the notion of vision;
> - Discuss goals and objectives;
> - Discuss organizational mission; and
> - Discuss contingencies.

Introduction

An ancient Sufi story said to have originated in India and dating back to the twelfth century has been used by many strategy experts in an attempt to define strategy.[2] This story is given as follows:

> "Beyond Ghor there was a city. All its inhabitants were blind. A king with his entourage arrived nearby; he brought his army and camped in the desert. He had a mighty elephant, which he used in attack and to increase the people's awe. The populace became anxious to see the elephant, and some sightless from among this blind community ran like fools to find it. As they did not even know the form or shape of the elephant, they groped sightlessly, gathering information by touching some part of it. Each thought that he knew something, because he could feel a part. When they returned to their fellow citizens, eager groups clustered around them. Each of these was anxious, misguidedly, to learn the truth from those who were themselves astray. They asked about the form, the shape of the elephant and listened to all that they were told. The man whose hand had reached an ear was asked about the elephant's nature. He said: 'It is a large, rough thing, wide and broad, like a rug.' And the one who had felt the trunk said: 'I have the real facts about it. It is like a straight and hollow pipe, awful and destructive.' The one who had felt its feet and legs said: 'It is mighty and firm, like a pillar.' Each had felt one part out of many. Each had perceived it wrongly. No mind knew all. (...) All imagined something, something incorrect."[3]

The majority of people today are not blind—to elephants—but leaders and strategists must consider perceptional blindness. This notion is expressed within the following notions:

---

[2] D. H. Meadows, *Thinking in Systems: A Primer* (White River Junction, VT: Chelsea Green Publishing, 2008).

[3] M. de Vries, *The Whole Elephant* (Deventer, The Netherlands: Ankh-Hermes BV, 2007).

"We are the blind people and strategy formation is our elephant. Since no one has had the vision to see the entire beast, everyone has grabbed hold of some part or other and 'railed on in utter ignorance' about the rest. We certainly do not get an elephant by adding up its parts. An elephant is more than that. Yet to comprehend the whole we also need to understand the parts."[4]

Resultantly, emergency management leaders "must be effective strategists" to ensure the successfulness of the organizations and for creating value for public benefit.[5] Emergency management strategy is more critical than ever before for many reasons, primarily related to enhanced response capabilities and impact. Modern natural or man-made disasters are not necessarily more powerful or destructive than in times past. The disasters of today are often more impactful resulting in greater destruction and death. Current technological advances offer more enhanced capabilities for disaster response, resulting in a greater responsibility upon the emergency responder and emergency response agency to deliver effective and innovative emergency response to any disaster.

Hurricane Katrina of 2005 offers an excellent example of more devastating disaster impact. Historical hurricanes such as Camille, a category 5 hurricane striking the gulf coast in 1969, presented comparable force; however, the impact paled in comparison primarily due to a less densely populated and built-out gulf coast. Another more recent example occurred in 2012 as Hurricane Sandy struck the United States' northeastern coastline. Sandy's devastation was complicated by widespread fires burning down entire communities.

Likewise, the 2011 15-meter tsunami resulted in Japan's Fukushima nuclear accident. Each of these natural disasters resulted in significantly increased impact based on compounded factors and related complexities. Emergency response communities are tasked to respond regardless of the complexity of the incident—fortunately, advanced technologies such as geographical information systems (GIS), WebEOC, etc. are available for planning as well as responding to the predictable and unpredictable incidents.

Thus, one may explore the issues associated with the attributes of who, what, when, where, how, and why regarding effective and innovative strategies for emergency managers, emergency responders, and all related stakeholders. These basic queries are essential aspects of most any level of strategic planning. The following figure presents these notions from the perspective of public involvement within the context of emergency management.

---

[4]H. Mintzberg, B. Ahlstrand, and J. Lampel, *Strategy Safari: A Guided Tour through the Wilds of Strategic Management* (NY: The Free Press, 1998), Kindle location 80.
[5]J. M. Bryson, *Strategic Planning for Public and Nonprofit Organizations: A Guide to Strengthening and Sustaining Organizational Achievement*, 4th ed. (San Francisco, CA: Jossey-Bass, 2011), Kindle location 307.

Category	Example(s)	Synopsis of Purpose and Benefit
Mass Media	Radio, television, internet, newspaper, magazines, and so forth.	Used to communicate with a mass audience. For instance, notification of incoming storms or evacuations.
Questionnaires	Surveys conducted via online, telephone, door-to-door, or mail modalities.	Used to gather data regarding public perceptions of emergency incidents. Maybe analyzed to determine public understanding of emergency events, to learn about the existence of emergency resources, and so forth.
Roundtables	Discussion among emergency response entities concerning hazardous conditions and situations that may affect the locality.	Identifies potential threats that may impact a locality and helps to clarify mutual understandings of these threats among participants.
Targeted Interviews	Specific individuals or groups may be queried to obtain their feedback concerning a specific question or item of interest.	Determines the perceptions of various demographic, individual, or organizational categories within a locality.
Town Hall Gatherings	Open, public forum between members of the public and emergency management entities.	Identifies areas of concern within a locality; facilitates public awareness and involvement; and provides a mutual exchange of information between emergency services and the residents of a locality.

Figure 6.2 Strategies for Enhancing Public Involvement.

Who

Emergency management strategy applies to more than the limited number of emergency managers and responders across the nation. Emergency management impacts every stakeholder including the American citizenry. A stakeholder map for government is shown within the following figure.[6]

This diagram shows a variety of perspectives regarding organizational attributes with respect to competition for limited resources. It also alludes to the mechanisms whereby organizations may collaborate and cooperate during incidents.

---
[6]Bryson, *Strategic Planning*, Kindle location 307.

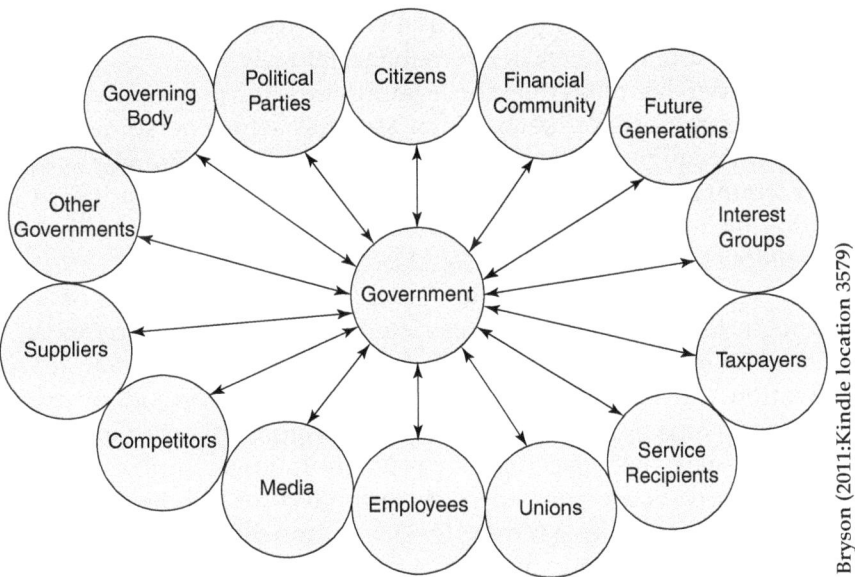

Figure 6.3 Bryson's (2011) Stakeholder Map for a Government

The variable of "who" involves the organization or stakeholder's lack of knowledge regarding organizational identity. It is unsurprising that strategy planning, development, and implementation are much more effective with informed stakeholders, sound policies, and efficient leadership.

Leadership

Modern and future leaders face unprecedented and complex challenges. These challenges are not only crossroads that our grandparents could not have imagined, but are junctures that we did not imagine even a decade ago. The challenges of emergency management and public safety, intertwined with constitutional and human rights issues, are now pivotal leadership dilemmas. As with any area of expertise, any aspects of human characteristics tend to be more an art than science. Thus, identifying potential emergency management personnel must accommodate some review of their historical performance as an individual.

With respect to Complex Leadership Theory, leadership is defined as an "emergent result of interacting individuals such that behavior & resource elements of the organization come together in useful ways"—certain social forces at play among actors, which may or may not include a formal leader.[7] In the absence of an agreed upon definition of leadership, there are accepted views of leadership including the

---

[7] B. B. Lichtenstein, M. Uhl-Bien, R. Marion, A. Seers, O. Douglas, and C. Schreilber, "Complexity Leadership Theory: An Interactive Perspective on Leading in Complex Adaptive Systems," *Emergence: Complexity and Organization* 8, 4 (2006): 6.

managerial and the symbolic or interpretive views. Traditional constructs, like the military, dictate positional power in the managerial view.

Traditional leadership often relies on assumptions that may be myths.[8] Assumptions, like the hierarchical approach which assumes senior personnel know more than others, may endanger effective leadership. Although a hierarchy is believed to accomplish accountability and control, it may be based on circular logic—a model suggesting that the informed strategic leader makes better decisions and the information is available based on organizational position.[9]

Often, followers define leaders. However, "historically, studies of leadership have either neglected followership or restricted their concern to a focus on followers' attributions of exceptional qualities to leaders through, for example, romanticism, idealization, and fantasy."[10] More important to this discussion may be the view of leadership as a "positive process of disproportionate social influence,"—one that distinguishes "between influence (where followership is voluntary) and power (where followers are coerced into compliance or obedience.)"[11] It is concluded that leadership studies need to develop a "much deeper understanding of follower identities and of the complex ways that these selves may interact with those of leaders."[12] The argument can be made that meta-leaders intuitively understand the complexity and integration of follower identities as that understanding may be one of the most powerful attributes of persuasion in this context.

Leadership is explained as innovative thinking, acting and learning—a requirement for emerging emergency management leaders who "work in a far less scripted fashion" than some.[13] Leadership is viewed as strategy—the art of seeing the sum as more than its parts. This art can be compared to Snowden's (2007) tools for managing complexity: interactive communication; creation of rules; stimulation; dissent and diversity. This perspective reveals that complexity science offers a new approach to leadership, one that appears to be well aligned with the charge of emergency management leaders in that "good leadership is not a one-size-fits-all proposition."[14]

---

[8]C. R. Paparone, "Deconstructing Army Leadership," *Military Review* (2004): 2, http://www.au.af.mil/.

[9]Ibid.

[10]D. Collinson, "Rethinking Followership: A Post-Structuralist Analysis of Follower Identities," *The Leadership Quarterly*, 17 (2006): 186.

[11]Ibid.

[12]Ibid., 180.

[13]L. J. Marcus, B. C. Dorn, and J. M. Henderson, Meta-Leadership and National Emergency Preparedness: Strategies to Build Government Connectivity (2006): 43, www.hks.harvard.edu/.

[14]D. J. Snowden and M. E. Boone, "A Leader's Framework for Decision Making," *Harvard Business Review* (2007): 7, http://www.mpiweb.org/.

Complexity science offers the Cynefin framework consisting of five contexts: simple, complicated, complex, chaotic, and disorder.[15] Complex and chaotic contexts are unordered. The complex domain is the realm of unknown unknowns where solutions cannot be readily attained as again, the whole is more than the sum of its parts.[16]

A chaotic context is where America's leadership and emergency response agencies often find themselves experiencing a variety of unknown variables. In this context, solutions are elusive as constants and manageable patterns do not exist. Therefore, "humans use patterns to order the world and make sense of things in complex situations."[17] This "pattern-making tendency" integrates our multiple experiences and provides confidence.[18] Complexity theory studies how patterns emerge through interaction of many agents. As patterns are sensed or discovered the opportunity to stabilize the desirable patterns and destabilize the undesirable patterns creates an environment for emergence of acceptable patterns. Leading in the future of emergency management may well require a "strategic process that incorporates the dynamic realities of complex adaptive systems" by "recognizing and managing systemic patterns."[19]

What

Strategy is "a pattern of purposes, policies, programs, projects, actions, decisions, or resource allocations that define what an organization is, what it does, and why it does it."[20] The root word for strategy is the "Greek word *strategos*, meaning 'a general'"—or a commander of the army.[21] The foundations of strategy were "discussed by Homer and Euripides, and many other early writers" and that the Greek word "stratego means to plan the destruction of one's enemies through effective use of resources."[22] The disadvantage of strategy is that "it has to be realized that every strategy, like every theory, is a simplification that necessarily distorts reality.

---

[15]Ibid.

[16]Ibid.

[17]C. F. Kurtz and D. J. Snowden, "The New Dynamics of Strategy: Sense-Making in a Complex and Complicated World," *IMB Systems Journal*, 42, 3 (2003): 466, http://www.research.ibm.com/.

[18]Ibid., 477.

[19]C. Bellavita, "Changing Homeland Security: Shape Patterns, Not Programs," *Homeland Security Affairs* II, 3 (2006): 1, http://www.hsaj.org.

[20]Bryson, *Strategic Planning*, Kindle location 1908.

[21]M. Lindgren and H. Bandhold, *Scenario Planning: The Link between Future and Strategy* (New York, NY: Palgrave Macmillan, 2009), 36.

[22]J. Bracker, "This Historical Development of the Strategic Management Concept," *The Academy of Management Review* 5, 2 (1980): 219–224, http://www.jstor.org/discover/.

Strategies and theories are not reality themselves, only representations (or abstractions) of reality in the minds of people. No one has ever touched or seen a strategy."[23]

Given these notions, most strategy represents a plan. However, it is cautioned that "strategy is one of those words that we inevitably define in one way yet often also use in another. Strategy is a pattern, that is, consistency in behavior over time."[24] Strategy exists in every organization as evidenced by "some sort of pattern across purposes, actions, resource allocations, and so on."[25] The clarity and quality of these patterns may vary. In addition to patterns, for some organizations "strategy is position" specifically within a particular market; for other organizations, "strategy is perspective" describing specifically the foundational way an organization does things; and, to some, "strategy is play, that is a specific 'maneuver' intended to outwit an opponent or competitor."[26]

Emergency management agencies commonly have difficulty identifying strategic patterns as well as strategic issues. Strategic misalignment is common and often discovered in conjunction with resource deficits. Strategy may be productive and effective or may be misaligned, ineffective, or counterproductive. Strategies also can vary by time frame, from fairly short-term to long-term, and by level. Four basic levels include:[27]

1. cumulative strategy for the whole of the organization;
2. strategy for individual components of an organization;
3. strategies involving the various programs, services, or business processes; and
4. functional strategies (such as financial, staffing, communications, facilities, information technology, and procurement strategies).

Strategy, as defined, may be planned or unplanned. Respectively, these notions represent the "intended strategy and the other realized strategy."[28] Strategy is further divided into "deliberate strategies, fully realized" and "unrealized" strategies with a third category of "emergent strategy—where a pattern realized was not expressly intended."[29] More importantly, "the important question thus becomes: must realized strategies always have been intended? (That intended strategies are not always realized is all too evident in practice)."[30]

---

[23]Mintzberg, Ahlstrand, and Lampel, *Strategy safari*, Kindle location 265.
[24]Ibid., Kindle location 173.
[25]Bryson, *Strategic Planning*, Kindle location 1908.
[26]Mintzberg, Ahlstrand, and Lampel, *Strategy safari*, Kindle location 212.
[27]Bryson, *Strategic Planning*, Kindle location 5544.
[28]Mintzberg, Ahlstrand and Lampel, *Strategy safari*, Kindle location 174.
[29]Ibid., Kindle location 188.
[30]Ibid., Kindle location 174.

Although strategic planning may involve the stating of intent organizationally, changes occur in reality that produce a combination of intent with practice. Therefore the emergency management environment must be vigilant understanding that the unknown, unknowable, or unforeseen describes incidents as well as responses and related strategies—however, there are opportunities in each. *Webster* defines "opportunity" as a favorable juncture of circumstances.

Knowing and understanding our strategy is critical to mission alignment and goal achievement. Given the maze of directives, policy documents, initiatives, and current challenges within the U.S. Homeland Security and the Federal Emergency Management Agency (FEMA), and given the absence of a grand strategy that guides these efforts, it becomes an emergency management task to discover strategically not only what works, but where, when, and how it works.

Strategy may exist in the absence of strategic planning, strategic thinking, or strategic learning. Strategic planning is defined as "a deliberative, disciplined approach to producing fundamental decisions and actions that shape and guide what an organization (or other entity) is, what it does, and why it does it."[31] Strategic planning is not a remedy for all ills and ironically offers "more ways to fail than to succeed."[32] Similarly, strategic learning involves discovering what is successful, and is often based on a current knowledge of what is unsuccessful. Strategic thinking, strategic acting, and strategic learning involve the following notions:

- Strategic thinking contextually involves the pursuit of goals and purposes.
- Strategic acting contextually involves future consequences regarding the achieving of purposes and learning.
- Strategic learning contextually involves accommodating change dynamically and improving systems and organizations.

Elements of strategy are demonstrated within homeland security pursuits of the federal government. For instance, the QHSR (2010:5) states that in the emergency management environment, the purpose is "largely one of leadership and stewardship on behalf of those who have the capabilities to get the job done." The QHSR states a vision; defines mission; prioritizes goals, objectives, performance measures, and strategic outcome statements for each mission—and together these components chart the course of action—the strategy. Specifically, the QHSR (2010:12) outlines five homeland security missions and associated goals:

1. Preventing Terrorism and Enhancing Security;
2. Securing and Managing our Borders;

---

[31]Bryson, *Strategic Planning*, Kindle location 321.
[32]Ibid., Kindle location 1023.

3. Enforcing and Administering Our Immigration Laws;
4. Safeguarding and Securing Cyberspace;
5. Ensuring Resilience to Disasters.

Mission five is related specifically to emergency management. The goals associated with mission five are:

Goal 5.1: Mitigate Hazards

"Mitigation seeks to break out of the cycle of disaster damage, reconstruction, and repeated damage. Mitigating vulnerabilities reduces both the direct consequences and the response and recovery requirement of disasters" (QHSR, 2010:59).

Goal 5.2: Enhance Preparedness

"Active participation by all segments of society in planning, training, organizing, and heightening awareness is an essential component of national preparedness. While efforts have traditionally focused on the preparedness of government and official first responders, individuals prepared to care for themselves and assist their neighbors in emergencies are important partners in community preparedness efforts" (QHSR, 2010:60).

Goal 5.3: Ensure Effective Emergency Response

"A resilient Nation must have a robust capacity to respond when disaster strikes. Such response must be effective and efficient and grounded in the basic elements of incident management (QHSR, 2010:62).

Goal 5.4: Rapidly Recover

"Major disasters and catastrophic events produce changes in habitability, the environment, the economy, and even in geography that can often preclude a return to the way things were. We must anticipate such changes and develop appropriate tools, knowledge, and skills to adapt, improve sustainability, and maintain our way of life in the aftermath of disaster" (QHSR, 2010:63).

The QHSR (2010:70) acknowledges that all disasters are local stating that "state, local, tribal, and territorial governments are on the front lines . . . and are the first responders to incidents of all types. It is suggested that every emergency operations center and manager know and understand the strategy for their organization.

FEMA's Strategic Plan for fiscal years 2011–2014 was published in February 2011. FEMA's 2011 strategic initiatives are to (1) "foster a Whole Community Approach to emergency management nationally; (2) build the Nation's capacity to sustain and recover from a catastrophic event; (3) build unity of effort and common strategic understanding among the emergency management team; (4) enhance FEMA's ability to learn and innovate as an organization."[33] This strategy is related to the "President's National Security Strategy (NSS), the Department of Homeland Security's Quadrennial Homeland Security Review (QHSR), DHS missions, and the priorities expressed by FEMA Administrator's Intent."[34] FEMA's strategic priorities, as cited in this strategic plan, are given as follows:[35]

1. FEMA must strengthen the Nation's resilience to disasters. FEMA must enable individuals, families, and communities to withstand disruption, absorb or tolerate disturbance, act effectively in a crisis, adapt to changing conditions, and grow stronger over time.
2. FEMA must build unity of effort among the entire emergency management team through clearly defined roles and responsibilities for communities and individuals, access to information, and a shared understanding of how risks are managed and prioritized.
3. FEMA must effectively support the needs of disaster survivors and the recovery of affected communities.
4. FEMA must work with its partners to address our most significant risks. Accepting that risk cannot be totally eliminated, it is essential to develop a common understanding of the risks responders and communities face, and to plan, understand preparedness gaps, and build capabilities together to address those of highest collective concern.
5. FEMA must build, sustain, and improve its workforce and develop its current and future leadership. People are the backbone of any organization and FEMA is no exception. FEMA staff must have the tools they need to accomplish the mission. FEMA's ability to develop its workforce is the single most important driver of the Agency's future success.

Strategic Issues

Strategic issues are "fundamental policy questions or critical challenges affecting the organization's mandates, mission and values, product or service level and mix,

---
[33] FEMA, *Strategic Plan: Fiscal Years 2011–2014* (Washington, DC, 2011), 1.
[34] Ibid.
[35] Ibid., 5.

clients, users or payers, costs, financing, organization, or management."[36] Strategic issues inherently involve conflicts arising from a variety of organization issues. For instance, debates and disagreements may involve methods, personnel, resources, implementation, activities, goals, objectives and so forth. Regardless, strategic issues generally emphasize the following:

1. Addressing the need for new or revised high-level rules for making rules; institutional redesign; or adaptations involving new knowledge exploration, new concepts, changes in basic stakeholders and/or stakeholder relationships, or radical new technologies;
2. Creating a process (for example, a strategic planning process) to develop mission, vision, goals and realize them in practice;
3. Producing programs, products, projects, and services;
4. Controlling strategy delivery in the present;
5. Developing future capabilities; and,
6. Maintaining and enhancing stakeholder relations.

When

As noted above, strategies are as likely to fail as to succeed. As the humorous Yogi Berra quote reminds, knowledge of a goal or destination point can be helpful in realizing attainment or arrival. However, achieving a goal or arriving at a destination involves a starting point and a journey. The first reason involves resources or cash flow necessary to accomplish the journey. The second reason exists if organizational resources are insufficient to craft strategy or if strategic designs are abandoned or ignored.

Where

Increased responsibilities for emergency management based on legislative mandates, national response and operating frameworks and guidelines, increased capabilities, and enhanced technologies make the federal, state, and local emergency management environment more relevant than ever before. Based on national framework and guidelines as well as professional expectations, strategic planning requirements for public organizations are applicable to the emergency management environment. Those organizational requirements are:

1. To think and learn strategically as never before.
2. To translate their insights into effective strategies to cope with their changed circumstances.

---

[36]Bryson, *Strategic Planning*, Kindle location 1820.

3. To develop the rationales necessary to lay the groundwork for the adoption and implementation of their strategies.
4. To build coalitions that are large enough and strong enough to adopt desirable strategies and protect them during implementation.
5. To build capacity for ongoing implementation, learning, and strategic change.

How

So how is this "work" accomplished? Experts offer recommendations and solutions yet often fail to include the prescriptive "how to." Can it be that the absence of such strategic, prescriptive instruction is due to the nature of crisis or emergencies within any context? Can it be that traditional strategic approaches are inadequate for such analysis and planning? Scenario planning is the appropriate tool in this context because "the use of scenario planning, as opposed to strategic planning, is an appropriate but at present underused means of anticipating future crises and coping with or even preventing them."[37]

Scenario planning is "a tool in the strategist's arsenal"—a tool that is most useful when prediction of the future is unlikely.[38] This tool relies upon "judgment" and "insight" and less on "formal analysis" or "figures."[39] If one cannot predict the future course, control the future course, or assume stability of the future course—one should engage in traditional strategic planning and its resultant inflexible product.[40]

It must be noted that "emergent strategy emphasizes learning."[41] This learning is best accomplished through scenario planning and in the context of emergency management. It is likely that employing scenario-based planning "as a cognitive device" offers opportunities to structure events and patterns.[42]

Primarily, the "purpose of scenario work" is to "think in terms of asking the why question, trying to find the causes of the causes of the causes. It therefore needs to be a process that uses all cognitive aspects, experience, existing insights, predetermines, predictable structures, uncertainty, doubts, expert knowledge, remarkable people and their original ideas, intuition, curiosity, courage (to suspend disbelief), invention, originality, emotion, intellectual sparkle. It takes time and it cannot be forced."[43]

---

[37]P. Bobbitt, *Terror and consent: The wars for the twenty-first century* (New York, NY: Alfred A. Knopf, 2008), 529.
[38]Mintzberg, Ahlstrand, and Lampel, *Strategy safari*, Kindle location 818.
[39]Ibid.
[40]Ibid., Kindle location 950.
[41]Mintzberg, Ahlstrand, and Lampel, *Strategy safari*, Kindle location 2777.
[42]Kees van der Heijden, *Scenarios: The art of strategic conversation* (West Sussex, England: John Wiley & Sons Ltd., 2005), Kindle location 812.
[43]Ibid., Kindle location 933.

Strategy Formation

Effective leaders utilize planners for determining the purposeful uses of strategies regarding goals, vision, and other strategic issues. A common failure of strategy formation or development is discovered through the countless consulting contracts issued for the purpose of an outsider conducting a strengths, weakness, opportunities, and threats (SWOT) analysis and subsequently setting a strategy for the organization. Some may "oversimplify strategy" and subsequently deny the rigors of strategy formation thereby contributing to root causes of organizational problems and strategic issues.[44]

Crafting strategy may be approached through the following ten specific schools of thought:[45]

1. The Design School—strategy formation as a process of conception.
2. The Planning School—strategy formation as a formal process.
3. The Positioning School—strategy formation as an analytical process.
4. The Entrepreneurial School—strategy formation as a visionary process.
5. The Cognitive School—strategy formation as a mental process.
6. The Learning School—strategy formation as an emergent process.
7. The Power School—strategy formation as a process of negotiation.
8. The Cultural School—strategy formation as a collective process.
9. The Environmental School—strategy formation as a reactive process.
10. The Configuration School—strategy formation as a process of transformation.

The ten strategy formation schools of thought are grouped into three categories. The first three—Design, Planning, and Positioning—"are prescriptive in nature—more concerned with how strategies should be formulated than with how they necessarily do form."[46] The next six schools of thought—Entrepreneurial, Cognitive, Learning, Power, Cultural, and Environmental—highlight the "process of strategy formation" with a focus how strategies are made.[47] The last category includes only one school of thought—Configuration. Mintzberg, Ahlstrand, and Lampel note that "it could be argued that this school really combines the others" as proponents "cluster the various elements—the strategy making process, the content of strategies, organizational structures and their contexts—into distinct stages or episodes."[48]

The first three schools may be summarized to outline basic premises. It should also be noted that emergent strategy "emerges" within the context of the learning school. By far, the design school represents the "most influential view of the

---

[44]Mintzberg, Ahlstrand, and Lampel, *Strategy safari*, Kindle location 565.
[45]Ibid., Kindle location 99.
[46]Ibid., Kindle location 106.
[47]Ibid., Kindle location 117.
[48]Ibid., Kindle location 120.

strategy-formation process . . . at its simplest, the design school proposes a model of strategy making that seeks to attain a match, or fit, between internal capabilities and external possibilities."[49] The founder of IBM, Thomas J. Watson, indicated that "Thought has been the father of every advance since time began—'I didn't think' has cost the world millions of dollars."[50] In the 1940s, the one-word slogan "THINK" appeared in large block-letter signs in IBM offices and plants throughout the company.

The planning school of strategy can be summarized as follows:[51]

1. Strategies result from a controlled, conscious process of formal planning, decomposed into distinct steps, each delineated by checklists and supported by techniques.
2. Responsibility for that overall process rests with the chief executive in principle; responsibility for its execution rests with staff planners in practice.
3. Strategies appear from this process full blown, to be made explicit so that they can then be implemented through detailed attention to objectives, budgets, programs, and operating plans of various kinds.

The positioning school of strategy is summarized as follows:[52]

1. Strategies are generic; specifically common, identifiable positions in the marketplace;
2. The marketplace (the context) is economic and competitive;
3. The strategy formation process is therefore one of selection of these generic positions based on analytical calculation;
4. Analysts play a major role in this process, feeding the results of their calculations to managers who officially control the choices;
5. Strategies come out from this process full blown and are then articulated and implemented; in effect, market structure drives deliberate positional strategies that drive organizational structure.

Strategy Maintenance

Not all strategies are successful. Strategies fail for four main reasons:

1. A basic strategy may be good but have insufficient resources devoted to its implementation, and therefore insufficient progress is made toward resolving the issue it was meant to resolve.

---

[49]Ibid., Kindle location 337.
[50]Ibid.
[51]Ibid., Kindle location 813.
[52]Ibid., Kindle location 1198.

2. Problems change, typically prompting a need for new strategies, on the one hand, and making what was once a solution itself a problem, on the other hand.
3. As substantive problem areas become crowded with various policies and strategies, their interactions can produce results that no one wants and many wish to change.
4. The political environment may shift.

Evaluating and maintaining strategy is essential for maintaining its relevancy and potency. Accomplishing these tasks involves the following attributes:

1. To maintain existing strategies, seek little change in current organizational (interorganizational or community) arrangements;
2. to maintain or marginally modify existing strategies, rely on implementers and focused input from consumers, and involve supportive advocates; and
3. to invest in distinctive competencies and distinctive assets necessary for the success of the strategies.

Planning Dilemma

The all-hazards approach to planning should be a rather simple activity. It fosters an idea of integrated planning. The planner could integrate events such as, acts of terrorism, earthquakes, and technological events into one inclusive plan. The plan could synergize the multi-disciplined, multi-jurisdictional preparedness needs while integrating federal, state and local governmental jurisdictions into the preparedness initiatives. Attempts are being made at creating such a planning environment. However, those attempts have not produced a model that provides remedy to the greater planning dilemma.

Planning is a capability. It is an organizational ability; an asset. The utility of that capability is dependent upon its mission. In other words, the planner may ask what type of plan is needed; what is the mission of the organization that the planning function serves or will be designed to serve; and, is the resultant plan a strategic, tactical, or operational product? The business interest may develop a business model. The military may develop a tactical plan. The emergency management agency may develop a strategic plan. Further questions may include considerations regarding a design model. For example, is the planning or plan scenario driven? Is it threat based? Or, perhaps a capabilities based planning product? Questions that emphasize the planning dilemma are: is this a contingency plan; a continuity plan; a response plan; or, as highlighted through the HSPDs, an "all-hazards" plan?

A discussion of all-hazards planning cannot occur beyond a superficial glance without examining the following areas of conflict inherent within the homeland security and preparedness community: the organizational challenge; the policy challenge; and, the planning challenge. Within the homeland security framework of policy, agencies, and national partnerships that include states and local governments, a formidable conflict exists. That conflict is ironically embedded in the preparedness initiatives. Preparedness is the prevailing, pre-imminent national goal of homeland security.

It is the priority. In the attempt to build an integrated preparedness mechanism that is national in scope as opposed to a federal initiative, the framers have created a dilemma that threatens the effectiveness of the very goal to be achieved. The dilemma is a planning dilemma that exposes the internal conflicts at the United States Department of Homeland Security (DHS), which is a critical organizational challenge; it highlights the challenges posed by Homeland Security Presidential Directives 5 and 8 in the planning environment; and, it reflects a potential policy meltdown.[53]

The subject purpose of HSPD-5 is management of Domestic Incidents. That purpose is stated in the first paragraph of the directive: "To enhance the ability of the United States to manage domestic incidents by establishing a single, comprehensive national incident management system."[54] The body of this directive states, "the Secretary will also provide assistance to State and local governments to develop all-hazards plans and capabilities, including those of greatest importance to the security of the United States, and will ensure that State, local and Federal plans are compatible."

In HSPD-8, President George W. Bush stated that, "this directive establishes policies to strengthen the preparedness of the United States to prevent and respond to threatened or actual domestic terrorist attacks, major disasters, and other emergencies by requiring a national domestic all-hazards preparedness goal, establishing mechanisms for improved delivery of Federal preparedness assistance to state and local governments, and outlining actions to strengthen preparedness capabilities of federal, state, and local entities." Paragraph (2)(a) states, "the term 'all-hazards preparedness' refers to preparedness for domestic terrorist attacks, major disasters, and other emergencies."[55]

Annex 1 to HSPD-8 was published in 2007 stating in paragraph (28) that, "this annex is intended to further enhance the preparedness of the United States by formally establishing a standard and comprehensive approach to national planning. It is meant to provide guidance for conducting planning in accordance with the Homeland Security Management System in the National Strategy for Homeland Security of 2007.[56]

---

[53]White House, Homeland Security Presidential Directive/HSPD-5 and HSPD-8, 2003.
[54]Ibid.
[55]Ibid.
[56]United States Department of Homeland Security, Homeland Security Strategy, 2007, http://www.hsdl.org/?view&did=479633.

To the casual observer, it may seem that these HSPDs are straightforward and offer clarity. However, to the practitioner and student of policy, preparedness, and planning, it is becomes a convoluted labyrinth of conflicted policy and strategy. It is the planning dilemma. The dilemma must be explored or the issues remain unresolved.

Prior to the 9/11 attacks and subsequent legislation and policy, it was typical to consider man-made incidents and natural disasters as unrelated in crisis and consequence management schemes. This does not mean that police, fire, emergency management technicians, and other responders did not respond to the same incidents in support of one another. It does mean that integrated planning was vague, if not absent in many federal, state, and local agencies. Two critical incidents illustrate the planning dilemma for preparedness planners. One is the attacks of September 11, 2001. The changes are obvious; the declaration of the War on Terror, unprecedented as a concept war rather than a war against opposing countries; the creation of the DHS; and, the immediate change at America's airports with the creation of the Transportation Security Administration.

The second illustration of the planning dilemma is Hurricane Katrina in 2005. These two incidents of national significance became the primary agents of change for public policy relative to national preparedness and emergency management strategy.[57] Hurricane Katrina swung the pendulum of focus from terrorism back to natural disasters. This ebb and flow of the planning dilemma between natural disasters and man-made incidents is easily examined in the chronology of the HSPDs and other relative documents published by the United States Department of Homeland Security and its Federal Emergency Management Agency.

The conventional protocol for planning was based upon the specific mission of the discipline. Law Enforcement planned according to its specific mission and concentrated on the man-made incidents. The Fire discipline planned according to its specific mission with a broad concentration on man-made consequence. The Emergency Management Discipline (Civil Defense) concentrated on the natural disasters. The response "landscape" changed with these two national incidents. That change is irrevocable.

Scenario Planning

Scenarios are "vivid descriptions of plausible futures."[58] Scenario planning "is predicated on the assumption that if you cannot predict the future, then by speculating upon a variety of them, you might open up your mind and even, perhaps, hit upon

---

[57]Ibid.
[58]Lindgren and Bandhold, *Scenario Planning*, 23.

the right one."[59] Scenario planning is useful in preparing for the contingencies. Scenarios are useful in strategic planning "to test for robustness and also to develop the responsiveness capability."[60]

Strategic Planning

Strategic planning may occur from the top-down or from the bottom-up within any organization. However, it cannot occur in a vacuum or be accomplished by an individual. Its successfulness is often dependent upon collaboration by individuals and organizations. Examples of factors contributing toward success are given as follows:

1. relying on teams;
2. focusing on network and coalition development
3. making leadership and followership development an explicit strategy
4. establishing specific mechanisms for sharing power, responsibility, and accountability

Strategic planning, subsequently combined with ongoing strategy implementation, defines strategic management within the context of governing organizations. Strategic management "has become an academic discipline" and leaving little room for innovation or emergent strategy, "strategic management has commonly been portrayed as revolving around the discrete phases of formulation, implementation, and control, carried out in almost cascading steps."[61] It is advised that "the 'grand challenge' to strategic management is to manage the balance between stability and flexibility."[62]

The Future

The Federal Emergency Management Agency's (FEMA) Strategic Foresight Initiative (SFI) of 2010 defined "drivers through identification of the underlying trends" and to build an "understanding of how the drivers and trends could impact the future of emergency management. The emergency management community through this initiative highlights well-known, commonly accepted drivers likely to affect the field of emergency management significantly in the near future. The nine

---

[59]Mintzberg, Ahlstrand, and Lampel, *Strategy safari*, Kindle location 813.
[60]Lindgren and Bandhold, *Scenario Planning*, 8.
[61]Mintzberg, Ahlstrand, and Lampel, *Strategy safari*, Kindle location 278.
[62]Lindgren and Bandhold, *Scenario Planning*, 7.

drivers defined by FEMA through the SFI include: "the changing role of the individual in society, climate change, critical infrastructure, the evolving terrorist threat, global interdependencies, government budgets, technological innovation and dependence, universal access to and use of information, and U.S. demographic shifts."

FEMA reports that the significance of the role of the individual is based on (1) empowerment of the individual primarily through communications technologies; (2) increased distrust in government officials to include emergency management; and, (3) expansion of the definition of 'community' in that individuals now belong to multiple virtual communities. Climate change is highlighted by FEMA as a driver with significant impact for emergency management. Impacts cited include (1) rising sea levels and more intense storms in coastal areas; (2) crop and livestock production; (3) stressed water resources; (4) new threats to human health; and, (5) social and environmental stresses; (6) increase and spread of wildfires.

Critical infrastructure is highlighted due to aging sectors including transportation, communication, energy and health. The FEMA report estimates $2.1 trillion over the next five years to sustain national critical infrastructure. Several terrorism trends offer significant challenge to emergency management: (1) terrorists' access to technological and scientific knowledge; (2) adaptively of terror organizations; (3) increase in domestic terror and transnational criminal organizations. Global interdependencies are defined as a driver related to (1) reduced American influence; (2) a shift in economic power from the West to the East; (3) disruptions in the global supply chain. These specific global drivers could "lead to the United States having a greater role in emergency management internationally."[63]

Government budgets as a driver is obvious. Budgetary discussions span local, state and Federal deficits and the report discussed Federalism, increased private-sector partnerships, and privatization of emergency management. Technological innovation and dependency is an impact driver due to (1) rapid innovation and adoption of enhanced technologies including "improvements in how we model and warn about disasters;"[64] (2) increased vulnerability to cyber attack; and (3) sustained public support based on performance.

Universal access to and use of information is a driver as (1) an environment where everyone is both producer and consumer of information; (2) more savvy public resulting in increased public accountability; and, (3) constant need to question and verify legitimacy and accuracy of information. This driver is relevant as multiple intelligence and information leaks headline national news. Finally, U.S. demographic shifts is identified as a driver based on (1) likelihood of internal

---

[63]FEMA, *Crisis response and disaster resilience 2020: Forging strategic action in an age of uncertainty* (Washington, DC: FEMA Office of Policy and Program Analysis, 2010) 4.
[64]Ibid., 5.

migration related to major disasters; (2) continued migration to metropolitan areas; (3) overall population growth with specific increases in the Hispanic, Asian and senior citizen populations.

Why

Why develop strategy? Why not let things continue on the current path? The planning school of strategy is notably susceptible to the hazards of predicting the future solely based on forecast developed by present or past realities. The following sample of famous quotes lends some insight into failure to think outside the box, failure to innovate, failure to explore disruptive technologies, and failure to maintain organizational openness to strategic opportunities.

> "I think there is a world market for maybe five computers." *Thomas Watson, chairman of IBM, 1943.*
>
> "There is no reason anyone would want a computer in their home." *Ken Olson, president, chairman and founder of Digital Equipment Corp., 1977.*
>
> "This 'telephone' has too many shortcomings to be seriously considered as a means of communication. The device is inherently of no value to us." *Western Union internal memo, 1876.*
>
> "The Americans have need of the telephone, but we do not. We have plenty of messenger boys." *Sir William Preece, chief engineer of the British Post Office, 1876.*
>
> "The concept is interesting and well-formed, but in order to earn better than a 'C', the idea must be feasible." *A Yale University management professor in response to Fred Smith's paper proposing reliable overnight delivery service. (Smith went on to found Federal Express Corp.)*
>
> "If I had thought about it, I wouldn't have done the experiment. The literature was full of examples that said you can't do this." *Spencer Silver on the work that led to the unique adhesives for 3-M "Post-It" Notepads.*
>
> "It will be years—not in my time—before a woman will become Prime Minister." *Margaret Thatcher, 1974.*
>
> "There is not the slightest indication that nuclear energy will ever be obtainable. It would mean that the atom would have to be shattered at will." *Albert Einstein, 1932.*
>
> "The bomb will never go off. I speak as an expert in explosives." *Admiral William Leahy, U.S. Atomic Bomb Project.*

Generally, champions of strategic planning find themselves in a position to convince stakeholders of the need for or organizational value in developing strategy. Derived from a list offered by Bryson (2011: Kindle locations 875), as examples of common reasons cited as justification for strategy development, the following items are relevant:

- "We face so many conflicting demands we need to figure out what our focus and priorities should be."
- "Our funders (or board of directors, or new chief executive) have asked us to prepare a strategic plan."
- "We know a leadership change is coming and want to prepare for it."

Enhanced and effective communications may be another purpose of or justification for effective strategy development. As related to command, control, and communications, the 9/11 Commission Report highlighted that "the attacks on 9/11 demonstrated that even the most robust emergency response capabilities can be overwhelmed if an attack is large enough. Teamwork, collaboration, and cooperation at an incident site are critical to a successful response. Key decisionmakers who are represented at the incident command level help to ensure an effective response, the efficient use of resources, and responder safety. Regular joint training at all levels is, moreover, essential to ensuring close coordination during an actual incident."[65]

Chapter Comments and Summary

This chapter introduces the notion of strategy. The discussions herein are merely introductory because a full discourse of strategic concepts is beyond the scope of this text. However, readers are encouraged to peruse books in strategic management or non-profit strategy for much deeper discussions of strategic concepts.

Strategy is a long-term endeavor that delineates the methods through which certain incidents may be managed through time. Generally, strategic planning envisions periods of five or more years. Throughout this period, adjustments to strategy may be made according to various contingencies that exist within the incident domain. It is impossible to imagine each of the contingencies that may affect strategic decisions, but it is possible to conceptualize, itemize, and prioritize the events that organizational leaders believe will occur through time and that influence organizational strategy.

Several questions must be answered within the context of strategic planning to fully accommodate the characteristics of a well-devised strategic plan. These queries involve: who, what, when, where, why, and how? Having an understanding of the appropriate answers to each of these questions influences the rendering of strategic decisions, the allocating of resources, and the implementing of strategic plans in due time.

---

[65] 9/11 Commission Report: 396–397.

Figure 6.4 Strategy is Important.

Terminology

Consequences
Contingency
Goals
Mission
Objectives

Scenario
Stakeholder
Strategy
Time
Vision

Thought and Discussion Questions

1. Perform some research regarding your state's emergency management strategy. Determine the five most probable incidents that are likely to occur and the strategies that are specified for each event. Write a brief essay that highlights your findings.
2. Strategy is a long-term planning endeavor that usually encompasses periods of at least five years. However, between the present time and the fifth year, other

aspects of planning can occur. Do some research regarding this period, and determine the types of planning that are appropriate. Write a brief essay that highlights your findings.
3. Review some of the emergency strategies that affect your locale. Within them, determine what types of events are anticipated. Given these events, identify the stakeholders that may be affected. Write a brief essay that describes how these stakeholders are expected to be affected by the delineated strategies.
4. Contingency planning is a relevant concern for strategic thinking. Based on your findings to the last question, identify the contingencies that may affect your identified strategies. Write a brief essay that highlights your findings.

References

1. Bellavita, C. 2006. Changing homeland security: Shape patterns, not programs. *Homeland Security Affairs* II (3): 1. http://www.hsaj.org. (accessed June 9, 2014).
2. Berra, Y. http://www.yogiberraclassic.org/quotes/ (accessed June 9, 2014)
3. Bobbitt, P. 2008. *Terror and consent: The wars for the twenty-first century*. New York, NY: Alfred A. Knopf.
4. Bracker, J. 1980. "The historical development of the strategic management concept." *The Academy of Management Review,* 5(2), pp. 219–224.
5. Bryson, J. M. 2011. *Strategic planning for public and nonprofit organizations: A guide to strengthening and sustaining organizational achievement*. 4th ed. San Francisco, CA: Jossey-Bass. Kindle.
6. Collinson, D. 2006. Rethinking followership: A post-structuralist analysis of follower identities. *The Leadership Quarterly* 17: 180-186.
7. de Vries, M. 2007. *The whole elephant*. Deventer, The Netherlands: Ankh-Hermes. http://www.marjadevries.nl/ (accessed June 9, 2014).
8. DHS. 2010. *Quadrennial Homeland Security review report: A strategic framework for a secure homeland*. Washington, DC: Department of Homeland Security.
9. DHS. 2012. *U.S. Department of Homeland Security strategic plan for fiscal years 2012–2016*. Washington, DC: Department of Homeland Security.
10. FEMA. 2011. *Strategic plan: Fiscal years 2011–2014*. Washington, DC: Federal Emergency Management Agency.
11. FEMA. 2012. *Crisis response and disaster resilience 2020: Forging strategic action in an age of uncertainty*. Washington, DC: FEMA Office of Policy and Program Analysis.
12. Kurtz, C. F. and D. J. Snowden 2003. The new dynamics of strategy: Sense-making in a complex and complicated world. *IBM Systems Journal* 42(3), pp. 462–483.

13. Lichtenstein, B. B., M. Uhl-Bien, R. Marion, A. Seers, O. Douglas, and C. Schreilber. 2006. Complexity leadership theory: An interactive perspective on leading in complex adaptive systems. *Emergence: Complexity and Organization* 8 (4): pp. 6–8.
14. Lindgren, M. and H. Bandhold. 2009. *Scenario planning: The link between future and strategy*. New York, NY: Palgrave Macmillan.
15. Meadows, D. H. 2008. *Thinking in systems: A primer*. White River Junction, VT: Chelsea Green Publishing.
16. Mintzberg, H., B. Ahlstrand, and J. Lampel. 1998. *Strategy safari: A guided tour through the wilds of strategic management*. NY: The Free Press.
17. Marcus, L. J., B. C. Dorn, and J. M. Henderson. 2006. Meta-leadership and national emergency preparedness: Strategies to build government connectivity, http://dspace.mit.edu/bitstream/handle/1721.1/55934/CPL_WP_05_03_DornHendersonMarcus.pdf?sequence=1 (accessed June 9, 2014).
18. Merriam-Webster Dictionary. n.d. http://www.merriam-webster.com/. (accessed June 9, 2014).
19. Paparone, C. R. 2004. Deconstructing Army leadership. *Military Review*. p. 2. http://www.au.af.mil/ (accessed June 9, 2014).
20. Snowden, D. J. and M. E. Boone. 2007. A leader's framework for decision making. *Harvard Business Review*. http://aacu-secure.nisgroup.com/meetings/ild/documents/Symonette.MakeAssessmentWork.ALeadersFramework.pdf (accessed June 9, 2014).
21. United States Department of Homeland Security. 2007. *National Strategy for Homeland Security*. http://www.hsdl.org/?view&did=479633 (accessed June 9, 2014).
22. U.S. Department of Transportation. 2013. Highway evacuations in selected metropolitan areas: Assessment of impediments. http://ops.fhwa.dot.gov (accessed June 13, 2013).
23. Van der Heijden, K. (2005). *Scenarios: The art of strategic conversation*. West Sussex, England: John Wiley & Sons Ltd.
24. White House. 2003. Homeland Security Presidential Directive/HSPD-5. http://www.fas.org/. (accessed June 9, 2014).
25. White House. 2003. Homeland Security Presidential Directive/HSPD-8. http://www.dhs.gov/. (accessed June 9, 2014).
26. White House. 2007. Homeland Security Presidential Directive/HSPD-8, Annex. http://www.dhs.gov/. (accessed June 9, 2014).

ROLES IN EMERGENCY MANAGEMENT: VOLUNTEER ORGANIZATIONS AND NGOS

"What I've always found is that people will respond to meet a need in a crisis if they know what to do. You give people the opportunity to be a part of something that will make a difference and they will step up."

James Lee Witt, Former FEMA Director

Figure 7.1 American Red Cross.

> **OBJECTIVES**
>
> The objectives of this chapter are to:
> - describe nongovernment organizations;
> - discuss the role of nongovernment organizations during an emergency;
> - describe the framework in which the nongovernment organizations operate;
> - describe the relationships among nongovernment organizations with stakeholders; and
> - describe functions of nongovernment organizations in emergency management.

Introduction

Fundamentally, emergency management is about preventing the loss of human life, the preservation of property, and preparedness in the event of a natural or man-made disaster. The practice of emergency management has steadily evolved into the necessity for all levels of government and nongovernment organizations (NGOs) to work together prior to, during, and after a disaster. The human element is paramount with respect to assessing, planning, responding, and recovering from any type of disaster. This has led to the significant development and improvement of service deliverables provided by both government agencies and NGOs, whose sole function is to participate, promote, and professionalize the process of emergency management.

Primarily, the resources of local, state, and federal government agencies are utilized to manage almost any event that may take place within our nation. Notwithstanding governmental resources, often times there are other NGOs that play an integral role and offer a unique or specific function that is significant to the effective management of a critical incident or disaster. These NGOs are positioned in the private sector (corporate entities), or as nonprofit groups (charitable support).

Specifically, this chapter discusses the purpose, role, and benefits of NGOs within the process of emergency management; identify key NGOs that play a role in emergency management in the private and nonprofit sectors; describe the concepts or frameworks suggested to coordinate efforts between all government and NGO entities during the emergency management process, and mention the potential problems that may surface during the disaster response or recovery phases when volunteers approach emergency managers or relief organizations and wish to be a part of the emergency response framework.

Nongovernment Organizations

Throughout the United States, there are literally thousands of organizations, both large/nationwide and small/local that play an integral role during the response and recovery phases of a disaster, by either filling the gaps that government agencies lack immediate resources for or complementing the dual response and recovery efforts of government agencies. Notwithstanding, every organization and agency plays a key role in emergency management from the public, private, and nonprofit sectors.

Given the diversity and purpose of humanitarian organizations, it remains a challenge to develop a standard categorization with respect to public involvement in disaster response and recovery. Thus, the most prolific and well-recognized moniker within the arena of emergency management is the term *nongovernment organizations*. Most importantly, it should be well emphasized that NGOs are clear stakeholders in the field of emergency management and partnerships should be developed and maintained during all four phases that include mitigation/prevention, preparation, response, and recovery (Federal Emergency Management Agency (FEMA), http://www.fema.gov/).

Essentially, the term nongovernment organization distinguishes groups that are voluntary organizations and private sector organizations (PSOs). Typically, NGOs are characterized as being nonprofit, community based, corporate sponsored, or religious, which are supported by private citizens who are motivated by and maintain a humanitarian disposition. Such voluntary organizations are strictly nonprofit and tax-exempt, whose funding consists exclusively of private donor sources that consist of both citizen and business contributions.

These particular NGOs may function due to their specific association with the PSOs to include members of the local chamber of commerce (various businesses), or alliance to the corporate sector (nationwide companies, industries, and businesses), which are willing to provide foodstuff, financial assistance, and other resources during a disaster. The voluntary contributions may also consist of work (labor), monetary (cash endowment), or provisions (foodstuff, clothing, hardware, etc.). In an effort to maintain their definitive commitment to disaster response and recovery, NGOs value their independence and neutrality. For the most part, NGOs prefer to be nonpolitical and stress free from bureaucracy that often impedes organizations. Instead, NGOs prefer to be well organized through greater field-level management and less hierarchy. Doing so marginalizes the possibility of civil conflict and perceived favoritism toward a particular agency or facet of the public.

Another significant characteristic of NGOs is their purpose-driven motivation to maintain a long-term commitment to recovery. Often times, volunteers will have a tendency to come and go, be in the game but not of it, and once the "at the scene" attention has dissipated, will themselves evaporate from the situation, whereas

NGOs maintain a passion for the welfare of the public to the extent that members of NGOs may put themselves in harm's way.

Within the context of discussing the assets and resources allocated to NGOs, it is important to understand the distinct differences between donations and grant funds. Primarily, a donation consists of cash or property, but may also include the expenditure of time, muscle, and energy during labor provided for reconstruction, security, medical, or caretaker business, whereas grant funding involves the discretionary appropriation of monetary funds designated exclusively to established NGOs for well-defined resources that may include personnel, equipment, utilities, technology, or other property, as well as for education or training purposes.

Generally, nonprofit voluntary NGOs exist within a local community. The most common and easily recognized NGOs include volunteer fire departments, homeless shelters and soup kitchens, Community Emergency Response Teams (CERTs), Civil Defense Groups, and church outreach programs. Other NGOs that function exclusively and are sponsored by the local citizenry include educational organizations (student groups and faculty/staff), occupational organizations (athletics, health care, hospitality, retail, technology), civic organizations (Rotary, VFW, American Legion, Lion's Club), or religious/denominational groups (Baptist, Catholic, Methodist, Presbyterian).

The advantage of a local, voluntary, nonprofit NGO is its familiarity of the culture within the community. During a disaster, local NGOs are a reliable source of information about at-risk persons and have connections with many local resources and community leaders. As a result, the victims and outside responding agencies are glad to see the commitment to, and knowledge of, the various neighborhoods and areas that the local NGOs maintain, which fosters strong collaborative continuity during the disaster. All of these groups are vital to the successful management of a disaster, and these groups must be included in all phases of the disaster planning process.

Various prominent NGOs, such as the American Red Cross, Habitat for Humanity, Salvation Army, and United Way, which maintain nationwide and global partnerships. They are involved in providing disaster assistance such as counseling, food and shelter, monitoring environmental hazards, lobbying for public safety regulations, and inherently involved in promoting the professionalization of emergency management. Such established NGOs are well organized, respected, and have the resources necessary to respond immediately to disasters with great effectiveness.

Because of their vast alliances within the corporate sector and local business communities, these specific NGOs have developed and maintained strong partnerships with state and local governments to the extent that interoperability and interdependency has become a by-product of the emergency management process. While not directly affiliated with federal, state, or local government, such renowned NGOs do receive grant funding from FEMA to support and sustain shelters, and to purchase or replace necessary equipment that is essential during the response and recovery phases.

Some of the larger, well-known NGOs are also recipients of support from reputable corporations and industries, such as Motorola, Dell, Verizon, AT&T, Microsoft, Exxon, Chevron, and the auto manufacturers Chrysler and Ford. The vast amounts of donations and grants contributed to NGOs by the corporate/industrial sector are expended toward supporting high-level disaster training institutes, safety education and training programs for first responders, and other necessary resources that are specifically designated for disaster preparedness, response, and recovery purposes.

Role of Nongovernmental Organizations

The role of an NGO is an integral part of response and recovery support regarding disaster victims. This support includes food, shelter, counseling services, and medical assistance. NGOs are often able to provide specialized services for assisting members of the public who may have special needs, including individuals with disabilities.

In addition, an important facet of NGOs is their ability to provide valued assistance that truly exemplifies commitment, competency, and caring. The value-driven purpose of each NGO remains the inherent operational foundation for prioritizing the service deliverables to disaster victims. Examples of NGO and voluntary organization contributions include the following:[1]

- The training of volunteers.
- Identifying of options to be used as possible shelters.
- Managing supplies and resources used for a disaster.
- Discovering the actual disaster victims who are in need of assistance, determining what their exact needs are, and deploying the necessary provisions as needed.
- Identifying and providing essential resources for post-emergency cleanup including cleaning, foodstuff, clothing, shelter, and labor needs.

Some NGOs are authorized as official support groups that are responsible for delegation and management of all disaster support resources during a disaster. This status arises because of their prolific ability to respond during a disaster. For example, the American Red Cross provides services in six major areas: domestic disaster relief; community services that help the needy; support and comfort for military members and their families; the collection, processing, and distribution

---

[1]FEMA, http://www.fema.gov/NRF.

of lifesaving blood and blood products; educational programs that promote health and safety; and international relief and development programs.[2]

In October 2010, the American Red Cross and FEMA signed a new memorandum of agreement (MOA), in which the Red Cross was authorized to co-lead mass care responses during emergencies. Specifically, the unprecedented agreement involves the Red Cross serving ". . . as a national leader tasked with supporting state governments and other non-profit organizations to build stronger disaster response plans and capabilities in mass care."[3]

Simultaneously, the Red Cross continues to provide shelter, food, emotional support, first aid services, cleanup supplies, and comfort items to needy people. These mass care responsibilities are part of the National Response Framework, a federal guide indicating how the country will respond to anything from local emergencies to large-scale terrorist attacks and catastrophic natural disasters.[4]

The Salvation Army, an international movement, is an evangelical part of the universal Christian church. Motivated by the love of God, as a leader in Christian faith-based human services, the Salvation Army is committed to serving the whole person, body, mind, and spirit, with integrity and respect, using creative solutions to positively transform lives.[5] The Salvation Army has responded to emergencies and disasters for over a century, with groups of trained and committed volunteers who are immediately capable of responding anywhere in the United States and virtually anywhere in the world.

The Salvation Army's disaster work began in the United States in Galveston, Texas, after the hurricane of 1900. Galveston was virtually destroyed and over 5,000 residents were killed. Since the Galveston Hurricane event, the Salvation Army has played an integral part in disaster response and recovery by maintaining a strong contingent of volunteer response teams, and considers it a special assignment to not only attend to the needs of victims during a disaster but also assist first responders.

A traditional and well-renowned part of the Salvation Army is its numerous response vehicles strategically located throughout the United States. These resources are known as canteens, and are a fleet of mobile feeding units stocked and ready to respond wherever and whenever needed. In addition to providing food, clothing, shelter, and hygiene kits, the Salvation Army response teams provide emotional support to residents. The Salvation Army also maintains highly trained volunteers who are prepared to assume incident command responsibilities, which is a significant dynamic of the response and recovery phases.

---

[2]American Red Cross, http://www.redcross.org/.
[3]Ibid.
[4]Ibid.
[5]Salvation Army, http://www.salvationarmyusa.org/.

Framework for NGOs

The evolution and necessity of emergency management in recent years has required the obvious need to prepare for future disasters and makes it imperative that the vast array of stakeholders involved in the response and recovery efforts include competent and committed people of the best quality, who are held to the highest standard of professionalism. The nearly immeasurable contingent of resources available to local, state, and federal government has enabled emergency management agencies to handle almost any incident that occurs in our nation.

Yet, despite the government's tremendous assets, often times there are other organizations or groups that have a unique or specific function that is critical to effectively handle a crisis. As previously mentioned earlier in the chapter, the host of public, private sector, and nonprofit sector NGOs, which range from worldwide recognized groups such as the American Red Cross or Salvation Army to local business owners, all play a key role in emergency management.

Emergency management in the United States has a history of being haphazard and ineffective in terms of professional organization and effective legislation. In fact, the term emergency management is relatively new, despite the long history of people and governments actually managing an emergency. Comprehensive emergency management can be defined as the process of developing the operational emergency functions necessary to mitigate, prepare, respond, and recover from all man-made or natural disasters.

The 1960s and early 1970s experienced massive disasters. Examples include Hurricane Carla-1962, Hurricane Betsy-1965, Hurricane Camille-1969, and Hurricane Agnes-1972, as well as the Alaskan Earthquake of 1964, and the San Fernando, California Earthquake of 1971. During that period, the management process of response and recovery operations was primarily the sole responsibility of the Federal Disaster Assistance Administration, which was organized within the Department of Housing and Urban Development (HUD).[6] Other respondents included the U.S. Army Corps of Engineers, the U.S. National Guard, and other various agencies.[7]

Basically, the emergency management process was fragmented and not aligned with a comprehensive framework. As such, the magnitude and nature of these events served as the impetus for increased legislative efforts to address disasters. Eventually, new legislation, such as the 1968 National Flood Insurance Act, offered flood protection to homeowners. During 1974, the Disaster Relief Act firmly established the process of presidential disaster declarations. Primarily, the United States

---

[6]FEMA, http://www.fema.gov/about.
[7]George D. Haddow, Jane A. Bullock, and Damon P. Coppola, *Introduction to Emergency Management* (Burlington, MA: Butterworth-Heinemann Publications, 2011).

began the professionalization of emergency management during the tenure of President Jimmy Carter in 1979.[8]

Then, after the September 11, 2001, attack, the Homeland Security Act of 2002 established the Department of Homeland Security, whose primary mission was to fight against terrorism. Within the context of the Homeland Security Act of 2002, the FEMA was transitioned to the newly created Department of Homeland Security. Thereafter, in February 2003, President George W. Bush issued Homeland Security Presidential Directive-5 (HSPD-5) that specifically developed an all-hazards national system for emergency preparedness and response, and required all federal agencies to cooperate in the effort to establish the National Response Plan and the National Incident Management System (NIMS), which are the operational foundation of all existing emergency management plans.[9]

Subsequently, in December 2003, another significant HSPD-8 was implemented, which directed FEMA to develop an emergency management process known as the National Disaster Recovery Framework (NDRF). For the first time, the NDRF created by FEMA outlined recovery principles, roles and responsibilities of stakeholders, and provided standards for effective and efficient coordination and collaboration among all stakeholders that also included NGOs.[10]

Eventually, the FEMA NDRF established six new recovery support functions that provide a comprehensive framework, in which all stakeholders could partner together and communicate, collaborate, coordinate, in an effort to improve resource allocation and foster an integrated approach to the response and recovery phases of emergency management.

NGO Partnership Principles

Responding to incidents frequently exceeds the resources of government organizations. Volunteers and donors support response efforts in many ways, and it is essential that governments at all levels plan ahead to effectively incorporate volunteers and donated goods into their response activities.[11] The old cliché of a government team coming to a disaster or emergency scene, in a cloud of glory, including helicopters, ships, SUVs, and massive mounds of round hounds makes good fodder for television, but may actually inhibit or be detrimental to the response mission.

The success of managing any emergency, critical incident, or disaster is contingent upon partnerships between all stakeholders in an emergency management situation that are grounded in the following principles: collaboration, coordination,

---

[8]FEMA, http://www.fema.gov/about.

[9]W. L. Waugh and K.Tierney, *Emergency Management: Principles and Practices for Local Government* (Washington DC: ICMA Press, 2007).

[10]FEMA, http://www.fema.gov/national-disaster-recovery-framework.

[11]FEMA, http://www.fema.gov./pdf/emergency/nrf/PartnerGuidePrivateSector.

integration, flexibility, and professionalism. Without these critical concepts and ideas, the wide array of law enforcement, fire, medical, legal, media, community leaders, volunteers, and the friends and families of the victims will quickly overwhelm and render any incident command ineffective.

Collaboration is a matter of building relationships and consensus. This method has been found more effective than a command and control approach. Collaboration is built on trust and open communication and is essential in today's network of emergency managers. The ability to work collaboratively with public, private, and nonprofit agencies is a crucial leadership trait for an emergency manager.

Integration is bringing all levels of government, private, nonprofit, and volunteer sectors together and making them active participants in all phases of the emergency management process. The American emergency management system is founded on the principle of voluntarism and community aid from agencies such as the American Red Cross, the Salvation Army, and other groups. It is vital that as an emergency manager that a relationship with these groups is formed and cultivated before a disaster strikes.

Coordination is of vital importance among agencies before, during, and after a disaster. Effective channels of inter- and intra-agency communications are essential to plan and conduct training exercises, perform risk and needs assessments, and for operational decision making during a crisis. The importance of coordination is seen in the activation of multiagency coordination system and unified command structures.

Flexibility is the ability to adapt plans to the dynamic circumstances of a disaster. Rigid plans are to be avoided to allow for innovation and improvisation. Any operational plan is merely a starting point or a guide, instead of a step-by-step disaster response or recovery. In an effective emergency management environment, responders and officials will have the capabilities to interpret plans and adjust to the circumstances on the ground. Flexibility does mean that there are no rules guiding or controlling actions. Understanding the boundaries and limitations of authority and actions requires an extensive knowledge of the legal, social, and political realm, which is only developed over a career of experience and education.

Professionalism is a critical key component for the effective coordination of an emergency management situation. Leadership, followership, problem solving, planning, evaluating, and budgeting are just a few of the necessary skills needed in an emergency manager. All of these traits contribute to a professional warrior and a professional scholar that will have the appropriate traits to not let ego or personal or political desires interfere with the accomplishment of the mission.

Citizen Corps Councils

Another common partnership within the NGO arena that brings together local government, civic leaders, and the private sector for the purpose of readiness and response to critical incidents, emergencies, and disasters is known as a Citizen Corps Councils. Following the tragic events that occurred on September 11, 2001, state and local government officials determined that promoting and maintaining a safe and secure society required citizen involvement regarding the support of local first responders.[12]

Eventually, as a result of an intensive and proactive agenda, federal and state officials agreed that creating a training blueprint was necessary to prepare citizens to be proactive participants in supporting first responders. Thus, in January 2002, President George W. Bush introduced Citizen Corps, which was coordinated and managed by FEMA's Individual and Community Preparedness Division.

Figure 7.2 President George W. Bush waves during his visit to the Knoxville Civic Center in Knoxville, Tenn., Monday, April 8, 2002. The president spoke about the value and need of community service: "And so for those of you out there who are interested in participating, I want you to call up this number, 1-877-USA-CORPS, or to dial up on the Internet, www.citizencorps.gov. This is a way where you can help America," said President Bush.

Citizen Corps is composed of federally sponsored programs and nonprofit affiliate programs and organizations that share the common goal of helping

---

[12]Citizen Corps, http://www.citizencorps.gov/.

communities prevent, prepare for, and respond to crime, disasters, pressing public health needs, and emergencies of all kinds. It encourages all Americans to take an active role in building safer, stronger, and better prepared communities.[13]

The purpose of Citizen Corps is to encourage citizen involvement through education, training, and volunteer service in an effort to make communities safer, stronger, and better prepared to respond to the threats of terrorism, crime, public health issues, and disasters of all kinds.[14] For example, several objectives of Citizen Corps include supporting the initiatives of local Neighborhood Watch programs by fostering a proactive effort to report suspicious activity to the authorities, increase training in local communities for disaster preparedness, and recruit and train retired physicians to assist first responders during emergencies. The mission of Citizen Corps is as follows:[15]

- Preparing the public for local risks with targeted outreach.
- Engaging voluntary organizations to help augment resources for public safety, preparedness, and response capabilities.
- Integrating whole community representatives with emergency managers to ensure disaster preparedness and response planning represents the whole community and integrates nontraditional resources.[16]

Essentially, the mission of Citizen Corps is accomplished through a nationwide network involving state, local, and tribal coalitions known as Citizen Corps Councils. The Citizen Corps Councils put together the leadership and resources within a community to implement the Citizen Corps training blueprint, which is meant to develop and direct education and training programs for local volunteers in an effort to maintain support for emergency management and first responders. Such ad hoc Councils are created by elected or appointed local officials with involvement from the area emergency managers.[17]

As duly noted, "Citizen Corps Councils build on community strengths to implement the Citizen Corps preparedness programs and carry out a local strategy to involve government, community leaders, and citizens in all-hazards preparedness and resilience."[18] Citizen Corps Council activities include the following:[19]

---

[13]Ibid.
[14]Ibid.
[15]Ibid.
[16]Ibid.
[17]Ibid.
[18]Ibid.
[19]Ibid.

- Educating the community about disaster preparedness measures
- Implementing public education and outreach efforts
- Providing training to improve citizen preparedness
- Promoting the importance of making personal, family, and work emergency plans
- Coordinating volunteer opportunities that support local emergency efforts
- Coordinating citizen participation in community disaster response activities

Citizen Corps Councils utilize FEMA's Emergency Management Institute which offers many online courses for citizens to be trained in and familiar with emergency preparedness, mitigation, the NIMS, the Incident Command System (ICS), and the disaster response process. Below are examples of independent study courses offered to citizens by FEMA through the Citizen Corps:[20]

- IS 10.a-Animals in Disasters: Awareness and Preparedness
- IS 11.a-Animals in Disasters: Community Planning
- IS 22-Are You Ready? An In-depth Guide to Citizen Preparedness
- IS 244-Developing and Managing Volunteers
- IS 288-The Role of Volunteer Agencies in Emergency Management
- IS 317-Intro to Community Emergency Response Teams
- IS 324-Community Hurricane Preparedness
- IS 394.a-Protecting Your Home or Small Business in Disaster

Certified Emergency Response Team

Another important volunteer NGO that plays an integral role during the response phase is known as a CERT. The purpose of a CERT is to provide training through Citizen Corps and FEMA in an effort to equip the public and volunteers with the ability to protect themselves and their surrounding communities in the event of a critical incident or disaster.

After a major disaster, emergency managers and first responders are significantly challenged to meet the demands for services to survivors. Challenges may include hazardous waste, roadblocks, communication failures, and number of traumatized victims in need of immediate medical assistance. Such issues may initially hamper the progress of emergency services and marginalize the public's ability to access needed deliverables. Thus, the bottom line is that ordinary citizens will have to rely upon one another for help in an effort to meet their immediate life-sustaining needs.[21]

---

[20]Ibid.
[21]Haddow, Bullock, and Coppola, *Introduction to Emergency Management*, 2011.

Basically, a CERT consists of homeowners organized as groups to perform emergency management tasks in their neighborhoods. CERTs may also be known as neighborhood emergency response teams, or other similar names, and are organized to train neighborhood volunteers to perform basic emergency response tasks to include search and rescue, first aid, fire suppression, and estimating costs of damage.[22]

Essentially, a CERT will offer volunteers education programs about potential hazards they may face and offer lifesaving skills training. Becoming an official member of a local CERT requires a citizen to complete FEMA training courses including disaster preparedness, disaster fire suppression, disaster medical operations, light search and rescue, disaster psychology and team organization, and disaster simulation.

Based upon the circumstances following a disaster, these citizen responders use their training as part of a neighborhood or workplace team to help others when professional responders are overwhelmed or not immediately available. CERT members are trained and equipped to provide immediate assistance to victims in their area, organize spontaneous volunteers who have not had the training, and collect disaster intelligence that will assist professional responders with prioritization and allocation of resources when they arrive.[23] Over the years, the public along with emergency managers and first responders have acknowledged the tremendous benefit of partnering with a CERT during the response and recovery stages of a disaster.

Voluntary Organizations Active in Disaster (VOAD)

Throughout the United States, there remains a vast conglomeration of NGOs that diligently serve the public by providing the basic necessities to survive and recover during a disaster or emergency situation that include food, shelter, clothing, and medical assistance. One of the most prominent, nonpartisan, nonprofit membership associations that serves as a forum where organizations share knowledge and resources throughout the disaster cycle—preparation, response, recovery and mitigation—to help communities prepare for and recover from disasters is known as the National Voluntary Organizations Active in Disaster, or NVOAD.[24]

Formed in 1970, NVOAD assists member organizations with the coordination and communication processes during a disaster in order to provide the most efficient and effective response and recovery efforts. National VOAD consists of a coalition of some 55 state/territories VOADs, which are considered to be some of America's most reputable organizations that include faith-based,

---

[22]Michael Lindell, Carla Prater, and Ronald Perry, *Introduction to Emergency Management* (Hoboken, NJ: John Wiley and Sons Publishing, 2007).

[23]Haddow, Bullock, and Coppola, *Introduction to Emergency Management*, 2011.

[24]NVOAD, http://www.nvoad.org/.

community-based, no-profit, humanitarian NGO groups. The foremost objective of NVOAD is to complement the capability of state and local VOADs by ensuring that they are adequately prepared for future disasters and equipped to effectively execute their disaster response and recovery plans.

NVOAD helps to coordinate any assistance that NGOs may need to facilitate the operational services necessary during all phases of a disaster, which may involve both short- and long-term recovery efforts. NVOAD, for lack of a better term, is nonoperational. NVOAD does not have resources such as canteens, shelters, clothing, etc. Instead, the principal role of NVOAD involves coordination. Basically, once a disaster strikes, NVOAD rings the bell and communicates to members, coordinates where and what needs to be done, and NVOAD member groups then take action by providing the actual disaster response services. The national VOAD motto remains exemplary evidence of its tremendous commitment:[25]

Our Core Principles

- **Cooperation**—By this we mean that we need each other; we recognize no single organization has all the answers for all the challenges that arise during disasters. We understand that our common goals for a community can be best achieved by working or acting together with a common purpose. Operationally, "cooperation" recognizes the value of working together on at least one specific delivery of service or event, distinct from the more comprehensive "coordination" principle listed below.
- **Communication**—We treat fellow members as partners and foster a climate of openness to promote the regular sharing of information about and between our member organizations—their capacities, accomplishments, limitations and commitments. Members must develop and maintain effective channels for sharing information, listen carefully to each other, and deal openly with concerns.
- **Coordination**—Member organizations commit to working together, in a non-competitive manner, toward the goal of effective service delivery throughout the disaster cycle. Through careful planning and preparation, National VOAD member organizations form tactical partnerships to work in a coordinated, predictive fashion to more effectively utilize resources to accomplish a set of tasks.
- **Collaboration**—Member organizations establish shared goals and actively work together to achieve specific goals and undertake specific projects throughout the disaster cycle. We form strategic partnerships throughout the disaster cycle.

Credit: NVOAD, http://www.nvoad.org/.

---

[25]Ibid.

The practical purpose-driven mission of National VOAD, "Recognizes that these 4C's are progressive and based on building and maintaining trusting, mutually capable relationships. We promote and aspire to whole community collaborative relationships and practices throughout the disaster cycle."[26]

Aligned within the emergency management continuum of prevention, preparation, response, and recovery, National VOAD members represent a powerful force of goodwill in America, and provide the leadership necessary to provide the resources that make communities stronger and more resilient.[27] The NVOAD is a coalition that aims to streamline planning efforts by many voluntary organizations responding to disasters.

The NVOAD works very diligently to minimize duplication of services by sharing information among NGOs prior to and during a disaster, which will efficiently allocate the necessary provisions and prevent loss of resources. Essentially, once a disaster occurs, NVOAD initiates their communication process and contacts other state VOADs to encourage members and other NGOs to assemble and deploy to the scene. NVOAD helps a wide variety of volunteers and organizations work together in a crisis. In time of need, members of NVOAD deliver hope for a more positive future.[28]

NVOAD has also crafted a practical and user-friendly blueprint that provides the public with opportunities to become a volunteer member of a VOAD. Below is an example of basic guidelines NVOAD provides citizens wishing to be a part of an NGO within a state VOAD:[29]

- "Volunteer with an official VOAD organization and be trained before the next event to find meaningful volunteer opportunities following a disaster.
- There are many types of VOAD organizations for you to find a meaningful way to serve. Check out our member list of community-based, faith-based and other types of nonprofit groups in your community that have active disaster programs and need volunteers."[30]

Credit: NVOAD, http://www.nvoad.org/.

Local Emergency Planning Committees

Local Emergency Planning Committees (LEPC) are another framework that can be utilized to coordinate efforts between stakeholders to include NGOs, first

---

[26]Ibid.
[27]Ibid.
[28]Haddow, Bullock, and Coppola, *Introduction to Emergency Management*, 2011.
[29]NVOAD, http://www.nvoad.org/.
[30]Ibid.

responders, chemical/industrial businesses, utility companies, the public, and numerous government agencies. The primary purpose of LEPC is to assist in the preparation of emergency response plans for chemical emergencies.[31]

Basically, LEPC evolved from the Superfund Amendments and Reauthorization Act (SARA Title III) of 1986, which made it a stipulation that each state develop and maintain a LEPC in order to prevent, prepare, and plan for local industrial accidents that would be chemical, biological, or gas (hazardous waste). Under SARA, each state is required to and responsible for establishing a State Emergency Response Commission (SERC) to plan for hazardous material emergencies (HAZMAT). One of the primary aspects of a SERC is to assist local governments in developing an LEPC.[32]

Since October 1, 2005, all states have been required to become NIMS compliant in order to qualify for (FEMA) grants and other assistance. As stipulated by FEMA per the NIMS Integration Center, which was created by HSPD-5: "LEPCs are nonprofit community organizations (NGOs) that must include in their membership, at a minimum, local officials including police, fire, civil defense, public health, transportation, and environmental professionals, as well as representatives of facilities subject to the emergency planning requirements, community groups, and the media."[33]

According to the NIMS Integration Center, the following required elements are to be a part of the community emergency response plan that is developed by an LEPC:[34]

- "Identify facilities and transportation routes of extremely hazardous substances;
- Describe emergency response procedures, on and off site;
- Designate a community coordinator and facility coordinator(s) to implement the plan;
- Outline emergency notification procedures;
- Describe how to determine the probable affected area and population by releases;
- Describe local emergency equipment and facilities and the persons responsible for them;
- Outline evacuation plans;
- Provide a training program for emergency responders (including schedules); and,
- Provide methods and schedules for exercising emergency response plans."[35]

---

[31]FEMA, www.fema.gov/emergency/nims.
[32]FEMA, http://www.fema.gov.
[33]FEMA, www.fema.gov/emergency/nims.
[34]Ibid.
[35]Ibid.

NGOs and the Private Sector

Government agencies, including federal, state, and local organizations, are responsible for maintaining the safety and welfare of the lives and property of their citizens through protective safeguards and supportive assistance. The government, however, cannot and should not work alone during a critical incident. The many facets of an emergency incident require that necessary collaborative partnerships are developed, and coordination sustained between the government, the private sector, and NGOs during the emergency management process.

Obviously, the degree of resources, size of the business (structure), number of employees, and type of business (retail or industrial) within the private sector will affect the emergency management continuity relative to the nature of the incident itself. As distinctly highlighted within the Department of Homeland Security's National Response Framework, there are five functional responsibilities that can serve as a major foundation for the ability of PSOs to be as prepared as possible, respond accordingly, and recover within reason as summarized in Table 10.1.

Table 7.1 Private Sector Response Role

Category	Role in this category
Impacted organization or infrastructure	Private-sector organizations may be impacted by direct or indirect consequences of the incident. These include privately owned critical infrastructure, key resources, and other private sector entities that are significant to local, regional, and national economic recovery from the incident. Critical infrastructure and key resources (CIKR) are grouped into 17 sectors that together provide essential functions and services supporting various aspects of the American government, economy, and society. Examples of privately owned infrastructure include transportation, telecommunications, private utilities, financial institutions, and hospitals
Regulated and/or responsible party	Owners/operators of certain regulated facilities or hazardous operations may be legally responsible for preparing for and preventing incidents from occurring and responding to an incident once it occurs. For example, federal regulations require owners/operators of nuclear power plants to maintain emergency plans and facilities and to perform assessments, prompt notifications, and training for a response to an incident

(Continued)

Table 7.1 Private sector response role (*continued*)

Category	Role in this category
Response resource	Private sector entities provide response resources (donated or compensated) during an incident—including specialized teams, essential service providers, equipment, and advanced technologies—through local public-private emergency plans or mutual aid and assistance agreements, or in response to requests from government and nongovernmental-volunteer initiatives
Partner with state/local emergency organizations	Private sector entities may serve as partners in local and state emergency preparedness and response organizations and activities
Components of the nation's economy	As the key element of the national economy, private-sector resilience and continuity of operations planning, as well as recovery and restoration from an actual incident, represent essential homeland security activities

Thus, PSOs can be important resources that may serve as a significant factor prior to, during, and after a disaster. In addition to ensuring the safety and welfare of their respective employees, PSOs must collaborate with local emergency managers regarding post-incident preparedness for a disaster or emergency. PSOs must maintain the attitude of preparing for the worst but hoping for the best, and be progressive minded regarding resource sustainability (water, power, communication networks, transportation, medical care, security, etc.) during the response and recovery process.

Chapter Comments and Summary

There remain numerous communities in the United States that still lack the formality of a response and recovery plan in the event of a disaster. Ideally, prior to a disaster, it is obviously incumbent upon local government leadership to plan and prepare accordingly. In reality, however, a lack of personnel, provisions, and complacency serve as the proverbial stumbling blocks. Such poor preparation may be problematic because of leadership, budgetary constraints, or the abhorrent attitude of "it'll never happen."

Consequently, informal citizen groups are becoming more commonplace within communities. Quite often, the visionary process for promoting and maintaining a safe and prepared community is often initiated by citizens of a community that pursue the resounding resources offered by NGOs. Such advocacy groups quickly learn of significant links to a key stakeholder, which then develops into a collaborative partnership with an NGO that is integrated with a state VOAD and NVOAD. The incredible resources provided by NGOs and the private sector for the sole purpose of addressing preparedness, resiliency, relief, and recovery truly resonates during and after disaster. Thanks to such commitment on behalf of the public's welfare; the vision of citizens can become a reality as they prepare for the worst and hope.

Terminology

Citizen Corps Councils
Community Emergency Response Team
Donation
Funding
Government organization
Grant
Nongovernment organization
Red Cross
Salvation Army
Voluntary Organizations Active in Disaster
Volunteerism

Thought and Discussion Questions

1. Perform some research concerning volunteer organizations within your locality. Select two organizations, and write a brief essay that highlights their respective functions within the context of emergency management.
2. Regarding your response to the preceding question, consider the financial statuses of the selected organizations. Determine the methods whereby these organizations obtain funding. Write a brief essay that highlights your findings. Within your response, consider how you believe the selected organizations may improve their abilities to attract financial sponsors.
3. This chapter introduces the Red Cross and the Salvation Army. Perform some research regarding the primary missions and functions of both organizations. Write a brief essay that compares and contrasts these entities.
4. Select a disaster that occurred within the last 20 years. Perform some research about the actions and contributions of NGOs regarding the selected incident. Write a brief essay that highlights your findings.

References

1. American Red Cross. n.d. About us. http://www.redcross.org/about-us (accessed February 3, 2013).
2. American Red Cross. n.d. FEMA, Red Cross to share mass care responsibility for U.S. emergencies. http://www.redcross.org/ (accessed February 12, 2013).

3. Blanchard, B. W. 2007. FEMA – Principles of emergency management. Emergency Management Higher Education Learning Project. http://training.fema.gov/EMIWeb/edu/ (accessed February 6, 2013).
4. Calhoun County, Alabama. n.d. Calhoun County Emergency Management Agency – Calhoun County Citizen Corps. http://calhouncountycitizencorps.weebly.com/index.html (accessed February 17, 2013).
5. Cattaraugus County, New York. n.d. Office of Emergency Services – Local Emergency Planning Committee (LEPC). http://www.cattco.org/emergency-services/lepc (accessed February 16, 2013).
6. Cravens, J. n.d. Volunteering to help after major disasters. *Cravens & Coyote Communications*. http://www.coyotecommunications.com (accessed February 8, 2013).
7. Citizen Corps. n.d. About Citizen Corps—Mission. http://www.citizencorps.gov/citizencorps/index.shtm (accessed February 16, 2013).
8. Citizen Corps. n.d. Serve on your local Citizen Corps Council. http://www.citizencorps.gov/getstarted/training/getinvolved.shtm (accessed February 17, 2013).
9. Comfort, L. K. and A. G. Cahill. 1988. *Managing disaster, strategies and policy perspectives.* Durham, NC: Duke University Press.
10. Department of Homeland Security. n.d. FEMA—Community emergency response teams. http://www.fema.gov (accessed February 17, 2013).
11. Department of Homeland Security. n.d. FEMA—Department components. http://www.dhs.gov/department-components (accessed February 3, 2013).
12. Department of Homeland Security. n.d. Individual and community preparedness division. FEMA – About Citizen Corp. http://www.citizencorps.gov/ (accessed February 14, 2013).
13. Department of Homeland Security. n.d. FEMA—History. http://www.fema.gov/about (accessed February 12, 2013).
14. Department of Homeland Security. n.d. FEMA—Fact sheet: NIMS compliance requirements for local emergency planning committees. www.fema.gov/emergency/nims (accessed February 13, 2013).
15. Department of Homeland Security. n.d. FEMA—National incident management system. http://www.fema.gov/ (accessed February 7, 2013).
16. Department of Homeland Security. n.d. FEMA—National response framework. http://www.fema.gov/national-response-framework (accessed February 8, 2013).
17. Department of Homeland Security. n.d. FEMA partners with the Salvation Army to assist disaster survivors in New Jersey. http://www.fema.gov/ (accessed February 12, 2013).
18. Department of Homeland Security. n.d. FEMA—Private-sector coordination support annex. http://www.fema.gov (accessed February 9, 2013).

19. Department of Homeland Security. n.d. FEMA—Private-sector partner guide. http://www.fema.gov./pdf/emergency/nrf/PartnerGuidePrivateSector (accessed February 10, 2013).
20. Department of Homeland Security. n.d. FEMA—Recovery support functions: Leading coordinating agencies and key aspects. http://www.fema.gov/recovery-support-functions (accessed February 18, 2013).
21. Department of Homeland Security. n.d. FEMA—Superfund amendments and reauthorization act of 2002 (SARA), Title III. http://www.fema.gov (accessed February 5, 2013).
22. Emergency Disaster Systems. n.d. CERT gear safety pack. http://www.edisastersystems.com/store/images/P/CERT-gear-pack-1-resized.jpg (accessed February 17, 2013).
23. Haddow, G. D., J. A. Bullock, and D. P. Coppola. 2011. *Introduction to emergency management*. Burlington, MA: Butterworth-Heinemann Publications.
24. Kim, S. 2002. What is NVOAD? *Disaster Network News*, March 25, 2002. http://www.disasternews.net/news/article.php?articleid=2948 (accessed February 19, 2013).
25. Kim, S. 2004. NVOAD releases recovery manual. *Disaster Network News*, May 17, 2004. http://www.disasternews.net/news/article.php?articleid=1768 (accessed February 19, 2013).
26. National Voluntary Organization Active in Disaster (NVOAD). n.d. Who we are – Our mission. http://www.nvoad.org/ (accessed February 17, 2013).
27. Salvation Army, The. n.d. The Salvation Army mission and vision statements. http://www.use.salvationarmy.org/ (accessed February 5, 2013).
28. Waugh, W. L. and K. Tierney. 2007. *Emergency management: Principles and practices for local government*. Washington DC: ICMA Press.

ROLE OF THE EMERGENCY MANAGER AND DEVELOPING AN EFFECTIVE EMERGENCY MANAGEMENT ORGANIZATION

"All practitioners, politicians, and the public share something in common. One must be prepared to not only choose from good options for the best outcome, but also have the resolve and discretion to choose from the worst outcomes as the best option."

—Jeffrey M. Van Slyke

Figure 8.1 Emergency Managers in the Field.

OBJECTIVES

The objectives of this chapter are to:
- Define the position of an emergency manager;
- Discuss the role and responsibilities of an emergency manager;
- Describe the organizational structure of emergency management;
- Discuss the concept of accreditation;
- Discuss organization's functions of emergency management;
- Discuss the development process of an effective emergency management organization; and
- Introduce cooperation and partnerships among emergency management entities.

Introduction

All communities around the world, at one time or another, have faced, or will face, the threat of disasters within their borders. Whether these disasters are man-made or natural in origin, the loss of life and property is of real concern to those in all levels of emergency management. In the United States, all local, state, and federal emergency managers are responsible with the daunting task of providing for the safety and well-being of the residents and community in which they serve. However, because of the various sizes, geographic locations, and available resources, some emergency managers may encounter difficulties in effectively accomplishing this task. Due to the anticipation of responses to natural and man-made events, as well as the increased risk of terrorism, being prepared for emergencies has never been more important.

The four phases of emergency management responsibilities, mitigation, preparation, response, and recovery, extend from first responders to government officials. All of which entails a plethora of policy, practices, principles, politics, and partnerships. Many of the professional failures within emergency management result from a lack of recognition regarding the role and responsibilities of an emergency manager. Such unfortunate situations often result in confusion, lack of participation, conflict, and the blame game.

This chapter identifies, defines, and discusses the position of an emergency manager, its roles, and responsibilities. Included in this discussion are the key characteristics of an individual that are necessary to fulfill such a role, the communication aspects of an emergency manager that are essential for maintaining operational integrity, and the outcomes expected by the public and politicians.

Position of Emergency Manager

The position of an emergency manager has evolved throughout history out of circumstance and necessity. Countless lives and untold billions of dollars lost in natural disasters and man-made catastrophes have served as the catalyst for the creation and evolution of the emergency manager position. The position of an emergency manager has often been the subject of critical discussion with respect to its purpose and professional responsibilities. Oftentimes, the position of an emergency manager is found to not have the necessary authority to accomplish the assigned tasks. An emergency manager is not an emergency responder, their job is not to charge into a building filled with smoke or gunfire; that role belongs to the fire services and law enforcement. At a local level, the job of an emergency manager is twofold: reducing vulnerabilities and losses, and dealing with the consequences or aftermath of an event.

An emergency management practitioner is defined as the decision maker responsible for the formation, implementation, and operational coordination of a community's emergency management program. Specifically, "Emergency managers are professionals who practice the discipline of emergency management by applying science, technology, planning and management techniques to coordinate the activities of a wide array of agencies and organizations dedicated to preventing and responding to extreme events that threaten, disrupt, or destroy lives or property."[1]

The exact title of an emergency manager may vary. The local emergency management position is referred to with different titles across the country, such as civil defense coordinator or director, civil preparedness coordinator or director, disaster services director, and emergency services director. Most, however, within the profession are commonly referred to as emergency managers.

Over the years, the position of an emergency manager was integrated into the responsibilities of either the fire chief or police chief. As of today, many individuals serving as an emergency manager have experience in either police or fire services, or other background that has provided experience in command and control, decision-making, problem solving, strategic planning, and working effectively with a community and the media. Succinctly, the local emergency manager has the daily responsibility of overseeing emergency management programs and activities.

In an effort to successfully lead and maintain a proactive and progressive department, an emergency manager must exhibit results-oriented leadership skills and experience that emphasizes the importance of being committed to achieving department goals, along with the ability to influence employees to accomplish those goals. A high-performing organization will seek to shift the focus of

---

[1]T. Drabek, *The Social Dimensions of Disaster. FEMA Emergency Management Higher Education Project College Course Instructor Guide,* 2nd ed (Emmitsburg, MD: Emergency Management Institute, 2002).

management and accountability from activities and processes to contributions that achieve results. Thus, the ability of an emergency manager to consistently demonstrate the willingness and ability to make decisions, manage programs, and instill best practices, is an important indicator of a leadership commitment to results-oriented management.

Accordingly, an emergency manager should strive to create an organizational culture that involves empowering employees to improve operational and program performance while ensuring accountability and fairness. Reciprocally, an emergency manager must also recognize, and not lose focus of the accountability expectation regarding the degree of decision-making authority that has been entrusted to them internally within the organization, and externally from the public, politicians, and fellow practitioners with whom there is a shared professional partnership.

Role of the Emergency Manager

Essentially, the role of an emergency manager has been a continual work in progress. As politicians and the public continue to become more familiar with and understand the need for such a position, as well as provide the necessary resources to fulfill such a role, the emergency manager position will continue to evolve. In a matter of time, along with more disasters, the position of emergency manager should eventually be as common as the fire chief and police chief.

Today, almost every level or government, service, or industry has an emergency management position, or similar equivalent. The role of the emergency manager is neither easily performed nor well understood, nor is it readily accepted in many jurisdictions. The role of the emergency manager has evolved over the years due to the integration of resources among communities and various first-responder departments, and also as the result of numerous disasters that have played havoc with field-level response management resulting in counterproductive response operations.

The complexities of coordinating personnel and provisions necessary to respond in a timely and efficient manner have now resulted in a shift of emphasis from field command responsibilities to a broader spectrum of responsibilities of the emergency manager. Primarily, the day-to-day responsibilities of an emergency manager encompasses developing and maintaining programs and activities that benefit the safety and welfare of the community.

In order to perform the tasks of emergency management, the emergency manager must call on the necessary resources of the community. To succeed, the emergency manager must carry out the task of developing and maintaining effective relationships with government, private, and voluntary sectors of the community. The objectives of these relationships involve mutual consultations, exchanges of information, and agreements for cooperative action. The emergency manager is responsible for coordinating all the components of the emergency management system in the jurisdiction. Among those agencies contributing efforts may include the following:

- Civil Defense
- Fire and Police
- Emergency Medical Service
- Public Works

Thus, it is the emergency manager's responsibility to make certain that the agencies involved in the emergency management system are aware of the threats to the community, know the plans for emergencies, and can conduct recovery operations after a disaster. An emergency manager's role must also include an ongoing assessment of the availability and readiness of local resources most likely required during an incident and being diligent to identify any deficiencies specific to those that may impede the operations of the response and recovery phases. An emergency manager is also characterized as both a doer and a planner, which essentially requires participating in leading the response and recovery efforts by being a prepared like a P.O.E.T. (plan, organize, execute, train). These attributes involve the following responsibilities:

- Coordinate the planning process and work cooperatively with other community agencies and private sector enterprises.
- Develop and execute public awareness and education programs.
- Oversee damage assessments during an incident.
- Advise and inform local officials about emergency management activities during an incident.
- Involve private sector businesses and relief organizations in planning, training and exercises.
- Develop mutual aid and assistance agreements.
- Conduct exercises to test plans and systems and obtain lessons learned.
- Coordinate resources from all sectors before, during, and after an emergency.

The primary responsibilities of an emergency manager include:

- planning for an emergency,
- operating effectively during emergencies,
- creating awareness of possible threats,
- participating in preventive activities and mitigation, and
- conducting effective recovery operations during the aftermath of an incident.

These primary responsibilities and functions also include coordinating all components of the emergency management system. Examples of these components include public works and services, emergency medical services, law enforcement, fire services, volunteer organizations, and other pertinent groups that have some association with emergency management.

The secondary responsibilities of emergency managers include:

- inventorying resources,
- coordinating the planning process,
- establishing and maintaining networks,
- facilitating educational programs and public awareness,
- advising and informing the chief elected officials,
- identifying and correcting resource deficiencies,
- establishing a public warning system,
- alerting and communicating with officials
- identifying and analyzing the potential effects of hazards, and
- evaluating and improving existing plans.

Fundamentally, the role of an emergency manager is one of coordinating all aspects of a jurisdiction's mitigation, preparedness, response, and recovery capabilities. The emergency manager must work with chief elected and appointed officials to ensure that there are unified objectives with regard to the community's emergency response plans and activities. Thus, the role of an emergency manager demands coordinating all aspects of a jurisdiction's mitigation, preparedness, response, and recovery capabilities.

Organizational Structure

The organizational model of emergency management should emphasize the coordination of resources to maximize the ability of the community to respond to major emergencies and disasters. Through the knowledge of positional tasks, subsequent mastery through experience and training, an emergency manager will acquire the professional savvy to fulfill the essential responsibilities of their position. Included within the scope of one's role as emergency manager is the prolific ability to develop, organize, and maintain a department's formal operational program with respect to the phases of emergency management.

Likewise, it is incumbent upon an individual serving as an emergency manager to develop a standardized template and guidance for continuity of operations involving planning, preparedness, response, and recovery. For example, the Emergency Management Accreditation Program (EMAP) outlines that the standards

CHAPTER 8 Role of the Emergency Manager   221

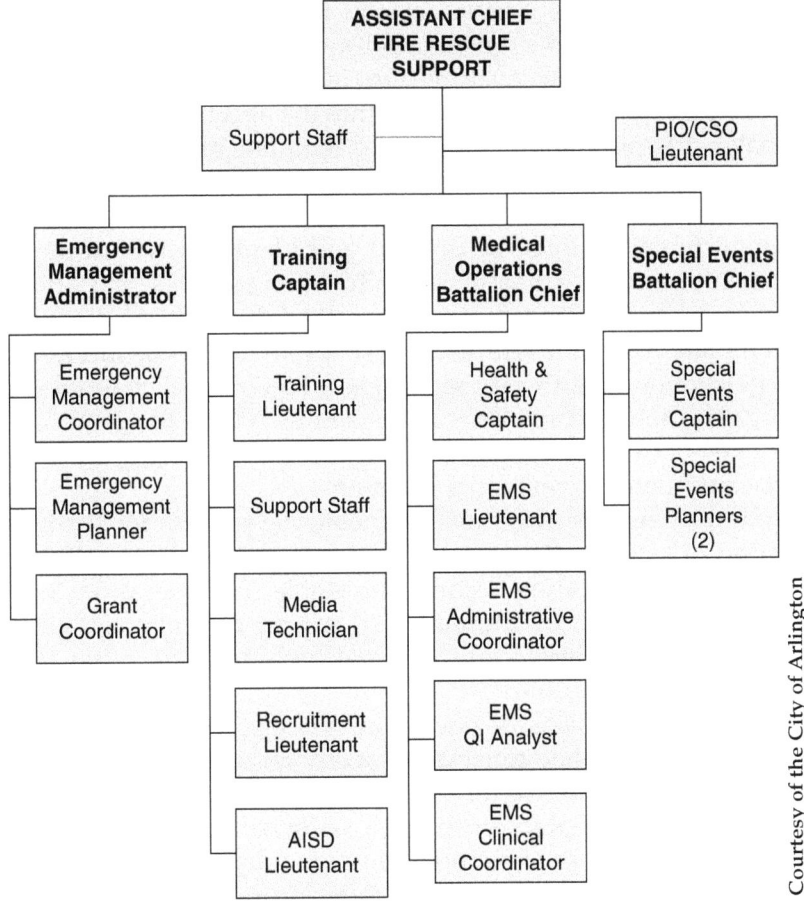

Figure 8.2 Arlington, Texas Organizational Structure (Fire Services)

and best practices with respect to the necessary steps for creating an Emergency Operations Plan (EOP) and Emergency Response Plan (ERP) should include:

- determining the hazards,
- carrying out hazards analysis,
- planning,
- testing the crafted plans, and
- maintaining all plans.

Certainly, the minimum objectives of both an EOP and ERP should include reducing incidence of injuries and deaths. These objectives also include minimizing any exposure to threats and hazards. They also involve minimizing both environmental and property damages while simultaneously protecting a variety of resources—equipment, people, facilities, documents, and so forth. Regarding any incident, these objectives also accommodate recovering from incidents quickly, effectively, and efficiently whereby services may be restored within the affected location.

Additionally, the continuity of operations template of an EOP and ERP should accommodate a variety of topics. It must define the scheduling of planning purposes. It must incorporate a contingency plan involving reduction in staffing levels. All essential functions must be defined and identified within the contexts of local government, businesses, health care, education facilities, and utilities. Within these contexts, continuity of government must exist and be expressed within the operations template. Thus, the template must specify orders of succession, and the appropriate delegation of authority regarding a hierarchical chain-of-command is model for organizational structuring. All positions within the chain-of-command must identify key personnel who are responsible for the execution of steps needed to ensure the continuation of essential functions.

The operations template also must delineate the process for the identification and preparation of facilities that can be used to accomplish essential functions if a department's primary location becomes unusable. The operations template also specifies the availability and redundancy of critical communications systems to support connectivity to internal and external organizations, other departments, essential partners, and the public.

Some chronological aspects of operations are also contained within the template. The operations template must also specify an operational plan for the first 12 hours of an incident, and also for the activation and relocation phases. It must also include an operational plan for the first 30 days for an incident that involves infrastructure, for a period of 18 months for an incident that necessitates a reduction of staff level, and for an alternate operations location. Reconstitution must also be a consideration of the operations plan. Thus, it must accommodate the activities that are necessary for returning to the original location.

Collaboration

Within the scope of planning and developing operational process and continuity of practices, it is also essential that an emergency manager understand the significance of collaboration, which is not synonymous with coordination. Essentially, collaboration is the operational dynamic consisting of partnerships that maintain a cooperative commitment to the operational group effort during an emergency incident. In essence, collaboration is distinctively about relationships, which involves the process in which people work together as a team on a common mission that requires:

- communication,
- decision-making,

- delegation, and
- resources.

Characteristics of a productive and effective collaborative team include:

- a common goal,
- identifying the tasks,
- a leader that provides direction and guidance,
- building trust through open communication and positive accountability,
- constructive conflict resolution, and
- acknowledgement of and respect for each individual and their contributions.

Ideally, the most efficient approach to constructing a collaborative team that will serve the purpose of communicating expectations and delegating tasks is to keep things simple by developing team roles to include:

- leader,
- task master,
- innovator,
- organizer,
- evaluator, and
- finisher.

Given the fact that every individual has different experience, training, education, personality, and competency levels, it is incumbent that emergency managers properly profile staff, or other readily available partners, to be sure that a person working within an emergency incident is as aligned as possible with what is prevalent in their background. Otherwise, the result may be the proverbial square peg in the round hole, which may compromise operations.

On the other hand, within the discretionary mindset of a leader, an emergency manager must also be a risk taker. There simply may be an opportunity during an incident when either due to lack of personnel, or in dire straits for something to be done, that a leader has to nurture the less experienced staff and prepare that person for the worst and hope for the best outcome.

Obstacles to collaboration include differences among agencies that may involve:

- terminology,
- operational directives,
- experience and/or training,
- lack of resources, and
- mission and culture.

Figure 8.3 Team Implementation is a Critical Aspect of Emergency Management.

Still, the benefits of collaboration include:

- elimination of duplication of services and resources expended,
- expansion of resource availability through sharing, and
- enhancement of problem solving and decision-making through cross-pollination of information, ideas, and experience

Elements and Functions of Emergency Management

For the most part, the emergency management profession in the United States continues to remain implacable in its progressive efforts to sustain an emergency management system that complements the freedoms our nation relish. Moreover, much to the chagrin of other countries, the success of our nation's efforts to maintain and enhance the emergency management programs is the result of endurance, perseverance, and lessons learned.

Basically, there are seven aspects of the emergency management profession that play an essential role in its development and operational function:

- Leadership
- Partnerships
- Communications
- Hazard mitigation
- Statutory and Budget Authority
- Implement progressive training and technology
- Public Awareness and Preparedness

Leadership

Fundamentally, leadership is the most critical aspect of any government, business, athletic, education, or other occupational function, and paramount to the emergency management profession. A successful leader is a progressive-minded individual committed to transforming the vision and mission of an organization into reality by focusing on the four P's of leadership:

- Public
- Personnel
- Processes
- Provisions

The successful leadership practice involves assessing the needs of the public, personnel development through training and policy expectations; improving the processes of service deliverables; and equipping the department or agency with necessary provisions to perform. In addition, essential to the success of an emergency management organization relative to its ability to provide the necessary functions of the four phases must involve government support at the local, state, and federal levels. Likewise, the leadership of nongovernment organizations (NGOs) plays an integral role in making communities safer by being better prepared for, equipped to respond, and capable of recovering from a disaster.

A model example of leadership in action can be attributed to the evolution and progressive efforts of the Federal Emergency Management Agency (FEMA), which can be attributed to its director, James Lee Witt, and the support provided by President Bill Clinton. Through their efforts, both Witt and Clinton demonstrated the value of emergency management with respect to the public's welfare.

Thereafter, the efforts of President George W. Bush continued to complement the emergency management operations through the creation of the Department of

Homeland Security per presidential directives. The dynamics of such leadership remains a significant aspect of emergency management profession in its progressive efforts to be prepared for the worst, but respond at its best.

Partnerships

The process of emergency management remains a complex dynamic involving the work relationships between agencies and individuals. In an effort to sustain effective interoperability, there must be expectations regarding who is doing what, where, when, how, and why. Doing so will maintain the necessary synergy for providing the service deliverables in an efficient manner. The working together of two or more people, organizations, or things, especially when the result is greater than the sum of their individual effects or capabilities, will reduce unnecessary expenditure of resources, and also minimize conflict and confusion. Thus, the term partnership is simply an effort of a group committed to working together as a team for the purpose of using similar objectives to reach the same goals during the response and recovery phases of emergency management.

Typically, partnerships have included first responders (fire, rescue, law enforcement, and emergency medical personnel), local, state, and federal government emergency management organizations, and NGOs such as the Red Cross, Salvation Army, and others. As partners, it is critical to the success of each function that these groups establish and maintain a conventional game plan on behalf of the safety and welfare of the public, as well as for maintaining sustainably safe operations on their behalf during the response and recovery operations.

In recent years, a new set of partners have become involved in emergency management including the business sector and the media. The new threat of biological and chemical weapons has resulted in a greater role that the public health system play in emergency management. Increasingly, the general public has become more involved in emergency management through Community Emergency Response Teams (CERT) and involvement in planning and implementing community preparedness and mitigation programs and actions. Additionally, private sector organizations are useful when disseminating information to the general public.

How well these partners work together defines how well the emergency management system functions. Agreements such as the Federal Response Plan and Mutual Aid Compacts define the roles and responsibilities of each partner. Community agreements and partnerships define risks and what to do to protect families, homes, and businesses from harm. Training and exercises allow partners to work together and to refine and enhance their roles.

In every facet of a successful emergency management system or function, partnerships leverage resources and technical skills, promote the exchange of accurate and timely information, and ensure that all the resources of the government, the community, and private sector are brought to bear on disaster issues.

Figure 8.4 Partnership Between FEMA and Commercial Providers. FEMA cooperates closely with the private sector to display disaster registration information in New York City's Times Square. Photo by Louis Eswood, Nov 14, 2012

Source: FEMA/Louis Eswood

Communication

The most significant aspect of good communications in emergency management is a leadership commitment to communicate with all stakeholders. Such commitment involves an organized process that includes:

- Having public information officers (PIOs), or the individual(s) responsible for public affairs, participate in all planning and operational activities.
- Establish a homogeneity line of communications with all stakeholders that allows the media access to key leadership for the purpose of communicating timely messages to the public.
- Ensure that those acting in the role of PIOs, or other public affairs duties, are trained, equipped, and have the discretionary ability serve as an effective communicator to be successful.

Being prepared to communicate timely and accurate information to stakeholders, to the public, and to decision makers is a critical element in any emergency management function. This is especially true during the response and recovery phases of emergency management. The C.A.T. principle of Communication + Accountability = Trust remains a critical component of the communication process, especially when lives are in danger due to the imminent threat of a disaster. Within the midst of whatever event is occurring, or is about to occur, the public's welfare is the highest priority. Thus, the public needs to have a peace of mind regarding what is being communicated to them. The benefits of good communication are as follows:

- Good communications relies on the collection, analysis, and dissemination of accurate and timely information.
- Good communications accurately defines the task and identifies how it will be accomplished and in what time frame.
- Good communications keeps all partners and the public informed, and maintains realistic expectations among all stakeholders.

Hazard Mitigation

Interest in hazard mitigation has grown in the past decade in response to the significant increase in the frequency and severity of natural and man-made disasters and the resulting loss of life and property and economic losses. Reducing the future impacts of disasters must be the foundation of any effective emergency management system or function.

Hazard mitigation efforts are most critical at the community level where land-use and development decisions are made and construction codes and standards adopted and enforced. It is time that community leaders incorporated hazard mitigation into their decision-making process not just for disaster issues but for everyday decisions on where and how a community grows and develops.

Business leaders and residents must also incorporate hazard mitigation information and techniques into their decisions on where to locate their business or home or how to retrofit their business or home to protect their lives and livelihoods from disasters. Promoting and investing in hazard mitigation efforts is the most direct way that families, businesses, and communities can reduce the human and economic losses from future disasters.

Statutory and Budget Authority

Statutory authority is critical to any emergency management system. The Stafford Act in the United States clearly establishes the authority for FEMA operations. It establishes the agency's principal mitigation, preparedness, response and recovery programs, and their eligibility requirements. It provides FEMA with the legal authority to function inside the federal government and in partnership with state and local emergency management agencies.

The authority would be meaningless without the budget appropriation provided to FEMA annually by the U.S. Congress. The regular appropriation of funds ensures that FEMA can fund its programs and activities as well as reimburse its federal, state, and local partners for actions taken as part of the Federal Response Plan. Regular and consistent budget appropriations have ensured that FEMA programs and activities are conducted efficiently and effectively.

No emergency management system anywhere in the world can properly function without statutory authority and consistent budget appropriations. Both must be in place for the system to function properly. In many countries throughout the world, statutory authority for an emergency management system is in place but regular and consistent budget appropriations are not. A successful emergency management system cannot have one or the other; it must have both statutory authority and regular budget appropriations to sustain operational consistency.

Training, Tools, and Technologies

Those staff and volunteers involved in emergency management cannot successfully do their jobs without the proper training, the tools, and technologies. Ensuring that all personnel receive appropriate and current training is central to their ability to fulfill the mission of their organization. With the new terrorist risks, proper training for first responders is critical to first protecting their lives and safety and the lives and safety of the public at large.

Well-trained personnel must also have the tools they need to do their job effectively and efficiently. For example, FEMA provided its home inspectors with Palm Pad computers for entering data concerning a home inspection that standardized the cost estimating process and was linked to a computer-based registration system for providing assistance to disaster victims. These tools allowed FEMA to reduce the time it took to get assistance checks to disaster victims from 30 days to 7–10 days.

New advances in technology occur daily and emergency management must embrace new technologies that enhance the ability to serve the public. Recent technological advances in tracking hurricanes have improved evacuation and warning protocols, new building design technology created the Safe Room to protect families and individuals from tornadoes, and the increased use of GIS has improved land-use and hazard mitigation planning.

Understanding the public's needs in a time of crisis is critical to developing effective disaster programs and activities. The timing and delivery of financial and technical assistance is driven by the customer's needs and schedule. FEMA surveys disaster victims and meets regularly with state and local emergency management

Figure 8.5 FEMA Training Exercise. Calvin Gurley, FEMA Office of the Chief Financial Officer (OCFO), extinguishes a gas fire as part of a training demonstration at FEMA Headquarters. The training and display, conducted by the Mt. Weather Fire and Rescue training division, is part of National Fire Prevention Week, October 5–11, 2008.

Source: FEMA/Barry Bahler

officials in order to understand their customer's needs. This information is used to refine existing programs and develop new programs to meet the expressed needs of the customers.

Public Awareness and Preparedness

The first and foremost attribute of a successful emergency management system, or function, is its focus on the needs of the public, which is the most obvious priority within the profession of emergency management. The process of public awareness and preparedness is an integral part of the responsibilities of an emergency manager. The categorization of a disaster victim can be an individual, a business, a community, elected officials, community leaders, volunteer and nongovernmental groups, and the media. Many of these groups are also considered stakeholders and partners. There are also internal customers within every emergency management organization that include fellow employees and other department staff within the organization.

An emergency manager should remain consistently focused and committed to being proactive and progressive minded regarding the following significant aspects of promoting and maintain the safety and welfare of the public:

- Public awareness and education programs
- Establish a system to alert officials and the public
- Establish and maintain networks
- Review current plans in place and make improvements

An emergency manager should make a concerted effort to promote preparedness through public awareness by providing simple, but informative, reminders regarding why to prepare, and suggestions for being prepared, as well as a notification process. An emergency manager should begin with a discussion as to what types of disasters are most likely to happen within the community, or in proximity to your residence in a rural area, and include the following basic family disaster plan reminders:

- Create and practice a disaster plan
- Create a disaster supply kit
- Complete a preparedness checklist
- Learn what to do in an evacuation
- Know what to do if a disaster strikes

Many local emergency management agencies have websites that provide information to assist the public in preparation and response to a disaster to assist with developing a family disaster plan. Contact information of the local emergency management department should be readily available to the public along with what the public needs to know about what community's warning signals, what they sound like, and what the public should do when activated.

Developing an Effective Emergency Management Organization

Managing people strategically within the profession of emergency management will continue to require maintaining a highly skilled and focused workforce that is empowered to focus on results that are critically important. The emergency management system within our nation has, for the most part, developed as a result of reactive measures to disasters rather than through proactive, meticulous, and thorough use of science, research, and a practical understanding as to what operable functions will work and not work.[2] Fundamentally, emergency management is the preparation for, and carrying out of all emergency functions, other than military functions, to minimize injury and to repair damage resulting from natural or man-made disasters, and to provide support for rescue operations for persons and property in distress.[3]

The modern concept of Emergency Management in the United States has developed circuitously from its infancy during the World War II era when President Franklin D. Roosevelt advised each city to organize its own "Civil Defense" system to plan and prepare for dangers looming on the horizon. Thereafter, in 1979 President Jimmy Carter brought the incessant turmoil and fragmentation of Civil Defense services at the national level to an end by creating the Federal Emergency Management Agency (FEMA). Subsequently, the terrorist attacks in September 2001 caused another a major shift in the role and focus of Emergency Management programs throughout the United States upon the formation of the Department of Homeland Security was formed at the national level.

Intuition would lead many to believe that the outcomes of hurricanes Katrina and Rita in 2005 served as a wake-up call for residents everywhere to begin making disaster preparations. A study, however, by the Council for Excellence in Government, and the American Red Cross following the hurricanes suggested otherwise. Surprisingly, the public was neither taking measures to be prepared nor developing an awareness of their community or state emergency alert plan. Although the results of the survey vary by region, the overall findings were startling and suggested that the emergency management profession needed to be more proactive in its efforts to change prevailing attitudes and behaviors.[4]

EMAP Accreditation Program

The preparedness of our communities for natural and human-caused disasters is of vital and growing importance to public health and safety, to the environment, and to the economy. State and local emergency management programs—the entities responsible for planning and coordinating disaster prevention, mitigation, preparedness, response and recovery—play a crucial role in creating safer communities and in reducing losses to residents, businesses, and important infrastructures. In

---

[2]D. G. Kamien, *Homeland Security Handbook: Strategic Guidance for a Coordinated Approach to Effective Security and Emergency and Management*, 2nd ed. (New York, NY: McGraw-Hill Companies, Inc., 2012).

[3]Emergency Management Accreditation Program (EMAP), *Basics*, n.d., http://www.emaponline.org (accessed May 1, 2013).

[4]EMAP, *Assessing Your Disaster Public Awareness Program. Council of State Governments Through Support from Alfred P. Sloan Foundation* (October 2006), http://www.emaponline.org (accessed May 1, 2013).

an effort to assure that state and local emergency management capabilities are as strong as they can be, a dozen national organizations have worked together to create an accreditation process for emergency management programs known as the Emergency Management Accreditation Program, or EMAP.[5]

Upon conclusion of a presentation on the need for emergency management standards and accreditation at the 1997 National Emergency Management Association (NEMA) conference, numerous emergency management organizations and other key stakeholders, began working on what is now EMAP. Thereafter, a group of national organizations, including the National Emergency Management Association, International Association of Emergency Managers, Federal Emergency Management Agency, U.S. Department of Transportation, Association of State Flood Plain Managers, Institute for Business and Home Safety, International Association of Fire Chiefs, National Association of Counties, National Association of Development Organizations, National Conference of State Legislatures, National Governors Association, National League of Cities, and the U.S. Environmental Protection Agency, created EMAP as a means of fostering continuous improvement of emergency management program capabilities.

According to its website, EMAP is a tax-exempt, nonprofit organization with headquarters at the Lexington, Kentucky, offices of The Council of State Governments (CSG), of which it is an affiliate. Many organizations collaborated on and supported the development of EMAP, including: NEMA, International Association of Emergency Managers (IAEM), U.S. Department of Homeland Security Emergency Preparedness & Response Directorate (EPR/FEMA), U.S. Department of Justice Office of Justice Programs, U.S. Department of Transportation, National Governors Association, National League of Cities, CSG, National Conference of State Legislatures, National Association of Counties, individual states, and others.[6]

The EMAP is governed by a board of directors. It consists of 10 members, referred to as the EMAP Commission, and it includes state, local, academic, and private sector representatives. The EMAP is an affiliate of The Council of State Governments and has a history of partnering with public and private organizations to improve the protection of the public by building stronger emergency management systems.[7] The EMAP recognizes the capabilities of a state or local government to bring together personnel, resources, and communications from a variety of agencies and disciplines in preparation for and in response to an emergency or disaster. The EMAP's process fosters benchmarking against consistent standards and continuous improvement in local and state government emergency management.[8]

---

[5]EMAP, *Basics*.
[6]EMAP, EMAP Standard—2011 Best Practices Document, n.d., http://www.emaponline.org.
[7]Ibid.
[8]Pennsylvania Emergency Management Agency (PEMA), *Governor Rendell Announces National Emergency Management Accreditation for PEMA*, n.d., http://www.pema.state.pa.us (accessed May 2, 2013).

The primary purpose of EMAP consists of developing standards of best practice, assessing operational programs of various organizations, and adding requirements for documentation and verification that neither standards nor self-assessment alone can provide. EMAP consists of:

- agreed-upon national standards *(Emergency Management Standard)* developed with input from emergency managers and state and local government officials;
- self-assessment and documentation;
- on-site assessment by a team of trained, independent assessors;
- committee review and recommendation; and
- accreditation decision by an independent commission.

The Emergency Management Standard of EMAP consists of 63 standards by which programs that apply for EMAP accreditation are evaluated. The Emergency Management Standard is designed as a tool for continuous improvement as part of the voluntary accreditation process for local and state emergency management programs. According to EMAP, it "fosters excellence and accountability in emergency management and homeland security programs by establishing credible standards applied in a peer review accreditation process."[9]

The accreditation process is based on compliance with collaboratively developed national standards, called the EMAP Standard, which is based on the National Fire Protection Authority 1600 Standard on Disaster/Emergency Management and Business Continuity Programs. Accreditation is available to U.S. state, territorial, and local government emergency management programs and anyone can register on EMAP's website to receive standards and guidance materials. The Emergency Management Accreditation Program (EMAP) is a program based on certain standards that assesses and accredits a state, regional, tribal, or local government and their emergency management programs. These programs are charged with the responsibility of coordinating prevention, mitigation, preparedness, response, and recovery activities for natural and human-caused disasters. The Emergency Management Standard covers:[10]

- Program Management
- Administration and Finance
- Laws and Authorities
- Hazard Identification, Risk Assessment, and Consequence Analysis
- Hazard Mitigation

---

[9]EMAP, http://www.emaponline.org.
[10]Ibid.

- Prevention and Security
- Planning
- Incident Management
- Resource Management and Logistics
- Mutual Aid
- Communications and Warning
- Operations and Procedures
- Facilities
- Training
- Exercises, Evaluations, and Corrective Action
- Crisis Communications, Public Education, and Information

For state and local emergency management programs, EMAP provides the opportunity to be recognized for compliance with national standards, to demonstrate accountability, and to focus attention on areas and issues where work or resources are needed. There is no requirement that a local program wait until its state program is accredited to go through the process. While EMAP does not accredit private sector programs, a private sector organization or an individual can register with EMAP and use the Emergency Management Standard, EMAP Program Assessment Tool, and other resources to benchmark and improve emergency preparedness activities.

Because gauging the capabilities of state and local programs before a disaster strikes is a major challenge for government and community leaders, the goal of EMAP is to provide a meaningful, voluntary accreditation process that can be used by state, territorial, and local programs. By offering consistent standards and a fair accreditation process, EMAP will strengthen a communities' ability to prepare for and respond to all types of hazards, from tornadoes and earthquakes to school violence and bioterrorism.[11]

An emergency management program will benefit from participating in the EMAP as follows:[12]

- Provides benchmarks for program management and operations.
- Focuses on comprehensive emergency management.
- Encourages collaboration of state- and community-wide programs. rather than individual agencies

---

[11]National Governors Association (NGA)—Center for Best Practices, *Emergency Management Accreditation Program Managing for Results,* n.d., http://www.nga.org (accessed May 3, 2013).
[12]EMAP, *Basics.*

- Validates professional capabilities.
- Recognizes program quality and individual effort.
- Demonstrates effective use of public resources and provides justification for resources.
- Encourages intra- and interagency communication and team-building through the assessment and accreditation process

The position of an emergency manager will also benefit as follows:[13]

- broadening perspective on the profession of emergency management and emergency preparedness activities;
- deepening the understanding of one's own program organization and operations;
- expanding experience and knowledge through networking and sharing of best practices; and
- contributing to individual skills to improve the nation's public safety and security by increasing emergency management program effectiveness.

Emergency management accreditation represents a significant achievement, as it is a means of demonstrating, through program assessment, documentation, and on-site assessment by an independent team, that a program meets national standards (EMAP Blunt). The EMAP accreditation is voluntary and not tied to funding. Its intent is to encourage examination of strengths and weaknesses, pursuit of corrective measures, and communication and planning among different sectors of government and the community-particularly in light of emerging problems such as changing climate and weather patterns and the threat of terrorism.

One of the major benefits of going through the certification is that each agency is graded on a standard set of requirements. The requirements are the same from agency to agency. With a common standard and level of training we begin to see some uniformity in accreditation. With common standards of mutually accepted best practices, our nation can have confidence in those agencies that have acquired the EMAP certification of accreditation. By going through the process of applying for and obtaining certification an emergency management agency can feel comfortable in their ability to prepare for and manage an incident. This certification is not limited to states only. It can be obtained all the way down to the local community level. As discussed earlier it is not necessary for a state to be certified before a local agency can achieve its certification.

---

[13]Ibid.

Chapter Comments and Summary

With contemporary concerns such as natural and man-made disasters, as well as the continual concern of the threat of terrorism, robust and broad-based emergency preparedness and planning are vital. The EMAP accreditation process will continue to be a catalyst for local, state, and territorial governments to bring their emergency management capabilities in line with recognized standards. EMAP enjoys broad support within the emergency management profession, and remains the prolific benchmark for organizational development in an effort to improve the preparedness, safety, and resiliency of communities.

Moreover, emergency managers and other key stakeholders that share mutual aid agreements will also have a greater peace of mind and confidence in other area agencies that also have acquired accreditation due to the recognized level of competency. Such alignment also creates stronger collaboration and generates a greater degree of compatibility between departments when two or more area agencies maintain accreditation certification.

Achieving accreditation also encompasses certain expectations. Any agency that holds EMAP certification should be expected to function at a certain level. As other agencies obtain this same certification, there becomes a common level of understanding and functioning between departments. Ideally, all agencies would achieve a minimum level of consistent certification based on the same standard. EMAP is a step in that direction.

Terminology

Accreditation
Civil defense
Contingency
EMAP
Long-term
Manager
Mitigation
Operations
Innovation
Partnerships
Planning
Preparedness
Short-term
Standard
Strategy
Volunteer services

Thought and Discussion Questions

1. This chapter introduces the basic notions associated with crafting an emergency organization. Visit with our local emergency services, and determine what management methods are used to embellish the effectiveness and the efficiency of the selected organization. Write a brief essay that highlights your findings.
2. This chapter introduces the concepts of planning from a variety of perspectives. Regarding the entity selected within the preceding question, determine what

methodologies it uses to craft, maintain, and evaluate its plans. How does its methodologies differ from the discussions herein? Write a brief essay that highlights your responses.
3. This chapter introduces the notion of accreditation. Do some research concerning the emergency services entities that are within your locality. What accreditations are associated with these organizations, their resources, or their activities? Write a brief essay that delineates your findings.
4. This chapter alludes to aspects of financial management that affect emergency service organizations. Talk with the leaders of your local emergency services organizations. Determine what responsibilities are associated with their managing of funding organizationally. Write a brief essay that compares and contrasts your findings.

References

1. Advancing Technology Across Oklahoma – ONENET. n.d. http://onenet.net/ (accessed May 2, 2013).
2. Boyd, A. and J. Caton. 1998. *Critical incident management guidelines*. Cambridge, MA: Volpe National Transportation Systems Center.
3. Drabek, T. 2002. *The social dimensions of disaster. FEMA emergency management higher education project college course instructor guide*. 2nd ed. Emmitsburg, MD: Emergency Management Institute.
4. Emergency Disaster Systems (EDS). n.d. The 72 hour survival kit–4 Person–3 Day emergency disaster kit. http://www.edisastersystems.com (accessed May 3, 2013).
5. Emergency Management Accreditation Program (EMAP). 2006. Assessing your disaster public awareness program. Council of state governments through support from Alfred P. Sloan foundation (October 2006). www.emaponline.org (accessed May 1, 2013).
6. Emergency Management Accreditation Program (EMAP). n.d. EMAP basics. http://www.emaponline.org (accessed May 1, 2013).
7. Emergency Management Accreditation Program (EMAP). n.d. Blunt announces Missouri's emergency management program one of few to achieve conditional accreditation. http://www.dps.mo.gov (accessed May 2, 2013).
8. Emergency Management Accreditation Program (EMAP). n.d. EMAP standard – 2011 best practices document. http://www.emaponline.org/ (accessed April 23, 2013).
9. Emergency manager. n.d. http://apocalypseprep.files.wordpress.com/2012/02/emergency-manager.jpg (accessed May 1, 2013).
10. International Association of Emergency Managers. 2010. IAEM code of ethics and professional conduct. http://www.iaem.com/about/IAEMCodeofEthics.htm (accessed September 4, 2011).

11. International Association of Emergency Managers (IAEM). 2007. Principles of emergency management supplement (September 11, 2007). http://www.iaem.com (accessed May 4, 2013).
12. Kamien, D. G. 2012. *Homeland security handbook: Strategic guidance for a coordinated approach to effective security and emergency and management.* 2nd ed. New York, NY: McGraw-Hill Companies, Inc.
13. National Governors Association (NGA) – Center for Best Practices. n.d. Emergency management accreditation program managing for results. http://www.nga.org/portal/site/nga/menuitem (accessed May 3, 2013).
14. Pena, A. Texas State Representative. n.d. A capitol blog. http://www.acapitolblog.com/2010_06_01_archive.html (accessed May 2, 2013).
15. Pennsylvania Emergency Management Agency. n.d. Governor Rendell announces national emergency management accreditation for PEMA. http://www.pema.state.pa.us (accessed May 2, 2013).
16. Waugh, W. L. and K. Tierney. 2007. *Emergency management: Principles and practices for local government.* Washington DC: ICMA Press.

PREPARING FOR THE INEVITABLE

Don't respond to what did not happen.

Bob Nations, Jr.

Figure 9.1 During 2011, Operation Lone Star (OLS), the "largest public health humanitarian effort in the country," an amalgamation of "state and local health and human services agencies, military forces, local service groups and other volunteers" participated in a "responder training exercise."[1]

---

[1] U.S. Department of Health and Human Services, "USPHS Officers Deploy as Part of Operation Lone Star," 2011, http://www.usphs.gov/newsroom/features/lonestar.aspx (accessed June 11, 2013).

OBJECTIVES

The objectives of this chapter are to:
- Discuss the concept of preparedness,
- Discuss preparedness for natural threats,
- Discuss preparedness for man-made threats,
- Discuss preparedness methods,
- Discuss organizations that advocate preparedness.

Introduction

Preparedness is the ultimate goal of all the tasks performed by emergency management and homeland security personnel. The threats posed by natural disasters and man-caused events, especially acts of terrorism and high profile crimes, are daunting when consideration is given to creating efforts that protect citizens and infrastructure in local communities.

Some historical influences must be surveyed as one begins a discourse in preparedness. James Witt, the Federal Emergency Management Agency (FEMA) Administrator during President Clinton's administration, is credited with institutionalizing the phrase, "all events are local." This phrase simply means that all disasters occur in local jurisdictions. That being the case, it is then necessary to examine by what authority do the local emergency management officials and senior elected officials act in preparedness activities. Preparedness is a readiness to anticipate and to respond to events and incidents that may or may not have devastating consequence to the nation, a community, neighborhood or individual. Preparedness engages the activities of planning, training, exercising, equipping, and evaluation which is the responsibility of the whole community but led through the professional emergency management and response personnel.

The National Incident Management System (NIMS) provides the mechanisms for emergency management/response personnel and their affiliated organizations to work collectively by offering a consistent and common approach to preparedness. Preparedness is achieved and maintained through a continuous cycle of planning, organizing, training, equipping, exercising, evaluating, and taking corrective action. Ongoing preparedness efforts among all those involved in emergency management and incident response activities ensure coordination during times of crisis. Moreover, preparedness facilitates efficient and effective emergency management and incident response activities.

Questions of Authority

In 1868, John Forest Dillon, a Supreme Court Justice in Iowa issued a ruling that municipal governments derive all of their authorities from the state and that they cannot act outside of that authority. His opinion further elaborates the idea that if the state so chooses they can revoke the recognition of a local government being an entity without engaging the will of the people. "Dillon's Rule has been a guiding doctrine in the constitutional relations between state and local government for more than a century. Simply stated, it declares that local jurisdictions are the creatures of the state and may exercise only those powers expressly granted them by the state."[2]

An opposing opinion to the Dillon Rule in support of Home Rule came from a Michigan Supreme Court Justice, Thomas M Cooley in 1871. Cooley's opinion stated, "In ruling for the city, Judge Thomas Cooley invoked a number of familiar themes, among them that local governments pre-dated the states and hence were entitled to self-government, a right they previously had enjoyed and had not relinquished under statehood. In addition, Cooley argued, local governments were essential to the decentralization of power, which in turn safeguarded citizens against the despotic tendencies of centralized state governments."[3]

These two opinions, along with a growing movement for Home Rule, eventually led to some limitations of state authority over local governments. Notably, another observation indicates that:

> "Of the formulas devised, the most successful in terms of adoptions has been the constitutional provision granting local control over the organization of municipal corporations. Although these provisions vary from state to state, generally they grant localities broad, though not exclusive, authority over their charters, including their design and adoption and the allocation of responsibility and authority among municipal officials"[4]

In most states within the United States, one finds state laws granting a Chief Local Elected Official (CLEO) emergency powers and authorities. Those powers and authorities are then generally delegated by the CLEO to the local emergency management agency.

In the national preparedness program as designed through the U.S. Department of Homeland Security (DHS), grants are provided for a vast array of preparedness efforts. Those grants generally are allocated to the states and then flow down to local governments. One reason to discuss authorities of the state and local

---

[2] J. G. Grumm and R. D. Murphy, "Dillon's Rule Reconsidered," *Intergovernmental Relations in America Today*, 416 (November 1974), 120.
[3] Ibid, 123.
[4] Ibid., 125.

governments is to gain insight into the relationship among the local, state, tribal, and federal government. The funding streams (grants or cooperative agreements) are the catalyst for state and federal intervention into local governments. In accepting federal funds, local public safety agencies—to include emergency management—agree to adhere to the programmatic and financial guidelines as adopted by the grantor. In partnering in the national preparedness efforts, local emergency management agencies receive and administer a wide spectrum of federal grants that are first received by the state and subsequently passed through to the local agency. It is postulated that:

> "In addition, Homeland Security Presidential Directives, several new federal laws, and a battery of new federal grant programs were introduced. Collectively, these measures dictated to local governments the exact steps they were expected to take. These measures placed terrorism preparedness above preparedness for all other types of disaster agents. Their effect was to move both emergency management and homeland security toward nation-centered dominance within an inclusive authority model. States and localities were told by DHS authorities that they would be actively consulted and welcomed as 'partners.' Yet, the profusion of top-down directions and the vast sums of federal money used to steer states and localities in various directions left little room for state codetermination and even less for local freedom."[5]

The local practitioner rarely focuses on the contingencies of authority. It is an important debate to be knowledgeable of and to appreciate. However, there is demand, if not duty, as a local manager to provide for the public safety and welfare of the local citizenry.

One other challenge necessary in a discussion of preparedness in the context of emergency management and homeland security is the difficulty in definitions. Some of the key words in the national preparedness efforts are preparedness, emergency management, homeland security, natural disasters, and terrorism. With each of the terms, the exception being natural disasters, there is an inherent difficulty in giving them concrete definition. It is much easier to illustrate descriptively their actions or activities—physical description—than to define them. It is sufficient for this text to circumvent an exhaustive discussion in an attempt to enumerate the various arguments and challenges in defining these terms and to simply examine their actions, intent, activities, and anticipated outcomes.

The DHS outlines five tasks that participating agencies must perform in aligning with the national preparedness paradigm: planning, organization, equipping,

---

[5]F. L. Edwards, "Federal Intervention in Local Emergency Planning: Nightmare on Main Street," *State and Local Government Review*, 39, no. 1 (2007), 36.

training, and exercise and evaluation. Further, there are five mission areas that guide the efforts toward preparedness: prevention, protection, mitigation, response, and recovery. These five mission areas are defined by the Federal Emergency Management Agency (FEMA) thusly:[6]

Prevention—Prevent, avoid, or stop an imminent, threatened, or actual act of terrorism.

Protection—Protect our citizens, residents, visitors, and assets against the greatest threats and hazards in a manner that allows our interests, aspirations, and way of life to thrive.

Mitigation—Reduce the loss of life and property by lessening the impact of future disasters.

Response—Respond quickly to save lives, protect property and the environment, and meet basic human needs in the aftermath of a catastrophic incident.

Recovery—Recover through a focus on the timely restoration, strengthening, and revitalization of infrastructure, housing, and a sustainable economy, as well as the health, social, cultural, historic, and environmental fabric of communities affected by a catastrophic incident.

Planning

One of the most important tasks undertaken by emergency management is planning. Planning is a comprehensive exercise that encompasses the total cycle of a crisis from prevention to recovery. In the planning activity the planner anticipates scenario's that are potential threats to the disruption of the routine. Anticipation is based upon various assessments that identify threats, risks, and consequence most likely to occur in a jurisdiction. For example, a jurisdiction that does not have "high rise" or "skyscrapers" will probably not anticipate a scenario involving a collapse of a tall building—if there are no tall buildings then there will not be the potential for the collapse of a nonexistent structure. However, if there is a major waterway in the jurisdiction and bridges that accommodate vehicular traffic on the bridge, there is a rationale to anticipate the collapse of the bridge(s). A foundational document in emergency management is the Emergency Operation Plan. FEMA aptly describes the Emergency Operations Plan (EOP) in the following:[7]

---

[6]FEMA, National Disaster Recovery Framework, 2013, http://www.fema.gov/national-disaster-recovery-framework-0.

[7]Ibid.

"The centerpiece of comprehensive emergency management is the emergency operations plan (EOP). Each jurisdiction develops an EOP that defines the scope of preparedness and incident management activities necessary for that jurisdiction. A jurisdiction's EOP is a document that:

- Assigns responsibility to organizations and individuals for carrying out specific actions at projected times and places during an emergency that exceeds the capability or routine responsibility of any one agency;
- Sets forth lines of authority and organizational relationships and shows how all actions will be coordinated;
- Describes how people and property are protected in emergencies and disasters;
- Identifies personnel, equipment, facilities, supplies, and other resources available—within the jurisdiction or by agreement with other jurisdictions—for use during response and recovery operations;
- Reconciles requirements with other jurisdictions; and
- Identifies steps to address mitigation concerns during response and recovery activities."[8]

As a public document, an EOP also cites its legal basis, states its objectives, and acknowledges assumptions. An EOP is flexible enough for use in all emergencies. A complete EOP describes the:

- purpose of the plan,
- situation,
- assumptions,
- concept of operations (CONOPS),
- organization and assignment of responsibilities,
- administration and logistics,
- plan development and maintenance, and
- authorities and references.

The EOP becomes a guide or framework for local and state operations that builds a consensus throughout the local jurisdiction to provide a comprehensive description of the results of assessments unique to the jurisdiction. The EOP format may be similar in jurisdictions, however, each EOP will have specific data based upon the needs of the respective jurisdiction. Threat and risk data will be collected, analyzed, and assimilated based upon the vulnerabilities and anticipated

---

[8]Ibid.

consequence within the jurisdiction. The EOP will also consider the resources and assets available within the jurisdiction. It is obvious that all jurisdictions will not have those same resources available to them. In that case, mutual aid agreements will be executed in the nonevent environment—and that becomes a component in planning. It is not uncommon for jurisdictions to have agreements with contiguous jurisdictions in sharing assets such as equipment and personnel.

Although the EOP may appear to be a static document, it is not. The community environment is a dynamic environment and the EOP and all other planning must remain dynamic. It is not the crisis that determines the effective, successful response and recovery—it is the management of the crisis that determines the long-term negative or positive consequence. Some events can be prevented and some cannot. Natural disasters—earthquakes, tornadoes, floods, hurricanes, and ice storms—are not preventable but they are survivable. Some man-caused events can be interrupted through diligent intelligence and information sharing—but, the reality is that some will be executed. In either case, comprehensive planning based on assessment data and risk analysis will be the key to an effective response and recovery.

Training

The old adage that states we do best what we do by rote is a reality in emergency management and homeland security. Training is the foundation to response. Training provides knowledge and skills necessary to develop core competencies in any endeavor but certainly in preparing to respond to emergencies, natural disasters, and man-caused incidents. Each jurisdiction assesses training needs based upon the size of their jurisdiction, the population, the critical infrastructure, and incidents most likely to occur in their community.

Further, each discipline—law enforcement, fire, emergency medical, health, emergency management, and other responders—has unique training needs from basic certifications to special competencies. One role of emergency management is to identify common training needs and offer those to the response community at large. Citizenry training is a critical effort for emergency management. Citizens are trained in basic first aid, CPR, and in the mechanics of turning off utilities, etc. as part of the organized Citizens Corp Program.

Training that combines all disciplines has proven beneficial to local communities. For example, in Threat Assessment courses it is advantageous to have the multidisciplines in the same class. Each discipline examines threats from their respective discipline which allows the group to create a more comprehensive assessment product. For example, police may survey a particular structure and examine it for physical security while the fire service may view the same structure for what it manufactures and determine how vulnerable it is for potential release of hazardous materials or other threats to the safety of citizens. Integrated training provides a mass of acquired knowledge for the practitioner that has practical applications for each discipline.

The following are some of the training venues supported by DHS/FEMA (2013):

Center for Domestic Preparedness—The Center for Domestic Preparedness (CDP) develops and delivers advanced training for emergency response providers, emergency managers, and other government officials from state, local, and tribal governments. The CDP offers more than 50 training courses at its resident campus in Anniston, Alabama, focusing on incident management, mass casualty response, and emergency response to a catastrophic natural disaster or terrorist act. Training at the CDP campus is federally funded at no cost to state, local, and tribal emergency response professionals or their agency.

Emergency Management Institute—The Emergency Management Institute (EMI) serves as the national focal point for the development and delivery of emergency management training to enhance the capabilities of federal, state, local, and tribal government officials, volunteer organizations, and the public and private sectors to minimize the impact of disasters.

National Training and Education Division—Training and Exercise Integration/Training Operations (TEI/TO) serves the nation's first responder community, offering more than 125 courses to help build critical skills that responders need to function effectively in mass consequence events. National Training and Education Division (NTED) primarily serves state, local, and tribal entities in 10 professional disciplines, but has expanded to serve private sector and citizens in recognition of their significant role in domestic preparedness. NTED draws upon a diverse group of training providers, also referred to as training partners, to develop and deliver NTED-approved training courses. These training providers include the National Domestic Preparedness Consortium (NDPC), the Rural Domestic Preparedness Consortium (RDPC), the Naval Postgraduate School (NPS), and Center for Domestic Preparedness (CDP), among others. NTED also provides oversight to the Competitive Training Grants Program (CTGP) which awards funds to competitively selected applicants to develop and deliver innovative training programs addressing high priority national homeland security training needs.

National Fire Academy—The National Fire Academy (NFA) is the nation's premier provider of leadership skills and advanced technical training fostering a solid foundation for local fire and emergency services stakeholders in prevention, preparedness, and response. The NFA employs resident and off-campus classroom, blended and distance learning options—including a national distribution systems of accredited state and metropolitan fire training systems and colleges and universities to reach America's first responders. All NFA courses receive college credit recommendation through the American Council on

Figure 9.2 Emergency Management Institute (EMI). According to the EMI, it "conducts a variety of courses for local and regional public agency emergency managers and interested members of the public. These courses are intended to improve the level of national emergency responsiveness to all forms of man-made and natural disasters."[9]

Source: Emergency Management Institute,
http://www.fema.gov/emergency-management-training-opportunities

Education and continuing education units through the International Association for Continuing Education and Training.

Equipping

Equipping first responders is one of the most critical activities undertaken. It includes equipping them in skills, knowledge, and abilities, as well as identifying and procuring the appropriate equipment necessary to respond to assessed threats and vulnerabilities, and in preparation for recovery efforts following an event. For example, police through local revenue and possibly Department of Justice grants acquire the basic equipment and training to perform their tasks. However, the threat of weapons of mass destruction—such as chemical introduction, or a bioterrorism scene, and certainly the use of or threat to use improvised explosive devices—have caused police agencies to "ramp up" beyond their traditional equipping and learn

---

[9]Emergency Management Institute, "Emergency Management Institute," 2013, http://www.fema.gov/emergency-management-training-opportunities (accessed June 11, 2013).

the use of specialized Personal Protective Equipment designed to work in chemical, biological, and explosive environments.

Public Health has the lead in bioterrorism response which has been a new equipping scheme since the anthrax attacks and threat in 2001 and continuing. Equipping integrates skills, knowledge, and abilities with learning and adapting to new equipment for the threats and risks present in today's response environment. DHS/FEMA join organization and equipping together to bring assessment results into decisions relevant to equipping responders in skills, knowledge, and abilities and in determining appropriate equipment needs.

Exercise

DHS/FEMA employs the Homeland Security Exercise and Evaluation Program (HSEEP) to guide national preparedness exercises at the local, state, tribal, and federal levels of government. Exercise provides the vehicle for testing training effectiveness, policy, procedures, equipment, operations, and the technologies to be employed. It indicates strengths and identifies gaps that can be improved. Exercises are designed on a complexity cycle which means it may start with a tabletop moving progressively over time to functional and full-scale exercises involving multidisciplined, multi-jurisdictional responders. The National Exercise Program publication provides this insight as follows:

> "Progressive Exercise Cycle: a discrete array of exercises that is anchored to a common set of objectives, built toward an increasing level of complexity over time, and involves participation of multiple entities . . . An effective National Exercise Program (NEP) is an essential component of our national preparedness. It validates our plans, tests our operational capabilities, maintains leadership effectiveness, and examines the ways we utilize our 'whole community' to prevent, protect from, respond to, recover from and mitigate disasters and acts of terrorism."[10]

Evaluation

The evaluation component allows for identification of smart practices and also guides jurisdictions in correcting any deficiencies identified through the exercise cycle. Evaluation examines the principal objectives and general objectives within the scope of the exercise(s) for optimal effectiveness. Evaluation is the culmination of the training and the cycle of exercises to support a state of preparedness. To facilitate the ability to develop a measurement of preparedness " . . . FEMA, in

---

[10]FEMA, National Exercise Program, 2011, http://www.fema.gov/pdf/prepared/ne_prog_base_plan.pdf, 3.

coordination with the Homeland Security Enterprise stakeholders, will develop a series of performance measures and metrics for each two-year cycle that will provide an assessment of how exercises are improving national preparedness."[11] Planning, training, equipping, exercise, and organization/evaluation continues to be a staple in the national preparedness efforts.

Five Mission Areas

The five mission areas are: prevent, protect from, mitigation, respond to, and recovery. "The National Planning Frameworks, one for each preparedness mission area, describe how the whole community works together to achieve the National Preparedness Goal. The goal is: "A secure and resilient nation with the capabilities required across the whole community to prevent, protect against, mitigate, respond to, and recover from the threats and hazards that pose the greatest risk.'"[12]

Prevention is the one mission area focused on acts of terrorism, specifically acts of terrorism that are threats within the continental United States and its territories. As one can imagine, this mission area is law enforcement driven because of the intelligence and information sharing demands. However, prevention is everyone's business and efforts to keep the public informed are a part of prevention strategies in local communities. In the National Prevention Framework, FEMA outlines the core capabilities in the prevention framework:

Intelligence and Information Sharing. Planning and Direction: Establish the intelligence and information requirements of the consumer.

Screening, Search and Detection. Locate persons and networks associated with imminent terrorist threats.

Interdiction and Disruption. Disrupt terrorist financing or prevent other material support from reaching its target.

Forensics and Attribution. Preserve the crime scene and conduct site exploitation for intelligence collection.

Planning. Initiate a time-sensitive, flexible planning process that builds on existing plans and incorporates real-time intelligence.

Public Information and Warning. Refine and consider options to release pre-event information publicly, and take action accordingly.

Operational Coordination. Define and communicate clear roles and responsibilities

---

[11] FEMA, National Exercise Program, 14
[12] FEMA, 2013.

Protection includes actions to deter threats, mitigate vulnerabilities, or minimize the consequences associated with an incident. Protection can include a wide range of activities, such as improving physical security, building redundancy, incorporating resistance to hazards in facility design, initiating active or passive threat countermeasures, installing security systems, promoting workforce surety, training and exercising, and implementing cybersecurity measures. Effective protection relies upon the close coordination and alignment of protection practices across the whole community as well as coordination with international partners and organizations.[13]

FEMA (2012) outlines the protection core capabilities in the National Protection Framework with the following:

1. Intelligence and Information Sharing
2. Interdiction and Disruption
3. Screening, Search, and Detection
4. Access Control and Identity Verification
5. Cybersecurity
6. Physical Protective Measures
7. Risk Management for Protection Programs and Activities
8. Supply Chain Integrity and Security

Mitigation efforts focus on reducing the loss impact of risks by engaging the community in awareness and creating efforts at reducing the potential consequence. It requires identification of the risks to a community and taking the necessary actions to diminish the loss impact. Further, mitigation focuses on preparing a community to be resilient—having the ability to bounce back effectively when disaster occurs. FEMA outlines the core capabilities of mitigation as:

Threats and Hazard Identification. Build cooperation between private and public sectors by protecting internal interests but sharing threats and hazard identification resources and benefits.

Risk and Disaster Resilience Assessment. Perform credible risk assessments using scientifically valid and widely used risk assessment techniques.

Planning. Incorporate the findings from assessment of risk and disaster resilience into the planning process.

Community Resilience. Recognize the interdependent nature of the economy, health and social services, housing infrastructure, and natural and cultural resources within a community.

---

[13]FEMA, National Protection Framework, 2012, http://ne-cipa.org/html/pdf/peo_nationalprotectionframeworkdraft_20120501[1].pdf.

Public Information and Warning. Target messages to reach organizations representing children, individuals with disabilities or access and functional needs, diverse communities, and people with limited English proficiency.

Long-Term Vulnerability Reduction. Adopt and enforce a suitable building code to ensure resilient construction.

Operational Coordination. Capitalize on opportunities for mitigation actions following disasters.

Response primarily focuses on responding when an eruption occurs interrupting the routine activity of a community. Response transitions into the recovery mission following the immediacy of acute response. Although there may be a response in the prevention and protection mission areas, the response mission area is more related to the activities during an eruption. FEMA outlines core capabilities in the response mission area as:

- Planning
- Public Information and Warning
- Operational Coordination
- Critical Transportation
- Environmental Response/Health and Safety
- Fatality Management Services
- Infrastructure Systems
- Mass Care Services
- Mass Search and Rescue Operations
- On-Scene Security and Protection
- Operational Communications
- Public and Private Services and Resources
- Public Health and Medical Services
- Situational Assessment

Recovery is a component of the comprehensive planning process. This mission area concentrates on what it takes to restore a community to its status prior to a disaster. Recovery planning can begin in the assessment of risk time frame as consequence can be anticipated although the jurisdiction will have to conduct damage assessments following the event to determine the more specific needs of the community to resume a sense of normalcy.

Summary—Response Case Analysis

In summary, there are many management tools and technologies applicable to the emergency management environment. Geographical Information Systems (GIS) is but one of those technological tools; however, it is an excellent decision-making tool applicable to both long term planning and innovative, response oriented decision-making in the emergency management. The GIS cases exemplified present this particular contrast. GIS integration into emergency management is extremely useful as oftentimes disparate data sets, outdated information, and conflicting or incomplete maps exists in and between governmental and private partner agencies. Efficient planning requires accurate data and coordination. The response phase of an emergency is not the time to seek accuracy or build capabilities—it is the time to make to good decisions based on accurate data coordinated across disciplines as well as jurisdictions.

Two Case Studies of GIS Integration into Disaster Management & Response

Case One: 2011 Floods of Shelby County, Tennessee

First, highlighting the historic Mississippi River flooding of 2011, as well as earthquake hazard/risk in the mid-south, we explore the integration of GIS into successful, real-time or near real-time disaster or crisis management, specifically highlighting critical components related to management or implementation beyond the technical aspects of GIS development. Integrating emergence and complexity principles coupled with disaster/crisis management through effective leadership, GIS technologies planning and implementation enabled innovation and improvision resulting in decision-making tools or processes for leaders.

> On April 27, 2011, with flood waters threatening the Shelby County, Tennessee area, Mayor Mark H. Luttrell, Jr., declared a state of emergency. At that time, there was a projection that 2,000 to 5,000 parcels of property were potentially vulnerable to damage from flood waters. This was significant because the Shelby County Emergency Operations Center had been operating since April 4, 2011, resultant of severe wind storms. Shelby County, Tennessee, had five major presidential disaster declarations before the end of May, 2011. The full-time operations/incident period ended on July 6, 2011. During this period the Mississippi River from North to South ravaged its path to southern Louisiana. Shelby County experienced most of its damage along its major tributaries from east to west. The challenge is: with surface rainfall the three major tributaries in Shelby County must "dump" into the Mississippi River. In the floods of 2011 the contingency was the inability for the tributaries to flow into the Mississippi due to

the flood stage of the Mississippi. As a result, the damaging flood waters "backed up" from west to east causing a significant threat to residents along the boundaries of tributaries. Early on, emergency response leadership recognized the threat and immediately employed the GIS developers from the University of Memphis. Their assignment was rather rudimentary, yet complicated. They were requested to provide a predictability mapping of what properties, at what predictable levels, would be impacted in two foot increments every 24 hours of operations. GIS and the developers—the information derived from their efforts—became the foundation for the decision-making model.

The second case describes a nationally and internationally recognized GIS initiative developed in 1995.

Case Two: MetroGIS, Minneapolis—St. Paul Region of Minnesota

MetroGIS is a voluntary collaboration of organizations in the Minneapolis-St. Paul Metropolitan Area that use GIS technology to carry out their business functions. The discussions that resulted in the establishment of MetroGIS began in the fall of 1995. Minneapolis-St. Paul Metropolitan Area's population is 2.8 million, covering 3,000 square miles, 187 cities and townships, 59 school districts, 39 watershed districts, and seven counties. The vision for the result of MetroGIS's efforts, or destination expected to be attained, is "organizations serving the Twin Cities Metropolitan Area are successfully collaborating to use geographic information technology to solve real world problems". The efficient use of geospatial information and shared knowledge of best practices benefit the region's citizens and their leaders:

- They are better able to solve real-world problems.
- In solving these problems, they make better decisions.
- Because better decisions are made, regional economies are strengthened.
- Citizens are better informed regarding geophysical and geopolitical objects and events.
- Because of all these factors, citizens and their leaders are more likely to reach community goals.

And, ultimately these outcomes play a substantive role in providing citizens a safe place to live and work; enhancing environmental systems and green space; and improving housing and transportation systems. As stakeholders use the enhanced capabilities available to them through MetroGIS, they better serve society's needs and, in the course of doing so, achieve effective cross-jurisdictional collaboration and substantive improvements to operational effectiveness and efficiency for their respective organizations.

Within any democratic society based on the rule of law, accurate, timely, geospatially referenced information is absolutely necessary for effective governance, planning, and coordination.[14] The Metropolitan Council for the Minneapolis—St. Paul Region of Minnesota (MC) had for years produced information, but it was often based on estimates and projections that did not take into account the carrying capacity of the land. MetroGIS grew out of the efforts of a group of public officials and managers, along with partners in other sectors, to remedy this shortcoming. They sought to create a shared GIS for the region that linked and made easily accessible business, government, and nonprofit databases of accurate, timely standardized, and needed geospatially referenced information; and acquired or developed the software applications to make use of the data to solve public problems.

Strengths and weaknesses or impediments to GIS implementation in the emergency response and management environment are primarily related to the complex and often chaotic nature of any particular incident. Recognizing limitations or weaknesses is critical to overcoming the same. It is noted that "the major constraint to utilizing GI Science technology is an understandable user interface and willingness to adopt new technologies."[15] Research needs to include "better temporal and spatial estimates" of resources; "better integration of physical processes;" and "better representations of risk and vulnerability, visual images that capture the spatial and temporal shifts in the risk and local vulnerability," and a greater degree of certainty related to the information presented such as "maps and other visualizations of data" that are "simple and easy to interpret by an educated public."[16] However, GIS, even in its current state of development and application, is an invaluable tool in disaster/crisis management for forward-thinking and innovative leaders.

Longley (1999: 846) highlights Cova's observation that "like many other GIS applications, emergency management is not isolated, and there are numerous related theoretical, management, application, and technical innovations that affect this application arena." Further complicating the landscape is the all-encompassing nature of the definition of emergency management. Longley (1999: 845) cites Hoetmer's definition of "emergency management as the discipline and profession of applying science, technology, planning, and management to deal with extreme events that can injure or kill large numbers of people, do extensive damage to property, and disrupt community life."

GIS development is a critical function of the prevention, protection, and mitigation phases, while GIS implementation is aligned best with the response and recovery phases. Exemplary elements of GIS development specific to

---

[14]J. M. Bryson, *Strategic Planning for Public and Nonprofit Organizations: A Guide to Strengthening and Sustaining Organizational Achievement*, 4th ed. (San Francisco, CA: Jossey-Bass, 2011), Kindle location 1369.

[15]S. L. Cutter, "GI Science, Disasters, and Emergency Management," *Transactions in GIS*, 7, no. 4 (2003), 442.

[16]Ibid.

preparedness/mitigation—as well as relational to today's re-categorization of terms to reflect prevention/protection/mitigation—are highlighted by Longley (1999: 849) inquiry related to (1) "the inherent spatial variation in the potential for natural hazard" through "natural hazard assessment and mapping;" (2) "the inherent spatial variation in human environmental vulnerability" through "vulnerability assessment and mapping;" (3) "the inherent spatial variation in risk" through "risk assessment and mapping;" (4) "which spatial strategy can be developed to reduce the effects of a particular hazard phenomenon" through "hazard mitigation" and (5) "which spatial strategy can be developed to reduce human vulnerability to a particular hazard" through "vulnerability mitigation."

Comprehensive Emergency Management (CEM) preparedness and response phases and current day response and recovery phases are well aligned with the benefit of GIS mapping applications for coordination among emergency entities; guiding the public; resource logistics and supply chains during emergencies; the characteristics of geographic locations; conceptualizing constraints and limitations of the incident location; and informing the public.

GIS application in these phases, involving response and recovery, are marked by the following queries: (1) What is the extensiveness of the incident location, including possible losses? (2) What logistics routes are feasible? (3) What evacuation routes and plans are feasible? (4) How can the location of a distressed individual be determined graphically? Although these queries were developed over a decade ago, they are pertinent for GIS development and application today. However, each emergency, crisis, or disaster requires unique queries designed with known data related to specific circumstances presented in the event. The following figure depicts a typical GIS software interface.

Specific to each unique incident, GIS—computer-based tools—is designed to support decision making. According to MacFarlane (2006:20) "GIS is a collection of tools that transform geographically-referenced data into information that is fit for purpose. However, without data that are suitable and sufficient to support the creation of the intended information, GIS can provide no effective part in the decision-making process." MacFarlane, author of the 2006 Guide to GIS Applications in Integrated Emergency Management, further defines GIS components:

> "Data and information are different things. Data are results, observations, counts, measurements, locations, attributes and other basic descriptive characteristics of things and places in a fairly 'raw' format. Information is more processed, ordered, summarized, selective and 'user-friendly' with the intention of assisting correct interpretation. Typically data are high-volume (a lot of records) whereas information is more tightly geared to the requirements of a specific end-use. One of the key strengths of tools such as spreadsheets, databases and GIS is their ability to transform, if appropriately used, data into information that can be appreciated and

acted on more readily. However, it is important to recognize that data are almost universally imperfect, therefore the decisions that are based on them may be misguided, and even when data and information are strong, decisions may still be misguided. Evidence is also widely used as a term and it is defined here as something that is created from information, through further sorting, selection, distillation or triangulation with other sources. In this respect it is similar to the term 'intelligence'; although specifically associated with the work of the security and intelligence services, the term is also widely used in contexts such as local government and regional development in a way that broadly equates with information and evidence. (2006:20)."

MacFarlane (2006:123–124) surmises that

"GIS applications in emergency planning and management have mirrored the development of GIS more broadly: early developments were carried out on single machines within separate agencies, and data and information sharing was partial and difficult to achieve . . . Future

Figure 9.3 GIS Interface. This type of interface not only is useful in providing graphical imagery but also facilitates emergency services routing.

developments will build on integration of data, systems and processes in the development of interoperable data, systems and processes that effectively remove the need for manual interventions to transmit data and information between organizations . . . It will now require vision and leadership to realize the gains."

GIS technologies, like most other technological products, are best tested and evaluated in the future, and as Bobbitt (2008: 545) notes, "the future is a land no one has visited."

Chapter Comments and Summary

This chapter introduces primarily the concept of preparedness. Admittedly, it is humanly impossible to identify and to predict instances of each and every threat that could possibly affect negatively a locality. Regardless, some measures of preparedness must occur to determine the primary threats that may endanger a locality. Failure to plan for disaster further invites the negative consequences that may result from both man-made and natural incidents.

Preparing for disaster necessitates a variety of considerations that vary widely. Everything from the feasibility of logistics routes to the protecting of incident areas must be accommodated during planning phases. Through planning, the safety of both responders and the general public are embellished. Although one may incorporate a variety of considerations when considering preparedness, no guarantee exists that preparedness efforts will be completely successful. Instead, because of unknown variables and the effects of Murphy's Law, events may occur that negate or impair preparedness initiatives.

Preparedness may be viewed from multiple perspectives. Some incidents have the potential of affecting the entirety of the nation. For instance, the events of September 11, 2001 affected the entirety of American society. Thus, a national perspective exists within the context of preparedness. Regional perspectives are also accommodated within preparedness. For instance, when hurricanes strike along the Gulf Coast, only those states that directly experience the effects of weather may suffer physical damage resulting from the storm. Thus, a regional perspective exists within the context of preparedness. Finally, a local perspective of preparedness exists. Emergencies arise daily among small towns and municipalities. For instance, a single building may be gutted by fire. Therefore, a local perspective exists regarding preparedness.

Any location that assumes it will never experience some emergency deludes itself. Disasters, whether small or large, shall occur in due time. Incidents may arise at any location, at any time, and with little to no warning. It is imperative that all localities acknowledge their relevant endangerments, and prepare accordingly. Without preparedness, the effects and consequences of disaster may be increased dramatically. Only through preparedness can a locality buffer itself against many of these negative effects and consequences.

Terminology

Center for domestic preparedness
Comprehensive emergency management
Emergency management institute
Equipping
Evaluation
Exercise
Geographic Information System
Hazard
Intelligence
Mitigation
National Fire Academy
National Training and Education Division
Planning
Preparedness
Public warning
Recovery
Response
Risk
Routing
Scenario
Training

Thought and Discussion Questions

1. This chapter introduces the notion of preparedness. Visit your local fire department, and determine what emergencies it envisions for your locality. Determine what preparedness measures it exhibits regarding these endangerments. Write a brief essay that highlights your findings.
2. Visit your local law enforcement entity. Determine what emergencies it envisions for your locality. Determine what preparedness measures it exhibits regarding these endangerments. Write a brief essay that highlights your findings.
3. Visit with your local government (e.g., mayor's office, commissioners, etc.). Again, determine what endangerments are envisioned for your locality. Determine what preparedness measures are exhibited regarding these endangerments. Write a brief essay that highlights your findings.
4. Based on your findings to the preceding questions, write an essay that compares and contrasts the identified endangerments and preparedness measures. Based on your findings, what recommendations do you have regarding the level of preparedness within your locality?

References

1. Bobbitt, P. 2008. *Terror and consent: The wars for the twenty-first century*. New York, NY: Alfred A. Knopf.

2. Bryson, J. M. 2011. *Strategic planning for public and nonprofit organizations: A guide to strengthening and sustaining organizational achievement.* 4th ed. San Francisco, CA: Jossey-Bass.
3. Cutter, S. L. 2003. GI science, disasters, and emergency management. *Transactions in GIS* 7 (4): 439–45.
4. Edwards, F. L. 2007. Federal intervention in local emergency planning: Nightmare on main street. *State & Local Government Review* 39 (1): 31–43.
5. Emergency Management Institute. 2013. Emergency Management Institute. http://www.fema.gov/emergency-management-training-opportunities (accessed June 11, 2013).
6. FEMA. 2008. *Producing emergency plans: A guide for all-hazard emergency operations planning for state, territorial, local, and tribal governments. INTERIM–Comprehensive Preparedness Guide 101 (July 11, 2008)*, 6. Cambridge, MA: U.S. Department of Transportation. http://www.hsdl.org/?view&did=487952. (accessed June 9, 2014).
7. FEMA. 2011. National Exercise Program. http://www.fema.gov/pdf/prepared/ne_prog_base_plan.pdf. (accessed June 9, 2014).
8. FEMA 2012. National Protection Framework. http://ne-cipa.org/html/pdf/peo_nationalprotectionframeworkdraft_20120501[1].pdf. (accessed June 9, 2014).
9. FEMA 2013. National Disaster Recovery Framework. http://www.fema.gov/national-disaster-recovery-framework-0. (accessed June 9, 2014).
10. FEMA. 2013. National Incident Management System (NIMS). http://www.fema.gov/preparedness. (accessed June 9, 2014).
11. FEMA 2013. National Mitigation Framework. http://www.fema.gov/national-mitigation-framework. (accessed June 9, 2014).
12. FEMA 2013. National Planning Framework. http://www.fema.gov/national-planning-frameworks. (accessed June 9, 2014).
13. FEMA 2013. National Prevention Framework. http://www.fema.gov/national-prevention-framework. (accessed June 9, 2014).
14. FEMA 2013. National Response Framework. http://www.fema.gov/national-response-framework. (accessed June 9, 2014).
15. FEMA 2013. Training. http://www.fema.gov/training-1. (accessed June 9, 2014).
16. Grumm, J. G. and R. D. Murphy. 1974. Dillon's rule reconsidered. *Intergovernmental Relations in America Today* 416 (November 1974): 120–32.
17. Cova, T. (1999) 'GIS in emergency management', in Longley, P. A., Goodchild, M. F., Maguire, D. J. and Rhind, D. W. (eds). Geographical Information Systems: Volume 2 - Management issues and Applications. John Wiley & Sons, New York. pp. 848–858.
18. MetroGIS. 2013. Minneapolis—St. Paul Region of Minnesota. http://www.metrogis.org/about/index.shtml. (accessed June 9, 2014).
19. U.S. Department of Health and Human Services. 2011. USPHS officers deploy as part of operation Lone Star. http://www.usphs.gov/newsroom/features/lonestar.aspx (accessed June 11, 2013).

CHAPTER 9 ADDENDUM

Derived from FEMA (2008), the following sections and their associated items characterize an EOP:

The Basic Plan

I. Promulgation Document/Signature Page
II. Approval and Implementation
III. Record of Changes
IV. Record of Distribution
V. Table of Contents
    A. Purpose, Scope, Situations, and Assumption
        1. Purpose
        2. Scope
        3. Situation Overview
            a. Hazard Analysis Summary
            b. Capability Assessment
            c. Mitigation Overview
        4. Planning Assumptions
    B. Concept of Operations
    C. Organization and Assignment of Responsibilities
    D. Direction, Control, and Coordination
    E. Disaster Intelligence
    F. Communications
    G. Administration, Finance, and Logistics
    H. Plan Development and Maintenance
    I. Authorities and References

Functional Annexes

    A. Direction and Control
    B. Continuity of Government/Operations
    C. Communications
    D. Warning
    E. Emergency Public Information
    F. Evacuation
    G. Mass Care
    H. Health and Medical
    I. Resource Management

Hazard-Specific Appendices (Note: This is not a complete list. Planning teams must define the annexes on the basis of their hazard analysis.)

 A. Earthquake
 B. Flood/Dam Failure
 C. Hazardous Materials
 D. Hurricane/Severe Storm
 E. Lethal Chemical Agents and Munitions
 F. Radiological Incident
 G. Terrorism
 H. Tornado

AGENCY/DEPARTMENT-FOCUSED EOP FORMAT

Basic Plan

 I. Promulgation Document/Signature Page
 II. Approval and Implementation
 III. Record of Changes
 IV. Record of Distribution
 V. Table of Contents
  A. Purpose, Scope, Situations, and Assumptions
   1. Purpose
   2. Scope
   3. Situation Overview
    a. Hazard Analysis Summary
    b. Capability Assessment
    c. Mitigation Overview
   4. Planning Assumptions
  B. Concept of Operations
  C. Organization and Assignment of Responsibilities
  D. Direction, Control, and Coordination
  E. Disaster Intelligence
  F. Communications
  G. Administration, Finance, and Logistics
  H. Plan Development and Maintenance
  I. Authorities and References

Response Agencies

 A. Fire
 B. Law Enforcement
 C. Emergency Medical

D. Emergency Management
   E. Hospital
   F. Public Health
   G. Others as Needed

Support Agencies

   A. Identify those agencies that have a support role during an emergency and describe/address the strategies they are responsible for implementing.

Hazard-Specific Procedures

   A. For any response or support agency, describe/address its hazard-specific strategies.

4.3 EMERGENCY SUPPORT FUNCTION EOP FORMAT

Basic Plan

   I. Promulgation Document/Signature Page
   II. Approval and Implementation
   III. Record of Changes
   IV. Record of Distribution
   V. Table of Contents
      A. Purpose, Scope, Situations, and Assumptions
         1. Purpose
         2. Scope
         3. Situation Overview
            a. Hazard Analysis Summary
            b. Capability Assessment
            c. Mitigation Overview
         4. Planning Assumptions
      B. Concept of Operations
      C. Organization and Assignment of Responsibilities
      D. Direction, Control, and Coordination
      E. Disaster Intelligence
      F. Communications
      G. Administration, Finance, and Logistics
      H. Plan Development and Maintenance
      I. Authorities and References

Emergency Support Function Annexes

  A. ESF #1 – Transportation
  B. ESF #2 – Communications
  C. ESF #3 – Public Works and Engineering
  D. ESF #4 – Firefighting
  E. ESF #5 – Emergency Management
  F. ESF #6 – Mass Care, Emergency Assistance, Housing, and Human Services
  G. ESF #7 – Resource Support
  H. ESF #8 – Public Health and Medical Services
  I. ESF #9 – Search and Rescue
  J. ESF #10 – Oil and Hazardous Materials
  K. ESF #11 – Agriculture and Natural Resources
  L. ESF #12 – Energy
  M. ESF #13 – Public Safety and Security
  N. ESF #14 – Long-Term Community Recovery
  O. ESF #15 – External Affairs
  P. Other Locally Defined ESFs

Support Annexes

  A. Financial Management
  B. Local Mutual Aid/Multi-State Coordination
  C. Logistics Management
  D. Private Sector Coordination
  E. Public Affairs
  F. Volunteer and Donation Management
  G. Worker Safety and Health

Incident Annexes

  A. Biological
  B. Catastrophic
  C. Cyber
  D. Food and Agriculture
  E. Nuclear/Radiological
  F. Oil and Hazardous Materials
  G. Terrorism
  H. Other Hazards as Required

THE EMERGENCY MANAGEMENT PROCESS: RISK PERCEPTION AND ANALYSIS, PREVENTION, AND HAZARD MITIGATION

"Be Prepared . . . the meaning of the motto is that a scout must prepare himself by previous thinking out and practicing how to act on any accident or emergency so that he is never taken by surprise."[1]

—*Robert Baden-Powell*

Figure 10.1 California wildfire. Wildfire that burned at least "6,500 acres of land along the U.S. 101 freeway near Camarillo and Newbury Park in Ventura County, suburban areas to the northwest of Los Angeles."[2]

---

[1] Robert Baden-Powell, "Emergency Quotes," http://www.brainyquote.com/quotes/keywords/emergency.html (accessed January 1, 2013).

[2] Adam Taylor, "Huge California Bushfire Spreads Over 6,500 Acres," *Business Insider*, 2013, http://www.businessinsider.com/california-wildefire-spreading-2013–5 (accessed June 9, 2013). *REUTERS/Gene Blevins*

OBJECTIVES

The objectives of this chapter are to:
- Introduce the concept of emergency management planning process;
- Introduce the concept of hazards analysis;
- Introduce the concept of vulnerability analysis;
- Introduce the concept of emergency domain assumptions.

Introduction

Numerous emergencies of gigantic scope and magnitude have affected the United States in recent years. These incidents represented both natural and man-made emergencies of regional and national proportions. Figure 10.1 shows the types of natural wildfire disasters that are experienced among the western states. The terrorist attacks against the United States, which occurred on September 11, 2001, are probably the most memorable man-made disasters on American soil after the turn of the century. The emergency response activities and events that transpired during the aftermath of the attacks were imperfect. For example, communications failed because "radio repeaters," that were atop the twin towers, and other fiber-optic networks, located near the incident site, were destroyed during the attack.[3] The remaining communications infrastructure was quickly overloaded with communication traffic, thereby causing a cascading effect of overloading the remaining communications infrastructure.[4] This overloading caused numerous failures and delays in communicating. These man-made attacks and the aftermath events showed the necessity of emergency preparedness, readiness, and planning.

During the last of the month in October 2012, Hurricane Sandy devastated the northeastern United States, and affected approximately "one-fifth" of the nation "directly."[5] The financial and economic damages resulting from Hurricane Sandy necessitated the debating of legislation to render approximately $60 billion in federal assistance.[6] The storm destroyed 25% of the "cell phone towers in an area

---

[3]Hüseyin Arslan, *Cognitive Radio, Software Defined Radio, and Adaptive Wireless Systems* (Dordrecht, The Netherlands: Springer, 2007), 407.
[4]Ibid.
[5]Susanna Kim, "Hurricane Sandy: Losses Estimated at $45 Billion," *ABC News*, 2012, http://abcnews.go.com/blogs/business/2012/10/hurricane-sandy-losses-estimated-at-45-billion/ (accessed January 3, 2013).
[6]Brian Faler, "Sandy Aid Worth $60 Billion Passes Senate as House Balks," *Bloomberg*, 2012, http://www.bloomberg.com/news/2012-12-29/senate-approves-60-billion-hurricane-sandy-aid-package.html (accessed January 3, 2013).

spreading across 10 states."[7] Communicating through the use of electronic devices was impossible in the area of "lower Manhattan."[8] Hurricane Sandy "killed over 120 people in 10 states," caused a loss of electrical power for approximately 8.5 million people, and caused approximately "40,000 people" to become homeless.[9] Although some storm warnings were issued before the impacting of the storm, "there were no hurricane watches or warnings issued by the National Hurricane Center for New Jersey ahead of Hurricane Sandy's landfall near Atlantic City."[10] Recovery activities are ongoing at the time of this authorship. This natural disaster also shows the necessity of emergency preparedness, readiness, and planning.

Both events were absolutely devastating and cataclysmic producing significant destruction and damage within their respective locations. Both events show the dangerousness of emergencies arising from man-made and natural origins. Both events show that emergencies may arise anywhere, anytime. Both events interjected different, varying periods of uncontrollable violence within their respective emergency domains. Both events involved the loss of human life. Essentially, although derived from different origins, both emergencies represented events involving different amounts of speed, surprise, and violence.

These events also involved complex emergency reactions and responses. In both cases, the related activities, processes, and procedures of emergency responders were neither haphazard nor random acts. Both events involved the executing of methodical, systematic endeavors to mitigate their effects and to employ response measures. The executing and adjusting of these endeavors occurred in accordance with actionable emergency plans. Such plans are derived from an emergency planning process that provides a foundational baseline regarding the basic operations that are to be performed during emergency situations.

No organization, group, or individual is insusceptible to the experiencing of emergency situations. Although a planning process may be identified through which emergency plans may be crafted, the resulting plan is unique for each organization. Organizations experience different dangers and have different resources. Therefore, any crafted plan must be expressed to uniquely satisfy the requirements of specific emergency domains among organizations. This chapter examines

---

[7]Jacob Urist, "Hurricane Sandy Exposes Faults of Modern Family Preparedness," *The Washington Post*, 2012, http://www.washingtonpost.com/blogs/on-parenting/post/hurricane-sandy-exposes-faults-of-modern-family-preparedness/2012/11/01/d488b106-2438-11e2-ac85-e669876c6a24_blog.html (accessed January 3, 2013).

[8]Ibid.

[9]Ryan Devereaux, "Hurricane Sandy: Two Weeks on, Coney Island Recovery far from Complete," *The Guardian*, 2012, http://www.guardian.co.uk/world/2012/nov/12/coney-island-struggles-sandy-recovery (accessed January 3, 2013).

[10]Henry Margusity, "No Hurricane Warning for What Could be the Most Expensive Storm in History," *Accuweather*, 2012, http://www.accuweather.com/en/weather-news/national-hurricane-center-no-a/839301 (accessed January 3, 2013).

the emergency planning process through which emergency plans are crafted, expressed, implemented, and maintained among organizations.

The Emergency Planning Process

Planning for emergencies is not a random endeavor. Instead, it involves meticulous planning to accommodate a variety of potential emergencies that may originate from man-made and natural disasters. A generic emergency planning process consists of five phases: (1) establish a planning group, (2) analyze hazards and capabilities, (3) develop the emergency plan, (4) implement the plan, and (5) feedback activities.

This process is iterative and reflexive. It may be repeated as necessary to accommodate new endangerments that exist within the emergency domain. It may also be repeated frequently and periodically to ensure that its contents are appropriate regarding present endangerments and organizational resources through time. Figure 10.2 shows these phases graphically.

The initial phase involves the forming of an entity whose specific goals and objectives are related to the crafting of an emergency plan. Different organizations have different titles for these entities. They may be labeled and designated as work groups, teams, task forces, committees, and so forth. The sizes and types of personnel vary depending upon the scope and magnitude of the problem domain and the availability of organizational resources and personnel.

After a team is established and solidified, it must be attentive to the perilous characteristics of the emergency domain. Such attentiveness occurs during the analyses of hazards and resources phase of the emergency planning process. It is

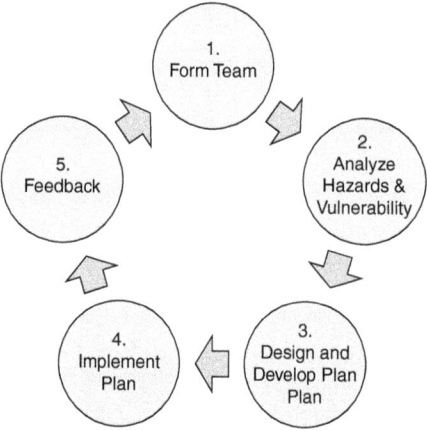

Figure 10.2 The emergency planning process.

during this phase that potential dangers are identified. The resources necessary for countering the effects of emergencies and responding to emergencies are identified within this phase. The outcome of this phase is a set of expressed requirements from which an emergency plan may be developed.

The third phase of the emergency planning process represents the developing of an emergency plan. Developing the emergency plan conforms to the contents of the requirements that were derived during the analysis period. The emergency plan is developed to express clearly and understandably all of the processes, procedures, and operations that occur during certain emergency situations. These activities are considered individually for each type of emergency that was stated within the requirements to produce specific emergency plans for each type of identified contingency that necessitates an emergency situation. The collection of individual activities is amalgamated to represent the cumulative emergency plan.

The fourth phase of the emergency planning process is the implementing of the emergency plan. The implementation of the plan is considered from two perspectives: practice and actual scenario. Practicing the emergency plan familiarizes personnel with their specific roles and the workings of the emergency plan. Such practice is essential for fielding any actual emergency response effectively and efficiently. Actual scenarios constituting live emergencies certainly necessitate the invoking of the emergency plan. If personnel have practiced specific scenarios, then the ability to respond to a live emergency is improved.

The fifth phase involves feedback and evaluation. After any experience of an emergency scenario, whether live or practice, feedback and evaluation are critical endeavors that contribute toward the improving of the emergency plan. These activities represent the determining of the "significance, worth, or condition of usually by careful appraisal and study."[11] They also represent the "transmission of evaluative or corrective information about an action, event, or process to the original or controlling source."[12] Essentially, during a period of evaluation, feedback is collected and analyzed to determine how the emergency plan may be improved (if necessary).

The emergency planning process contributes toward the crafting of an initial emergency plan and maintaining it through time. Any changes, such as personnel changes, resource changes, threat changes, and so forth, must be reflected within the emergency plan. Therefore, the crafting of an emergency plan is not static; instead, it represents a continuous endeavor through which initial plans are generated and maintained, and then are improved through time, change, and experience.

---

[11]"Evaluate," *Merriam-Webster Dictionary*, 2013, http://www.merriam-webster.com/dictionary/evaluation (accessed January 3, 2013).

[12]"Feedback," *Merriam-Webster Dictionary*, 2013, http://www.merriam-webster.com/dictionary/feedback (accessed January 3, 2013).

Emergency Planning Team

The first phase of the emergency planning process represents the forming of a team to craft the emergency plan. A team is a small group of individuals who have "complementary skills," and are all "committed to a common purpose, common performance goals, and a common approach for which they hold themselves mutually accountable."[13] Organizations and their outputs are no better than the individual people that comprise them. Therefore, team members may represent interests that are both internal and external to the organization and have some contributing perspective of the emergency domain. Team leadership and chain-of-command should be established and identified, and a "clear line of authority" must exist between the team leader and the peer team members.[14]

This phase also accommodates the mandating of a team mission. A mission statement is the expressing of purpose and existence that "communicates legitimacy to internal and external stakeholders."[15] The mission statement also defines parameters that characterize the emergency domain, thereby affecting the scope and magnitude of emergency planning. The team must be mindful of the stated mission throughout the entirety of crafting the intended emergency plan. Any consideration of mission also involves goals and objectives.

Goals represent the qualitative, intangible aspects of what the team hopes to accomplish, whereas objectives represent the quantitative, measurable attributes. Goals represent open statements regarding "what one wants to accomplish with no quantification of what is to be achieved."[16] Objectives are a conversion of mission "into specific performance targets" through which progress may be tracked and measured.[17]

Few things in life are free; emergency planning is not one of them. During the initial stage of the emergency planning process, a project schedule and budget must both be delineated to govern the crafting of the intended emergency plan.[18] Scheduling involves the determining, allocating, and assigning of resources and events

---

[13]Ricky Griffin and Gregory Moorhead, *Organizational Behavior: Managing People and Organizations*, 10th ed. (Mason, OH: South-Western Cengage Publishing, 2008), 286.

[14]Thomas Wahle and Gregg Beatty, "Emergency Management Guide for Business and Industry: A Step-by-Step Approach to Emergency Planning, Response, and Recovery for Companies of All Sizes," 2007, http://www.fema.gov/pdf/business/guide/bizindst.pdf (accessed January 4, 2013), 10.

[15]Richard Daft, *Organization Theory and Design*, 9th ed. (Mason, OH: Thomson South-Western Publishing, 2007), 58.

[16]C Appa Rao, B. Parvathiswara Rao, and K. Sivaramakrishna, *Strategic Management and Business Policy: Text and Cases* (New Delhi, India: Excel Books, 2008), 86.

[17]Ibid., 88.

[18]Ronald Perry and Michael Lindell, *Emergency Planning* (Hoboken, NJ: John Wiley and Sons Publishing, 2007).

chronologically that are necessary for the completion of an endeavor.[19] Establishing schedules and budgets provides a method of controlling the emergency planning process.

The first phase of the emergency planning process is foundational. It instantiates the personnel, activities, resources, functions, authorities, responsibilities, chains-of-command, and thrusts of the planning team. This phase establishes the boundaries and expectations of team performance quantitatively, qualitatively, and chronologically. From an integrative perspective, it also establishes the relationships that affect the functioning of the team both externally and internally and establishes legitimacy for the planning team.

Analysis of Hazards and Resources

The second phase of the emergency planning process involves considering both hazards and resources. A hazard represents "a situation in which danger exists and it is possible for an accident or incident to occur."[20] Hazards may arise from both man-made and natural origins and represent a wide array of endangerments. Examples range from terrorist attacks and automobile accidents to blizzards and hurricanes. Resources are those assets that are leveraged for problem solving and the mastering of challenging situations.[21] Resources can be tangible (e.g., humans) or they can be intangible (e.g., knowledge). In any case, the most important resource of any emergency planning initiative is people. Any crafted emergency plan is no better than the humans who produced it via the emergency planning process.

From a systems perspective, various types of techniques exist through which hazards analysis occurs. These are itemized as follows:[22]

> *Conceptual*—This category is the precursor for all succeeding analysis types.[23] Its primary function and purpose are to identify a listing of "top-level hazards that can be recognized."[24] The identification of hazards may involve experience with "similar systems, hazard checklists, mishap/incident hazard tracking logs, safety lessons learned, and so forth."[25]

---

[19]Kathy Schwalbe, *Introduction to Project Management* (Boston, MA: Course Technology, 2006).

[20]Ian Wallace, *Developing Effective Safety Systems* (Warwickshire, UK: Institution of Chemical Engineers, 1995), 85.

[21]Jeffrey Greenhaus, Gerard Callanan, and Veronica Godshalk, *Career Management*, 4th ed. (Thousand Oaks, CA: Sage Publishing, 2010), 292.

[22]Clifton Ericson, *Hazard Analysis Techniques for System Safety* (Hoboken, NJ: John Wiley and Sons Publishing, 2005), 32.

[23]Ibid.

[24]Ibid.

[25]Ibid.

Preliminary—This category involves an initial consideration of risk.[26] It involves considerations of chances regarding whether an event will or will not occur that poses some form of danger.

Detailed—This category extends the risk considerations that commenced within the preceding stage of preliminary analysis. It accommodates considerations of "hazard causal factors."[27]

System—This category further extends the implications of risk with respect to "hazard severity and probability." [28] The evaluation of potential hazards occurs to "identify all causal factors" including "interrelated fault events."[29]

Operations—This category accommodates analyses of any pertinent "operations and support functions."[30] It also considers "operational hazards" that may be averted or "mitigated" via modifications of "operational procedures when necessary."[31]

Health—This category focuses upon "human health hazards."[32] It involves considerations of the "human environment" involving "potential mishaps, causal factors, risk and safety critical factors," and any pertinent "safety" issues.[33] Examples include chemical, radiological, and biological hazards.[34]

Requirements—This category encompasses considerations of validation and verification to ensure the generating of sufficient requirements that affect emergency planning. It attempts to avoid the presences of any "safety gaps."[35] It interjects attributes of safety within the requirements specification.[36]

Contemplation of these categories may be facilitated through a review of the current organizational status regarding its ability to experience an emergency incident. Such consideration may be accomplished through the conducting of an analysis of strengths, weaknesses, opportunities, and threats (SWOT) to query facets of where the organization stands regarding its current capacities. Enacting the SWOT

---

[26]Ibid.
[27]Ibid., 38.
[28]Ibid., 39.
[29]Ibid.
[30]Ibid., 40.
[31]Ibid.
[32]Ibid., 41.
[33]Ibid.
[34]Ibid.
[35]Ibid., 42.
[36]Ibid.

analysis may involve a variety of both internal and external activities to provide sufficient information regarding organizational characteristics.

Internal organizational considerations may involve examining and evaluating the following items:[37]

- Evacuation plans
- Fire plans
- Safety and health programs
- Environmental policies
- Security procedures
- Insurance programs
- Finance and purchasing procedures
- Organizational policy
- Employee manuals
- Hazardous materials plans
- Process safety assessments
- Risk management plans
- Capital improvement programs
- Mutual aid agreements

External resources may involve considerations of the following resources and items:[38]

- Community emergency management office
- Mayor and municipal administrators
- Local emergency planning committees
- Fire department
- Police department
- Emergency medical services
- American Red Cross
- National Weather Service
- Public Works Department
- Planning Commission
- Telephone companies

---

[37]Thomas Wahle and Gregg Beatty, "Emergency Management Guide for Business and Industry," 11.
[38]Ibid.

- Electric utilities
- Hospitals
- Insurance Agencies
- Various businesses
- Various other agencies and organizations

Many more internal and external activities are necessary within this phase of the emergency planning process. Considerations of pertinent regulatory codes include the identifying of "applicable federal, state, and local regulations" ranging from "transportation regulations" to "zoning regulations."[39] Evaluating the critical products, services, and operations of the organization ranges from examining "lifeline services, such as electrical power, water, sewer, gas, telecommunications, and transportation" to facets of "operations, equipment, and personnel" that are essential for the continuity of operations.[40] Other evaluations range from examinations of personnel and equipment to "backup systems" and "organizational capabilities."[41]

These internal and external resources represent a diverse array of sources of information from which data regarding potential hazards may be obtained. The primary focus of these internal and external queries and analyses provides materials for assessing "current capabilities and about possible hazards and emergencies."[42] These activities provide an initial understanding of the external and internal endangerments that may pose risk to the organization and facilitate a basic understanding of the emergency domain. Through conducting such an analysis of potential hazards and their associated risks, an understanding of community implications may be fashioned from the perspectives of hazard and risk.[43]

This phase of the emergency planning process also incorporates vulnerability analysis. Through the use of vulnerability analysis, organizations gain a perspective of the "many types of emergencies" that may be experienced, the "projected impact of the emergency," and the availability of resources that are necessary to "respond to the emergency."[44] The vulnerability analysis incorporated the findings of the hazard and risk analyses to embellish this organizational perspective.

---

[39]Ibid.
[40]Ibid.
[41]Ibid.
[42]Ibid.
[43]U.S. Environmental Protection Agency, *Technical Guidance for Hazards Analysis: Emergency Planning for Extremely Hazardous Substances* (Washington, D.C.: U.S. Government, 1987), 1–3.
[44]Andrew Hiles, *The Definitive Guide to Business Continuity Management,* 2nd ed. (West Sussex, England: John Wiley and Sons Publishing, 2007), 264.

Vulnerability analysis involves a consideration of emergencies that impact directly an organization, the chance of the event occurring, and organizational risk associated with the occurrence of an event.[45] Vulnerability analysis necessitates a formal listing of emergencies and risk probabilities that may impact the organization from both man-made and natural origins. Risk probabilities are associated with attributes of "an immediate threat to life or health, the disruption of services (including information technology services), damage to critical infrastructure, the loss of goodwill or trust, financial impact, or legal impact on the organization."[46] Determining these probabilities involves considerations of "historical occurrences, known threats, and any intelligence provided by the law enforcement community."[47] It also involves considerations of the information that was derived from the analysis of hazards, possible geographic emergencies, and emergencies that may arise because of "human error."[48]

Vulnerability analysis also necessitates a consideration of organizational robustness with respect to the listed endangerments. This component of emergency planning evaluates the organizational "level of preparedness to respond effectively to each specific hazard," thereby further defining the current state of the organization and its abilities regarding emergency situations.[49] The evaluation of probabilities associated with these listed emergencies occurs through the use of a "matrix format" in which hazards are rated through the use of numerical values.[50] Probabilities may also be expressed through associating them with alphabetic descriptors. For example, the letter "L" may indicate a low risk, the letter "M" may represent a medium risk, and the letter "H" may represent a higher risk. Lower scores associated with potential emergencies indicate a greater level of organizational preparedness regarding the possible emergency.[51]

Figure 10.3 shows an example of the matrix that is generated from vulnerability analysis.[52]

---

[45]Rita Hammer, Barbara Moynihan, and Elaine Pagliaro, *Forensic Nursing: A Handbook for Practice*, 2nd ed. (Burlington, MA: Jones and Bartlett Publishing, 2013), 424.

[46]Ibid.

[47]Ibid.

[48]Thomas Wahle and Gregg Beatty, "Emergency Management Guide for Business and Industry," 14.

[49]Rita Hammer, Barbara Moynihan, and Elaine Pagliaro, *Forensic Nursing*, 424.

[50]Ibid.

[51]Ibid.

[52]Utah Department of Public Safety., 2011. "Vulnerability Analysis Methodology," 2011, https://www.google.com/url?q=http://publicsafety.utah.gov/emergencymanagement/documents/VulnerabilityAnalysisMethodologyMarch2011.pdf&sa=U&ei=XOHnUKuxMYz-8ASRlIHYBg&ved=0CAcQFjAA&client=internal-uds-cse&usg=AFQjCNGl5Y8vTxNw51h14sr1KpnQekNcsw (accessed January 5, 2013), p. 54.

Hazard Consequence and Impact Analysis Matrix

Hazard	Frequency of Occurrence	Public	Responders	COOP	Delivery Services	Property, Facilities, Infrastructure	Environment	Economics and Financial Conditions	Public Confidence in Governance
Severe Weather	4	H	H	H	H	M	M	H	*
Earthquake	2	C	H	H	H	E	M	C	*
Building Fire	3	C	H	M	L	C	H	M	*
Drought	3	L	L	L	L	L	H	H	*
Tornado	2	H	M	M	M	M	M	M	*
Landslide	2	H	M	L	M	M	H	L	*
Flooding	3	H	M	M	M	H	H	H	*
Dam Failure	1	E	M	M	M	H	H	H	*
Wildfire	5	H	H	L	M	H	C	M	*
Volcano	1	H	M	L	M	H	H	H	*
Problem Soils	5	L	L	L	L	L	M	L	*
Radon Gas	5	H	L	L	L	L	L	M	*
Technological Man-made (Overall)	5	C	C	E	E	E	H	E	*
WMD	5	H	H	H	H	H	H	H	*
Cyber-Terrorism	1	M	L	H	M	H	L	H	*
Agri-Terrorism	5	M	L	L	M	M	H	M	*
HazMat Transportation	1	H	H	M	H	H	H	M	*
HazMat Fixed Sites	1	H	H	M	H	H	H	H	*
Civil Unrest	2	M	M	M	M	M	M	H	*
West Nile Virus	1	H							*
Influenza	5	H	L	L	L	L	H	L	*

Frequency of Occurrence: Numerical Value		*Vulnerability Factor: Numerical Value*	
Annual Event	1	Low	L
Every 5 years or less	2	Moderate	M
Every 10 years or less	3	High	H
Every 30 years or less	4	Extensive	E
Greater than 30 years	5	Catastrophic	C

Courtesy of the Utah Department of Public Safety

Figure 10.3 Vulnerability analysis matrix.

This figure represents the resulting matrix that was produced by the Utah Department of Public Safety. An examination of this figure yields the existence of the characteristics of both hazards analysis and vulnerability analysis. It accommodates the potentials of both man-made and natural emergencies. The events associated with terrorism represent man-made emergencies, whereas earthquakes and drought represent natural emergencies. Therefore, it incorporates a consideration of endangerments that pose different amounts of risk and that arise from both categories of emergencies.

The matrix also incorporates facets of historical, geographic, organizational, societal, financial, economic, and environmental characteristics that affect the emergency domain. The historical attributes are associated with the quantifying of occurrence frequencies representing the known instances of each event. The geographic attributes are reflected in the hazards listing because it references the risk of dam failure, landslide, flooding, HazMat fixed sites, and so forth. Both organizational and societal characteristics are reflected in the expressions of risk levels associated with the public and responders categories as well as the hazard of civil unrest. Both financial and economic attributes are affiliated with the chance of each emergency occurring. The potential of environmental impact is also associated with the chance of each emergency occurring.

Interpreting the matrix is accomplished rather easily and understandably. Rows represent the identified types of emergencies that may occur, whereas columns represent the different characteristics of the emergency domain. The intersection of rows and columns signifies the chance of an event occurrence for each identified hazard with respect to different characteristics of the emergency domain. It primarily uses English letters to denote the potentials of risk that are associated with each type of identified hazard. Below the matrix, the legend describes each letter and its associated meaning within the context of risk probability. From a historical perspective, the use of numerical values is associated with the frequency distributions of hazardous events that were observed historically thorough time.

The overall design of the matrix is simple and straightforward. It communicates pertinent information quickly, and minimizes the potential of confusing readers through the use of legends to denote the meaning of symbols contained within the rows and columns. Further, its design constrains the matrix to a single page of information, thereby enhancing the simplicity of conveying information.

Although the producing of a matrix may seem rather easy to accomplish, there exists a myriad of factors that may impede the progression of vulnerability analysis. Conflicting personalities among team members may pose various challenges. Misunderstandings may occur when working with other organizations. The internal and external politics of organizations may impact team functioning. In some cases, only partial information may be obtainable thereby increasing the amount of uncertainty within the hazards and vulnerability analyses. Human error also may impact negatively analytical functions thereby interjecting defects and inaccuracies within both forms of analysis.

Regardless, all organizations must perform some type of analysis to determine the characteristics of potential endangerments that exist within the emergency domain. Identifying and expressing these potential endangerments is a critical and significant component of the emergency planning process because it establishes the foundational perspectives of emergencies that may be experienced through time. Careful attention must be devoted to the analyses of hazards and vulnerabilities.

Emergency Plan Design and Development

The third phase of the emergency planning process involves the crafting of an emergency plan. An emergency plan represents the specific actions that are "to be taken immediately after a disaster strikes."[53] This stage is entered only upon the completion of the hazards and vulnerability analyses and the specification of emergency domain requirements. The focus of this phase involves the producing of an expressed, written document that delineates the emergency processes and procedures of the organization. Essentially, the final output of this phase represents the existence of a realistic, viable emergency plan for the organization.

The basic components of a generic emergency generally consist of: (1) executive summary, (2) emergency management elements, (3) emergency response procedures, and (4) support documents.[54] The executive summary represents a succinct, understandable, and straightforward "summary of the planning issue."[55] The emergency management elements consist of items that are the "foundation for the emergency response procedures" that must be implemented in order to "protect personnel and equipment and resume operations."[56] The emergency response procedures comprise the specific processes, procedures, activities, and tasks that are implemented during certain emergency situations. Supporting documents may consist of any items that embellish the emergency plan ranging from telephone number contact lists to geographic maps of danger zones within the incident area.

Developing the emergency plan is methodical. The plan must incorporate the outputs of the hazards analysis and the vulnerability analysis. It must also conform to the specification of requirements that was delineated within the preceding phases of the emergency planning process. The emergency plan must also

---

[53]Gary Shelly and Misty Vermaat, *Discovering Computers: Fundamentals*, 5th ed. (Boston, MA: Course Technology, 2008), 479.

[54]Thomas Wahle and Gregg Beatty, "Emergency Management Guide for Business and Industry."

[55]Eric Berkowitz, *Essentials of Health Care Marketing*, 3rd ed. (Sudbury, MA: Jones and Bartlett Publishing, 2010), 453.

[56]Thomas Wahle and Gregg Beatty, "Emergency Management Guide for Business and Industry."

CHAPTER 10 The Emergency Management Process   281

conform to the tenets of organizational policies and any pertinent government regulations.

Based on the outputs of the hazards and vulnerability analyses, sets of processes, procedures, and activities are expressed for each potential endangerment that is anticipated by an organization. Any associated information, such as pertinent contact information for emergency personnel and organizations, is also specified for each potential endangerment. All of these items are amalgamated and contained within the emergency plan document.

Depending upon the complexity of the organization or the emergency situation, the high-level directions associated with each specific endangerment may be stated generally within one to two pages of writing within the master emergency plan document. Any supplemental information or enhancement of details may be contained within the supporting documents section of the emergency plan (e.g., appendix). Within the master document, the location of specific directions for each possible endangerment must be easily accessible to facilitate an ease of access during emergency situations. Therefore, emergency plans may contain a table of contents or an index that both show pagination appropriately.

The expressing of processes, procedures, activities, and any supplemental information must be understandable and unambiguous. It must communicate clearly directions that are to be followed during inclement circumstances and situations. Readers should know exactly what actions to perform and should be provided with sufficient information to successfully execute the prescribed processes, procedures, and activities. The language of the plan must be simple to accommodate a wide variety of readers whose vocabularies may differ significantly. Figures 10.4 and 10.5 show emergency plan excerpts regarding bomb threat directions.[57]

These figures are located within the *Moscone Center Emergency Preparedness Plan*. The Moscone Center is a convention center in San Francisco, California, which may be used by a variety of organizations and that accommodates a mass audience.[58] Ensuring the safety of event attendees is of paramount importance to the Moscone Center. Therefore, it maintains and periodically updates an emergency plan that encompasses the potentials of both man-made and natural hazards.

These figures show the elements of thoughtful emergency plan design. They both represent the characteristics of good communication through which process writing and informational content are used extensively to provide detailed directions and to solicit personnel knowledge regarding the instance of a bomb threat incident. Both figures communicate succinctly and clearly their intended messages

---

[57]Moscone Center, *Moscone Center Emergency Preparedness Plan* (San Francisco, CA: Moscone Center, 2012), 18–19.
[58]Ibid.

Bomb Threat

Bomb Threat Safety Guidelines

MAINTAIN RADIO AND ELECTRONIC SILENCE

1. Alert all personnel that radio silence is required via the PA system.

> **PA SCRIPT:**
>
> "May I have your attention, please observe radio silence until further notice. I repeat, please observe radio silence until further notice."

2. By telephone, notify the General Manager, Assistant General Manager, Security Manager, Event Manager, and Director of Operations.
3. Using the PA system, recall all SFCF personnel to the Exhibit Level Security Control Room.

> **PA SCRIPT:**
>
> "Your attention please, all Moscone Center Employees report to the Security Control Room. All Moscone Center Employees report to the Security Control Room."

4. Dispatch a Rover to meet the Police Bomb Squad. Direct them to the Security Control Room.
5. Complete the attached Bomb Threat Checklist as soon as possible.

Figure 10.4 Emergency plan bomb threat excerpts #1.

Credit: Moscone Center

From "Emergency Preparedness Plan" By Moscone Center. Copyright © 2012 by Moscone Center. Reprinted by permission.

💣 BOMB THREAT CHECK LIST 💣

Callers Voice:

- ❑ Accent
- ❑ Angry
- ❑ Calm
- ❑ Clearing Throat
- ❑ Coughing
- ❑ Cracking
- ❑ Crying
- ❑ Deep
- ❑ Deep Breathing
- ❑ Disguised
- ❑ Distressed
- ❑ Excited
- ❑ Familiar
- ❑ Laughter
- ❑ Lisp
- ❑ Loud
- ❑ Normal
- ❑ Nasal
- ❑ Raspy
- ❑ Ragged
- ❑ Slow
- ❑ Slurred
- ❑ Soft
- ❑ Stutter
- ❑ Other: _____
- ❑ Familiar?
- Who: _____

Background Noise:

- ❑ Animals
- ❑ Baby
- ❑ Bar
- ❑ Cellular
- ❑ Clear
- ❑ Cordless
- ❑ Factory
- ❑ Home
- ❑ Long Distance
- ❑ Motor
- ❑ Music
- ❑ Office
- ❑ PA System
- ❑ Pay Phone
- ❑ Plane
- ❑ People
- ❑ Static
- ❑ Street
- ❑ Voices
- ❑ In-house
- Other: _____

Threat Language:

- ❑ Educated
- ❑ Foul
- ❑ Irrational
- ❑ Incoherent
- ❑ Message read
- ❑ Taped
- ❑ Other: _____

Caller Description:

- ❑ Male
- ❑ Female
- Age: _____
- Race: _____
- ❑ Other: _____

Call Received by:

Name: _____
Position: _____
Phone #: _____

Questions To Ask:

When is the bomb going to explode?

Where is the bomb?

What does it look like?

What kind of bomb is it?

What will make it explode?

Did you place the bomb?

Why?

What is your name?

Where do you live?

Is there a way to contact you?

Exact wording of threat:

Date: _____
Time: _____ am/pm
#Threat Received At: _____

Page 19

Figure 10.5 Emergency plan bomb threat excerpt #2.
Credit: Moscone Center

using terms and phrases that are unambiguous and simple. These figures also are representative of threat-specific processes, procedures, and activities that must be followed when a specific type of emergency incident arises. Within these figures, the specific emergency incident involves a bomb threat.

The directions that detail the specific activities that must occur during bomb threat incidents are both accurate and complete. They are listed in the chronological order in which they must occur to be effective and to embellish the efficiency of emergency activities during any instance of an incident. They also provide the exact wording of phrases that are to be announced via the public address system during inclement situations.

Both internal and external organizational contexts are referenced within these figures. The directions reference specific organizational positions that are to be contacted during the incident and how they are to be contacted. For example, internal to the organization, the "General Manager, Assistant General Manager, Security Manager, Event Manager, and Director of Operations" are to be contacted through telephone calls.[59] External to the organization, the plan dictates the contacting of the "Police Bomb Squad" and the directing of their personnel to the "Security Control Room."[60]

Timely and accurate information is essential during any emergency incident. The Moscone Center example also facilitates the transcribing of information that may assist responders when diffusing the incident and mitigating its effects. Within these figures, the checklist collects information about the emergency incident through manual writing. Collecting this information manually ensures that it is not forgotten or misstated, and provides a means of accessing the pertinent details of the emergency situation. Also, collecting the details of the emergency domain also provides information that influences the actions of emergency responders. Through the use of such information, the tactics and resources necessary for an effective, efficient response may be better understood, crafted, and implemented.

The checklist provides sections to record details about the perpetrator, the situation, the weapon, and the background environment. It also facilitates the transcribing of information that is collected during conversations with the perpetrator. Such information may be used to craft a profile of the perpetrator and to evaluate the scope and magnitude of the emergency situation.[61] It may be shared with multiple responding organizations and personnel to facilitate a mutual understanding of the emergency situation among these entities.

---

[59]Ibid, 18.

[60]Ibid.

[61]Ronald Decker, *Bomb Threat Management and Policy* (Woburn, MA: Butterworth-Heinemann Publishing, 1999).

The checklist also accommodates the personnel and chronological attributes of information regarding the reporting of the emergency incident. Knowing the reporting individual helps to facilitate the flow of information within the chain-of-command and to also assign initial responsibility and authority. Knowing the exact date and time of the report is especially important if the perpetrator attempts to impose any time constraints regarding any stated demands.

These figures incorporate meticulous details regarding the potential of a bomb threat. The specification of a bomb threat within the emergency plan represents the expressing of an endangerment that was considered during the preceding phases of hazards analysis and vulnerability analysis within the emergency planning process. This consideration of a bomb threat is one of many endangerments that are considered within the emergency plan cumulatively.

With respect to its entirety, the Moscone Center plan represents the culmination of the plan development phase of the emergency planning process. The document presents a logical, organized structuring through which all potential endangerments are summarized and itemized. The ability to access specific incident types is accommodated through the use of a table of contents. Any supporting materials, such as telephone numbers for law enforcement and medical responders, are also contained within the master planning document. Cumulatively, the plan exhibits the characteristics of unique emergency planning design that is appropriate for the organization.

The producing of an emergency plan is not sporadic. It adheres to a methodical process through which emergency plans are crafted, finalized, and disseminated throughout an organization. The process of generating an emergency plan also must conform to any stated requirements that resulted from the preceding phases of the planning process, must incorporate the outcomes of vulnerability and hazards analyses, must incorporate document structuring to effectively and efficiently communicate information to readers, and must conform to organizational policy and any pertinent government regulations.

A generic process for generating an emergency plan is stated as follows:[62]

1. Identify challenges and prioritize activities.
2. Write the plan.
3. Establish a training schedule.
4. Continue to coordinate with external entities.
5. Maintain contact with peer entities.
6. Review, conduct training, and revise.
7. Seek final approval.
8. Distribute the plan.

---

[62]Thomas Wahle and Gregg Beatty, "Emergency Management Guide for Business and Industry," 19.

The salient characteristics of these process steps are highlighted as follows:

1. *Identify challenges and prioritize activities*—This step involves a consideration of milestones and goals regarding the drafting of the plan document.[63] Responsibilities should be delegated to specific individuals regarding anticipated activities and accomplishments.[64] Resources and impediments concerning the hazards analysis and vulnerability analysis should be considered within this step. Deadlines should also be identified.
2. *Write the plan*—This step involves each member of the team authoring a section of the emergency plan.[65] The writing of the plan should adhere to an expressed schedule of drafts, edits, and revisions.[66]
3. *Establish a training schedule*—A training schedule should be specified concurrently with the writing of the plan. After an initial training session occurs, any defects may be identified within the plan and rectified accordingly.
4. *Coordinate with external entities*—Information regarding external entities (e.g., fire, medical, and police organizations) may be incorporated within the plan. Questions to consider include: "Which gate or entrance will responding units use? Where and to whom will they report? How will they be identified? How will facility personnel communicate with outside responders? Who will be in charge of response activities?"[67] Certainly, many more necessary questions may be identified and included within the contents of the plan.
5. *Maintain contact with peer entities*—This step accommodate internal communication and coordination within an organization. It ensures that all components of the organization are informed and have some feedback regarding the crafting of the emergency plan. Within this step, concepts to consider include the methods and conditions of internal organizational collaboration, contact information for appropriate personnel, notification procedures, and so forth.[68]
6. *Review and revise*—This step involves disseminating the initial draft of the emergency plan to appropriate personnel for mutual commentaries. Training may occur during this step to provide feedback through which the plan may be altered to eliminate defects. Any identified "areas of confusion and overlap" may be corrected within the emergency plan.[69]
7. *Seek final approval*—Briefings should be arranged to inform organizational leadership regarding the salient aspects of the emergency plan. This step

---

[63] Ibid.
[64] Ibid.
[65] Ibid.
[66] Ibid.
[67] Ibid.
[68] Ibid.
[69] Ibid.

also accommodates the obtaining of "written approval" of organizational leadership.[70]

8. *Distribute the plan*—After the plan is approved, it may be disseminated throughout the entirety of the organization. Personnel should be familiarized with the emergency plan and its contents. Distribution of the plan may occur both internally and externally, and certain personnel may be required to maintain a copy of the plan in their personal residences. In some cases, some information contained within the plan may be confidential (e.g., corporate secrets, contact information).[71] Therefore, discretion must be exercised when determining whether some or all of the components of the plan are distributed both internally and externally.[72]

Cumulatively, this phase of the emergency planning process is twofold: plan design and plan development. Designing the plan is consistent with the attributes of presenting and communicating information accurately, efficiently, and effectiveness to bolster the soundness of an emergency response. Developing the plan is a methodical process through which drafting, editing, revising, reviewing, disseminating, and distributing of the emergency plan occurs chronologically. Both concepts are complementary; design and development each have specific outputs that culminate in the completing of the design and development phase of the overall emergency plan development process. This phase closes when the crafted emergency plan is approved, distributed, and familiarized appropriately both internally and externally.

Plan Implementation

The fourth phase of the emergency plan development process represents the implementing of the emergency plan. This phase occurs only when an emergency plan exists and permeates the appropriate organizational structures both externally and internally. Implementation represents the instantiating of the plan throughout the organization, and exercising its processes, procedures, and activities commensurately with respect to any emergencies that may arise through time or any practice emergency drills that may be mandated periodically.

This phase of the emergency planning process may incorporate various types of testing exercises and drills. Examples include tabletop exercises, functional exercises, or full-scale exercises as testing methods in which the emergency plan is exercised. A tabletop exercise represents testing that involves "classroom or conference"

---

[70]Ibid.
[71]Ibid.
[72]Ibid.

settings to orally explain potential responses to emergency scenarios.[73] Generally, this form of exercise involves a "limited scenario."[74] A functional exercise involves the realistic testing of "one or more" aspects of the emergency plan within a "field setting" that approximates "disaster conditions."[75] A full-scale exercise involves the testing of "all aspects and all organizational participants" using a "realistic field setting."[76]

Respectively, these categories of testing represent hypothetical, selected, and complete testing methodologies and perspectives through which the emergency plan is exercised. Testing reveals shortcomings in the emergency plan and confirms aspects of the emergency plan that function well. It also strengthens the ability of multiple organizations and different people to collaborate efficiently and effectively during both simulated and actual emergencies. Testing also familiarizes personnel and organizations with the resources and methods that they may experience during actual emergencies.

Simulated emergencies may be exercised through the use of various scenarios and test suites. A test scenario is defined as a "model of conditions and circumstances" that is used to "illustrate the connection between" with respect to "how conditions influence circumstances and how circumstances alter conditions."[77] Basically, it mimics the conditions of reality to emulate various facets of emergencies, including both natural and man-made incidents, to necessitate the rendering of human decisions and commensurate activities.

Test scenarios are used to examine and to evaluate the ramifications of "disaster impacts, of mitigation decisions, or of post-disaster search-and-rescue strategies."[78] Scenarios generally involve a "what if" question regarding the "hypothetical progression of circumstances and events designed to illustrate the consequences of some decision, action, or impact."[79] Test scenarios may be implemented separately or they may be implemented as a component of a test suite. A test suite represents a collection of related test scenarios that are designed to test a specific characteristic of the emergency plan.

During any form of testing, various observers may be used to record data and to assess and evaluate the exercise objectively. Observers do not participate directly during the progression of a testing exercise. Their role is observational,

---

[73]Perry and Lindell, *Emergency Planning*, 112.
[74]Ibid.
[75]Ibid.
[76]Ibid.
[77]David Alexander, *Principles of Emergency Planning and Management* (Hertfordshire, England: Terra Publishing, 2002), 42.
[78]Ibid.
[79]Ibid.

and it adheres to a methodical, "step-by-step observation procedure."[80] Such procedures may be unique among organizations and scenarios. Multiple observers should coordinate their activities to improve the effectiveness and efficiencies of their reviews.[81] Observers may use a variety of worksheets, record management systems, checklists, and other tools to record observations of the intended exercise.[82] Such data sets are evaluated and analyzed for review after the completing of test exercises.

Observations regarding the implementing of the emergency plan are not constrained solely to test exercises. Instead, observations may be made during actual exercises. By doing so, organizations may record the attributes of their live responses for review during the aftermath of the incident. Such reviews introduce the notion of feedback regarding the emergency plan.

Implementation of the emergency plan occurs from the perspectives of both actual and practice incidents. Regardless of the instance, any implementing of the emergency plan shows both the weak and the strong characteristics of the intended response and of the plan itself. During the implementing of the emergency plan, one or more observers may be present to record attributes of the emergency response, thereby establishing a basis for improving the emergency plan and the associated quality of response for any future incidents.

Feedback

Feedback represents the final phase of the emergency management planning process before any new iterations of the process occur. The emergency management planning process culminates in its final stage of feedback during which multiple factions communicate regarding their findings from analyzing observations, conclusions, and recommendations concerning the emergency plan. Based upon the findings gained from the observing of both simulated and actual emergencies, various conclusions and recommendations can be offered regarding the emergency plan. Feedback can also be gained from interviewing the participants of the emergency incident including both victims and responders. Feedback may also be gained from expert opinion.[83]

Soliciting feedback occurs through a number of methods. Feedback may be gained from the use of manual, Internet, or intranet surveys through which respondents may rate various attributes of the emergency plan and the associated

---

[80]Michael Karmis, *Mine Health and Safety Management* (Littleton, CO: Society for Mining, Metallurgy, and Exploration, 2001), 220.

[81]Ibid.

[82]Ibid.

[83]Robert McCreight, *An Introduction to Emergency Exercise Design and Evaluation* (Plymouth, UK: Government Institutes—Scarecrow Press, 2011).

response. Personal interviews may be used to talk directly with people to gain their opinions and perceptions of the emergency plan and the associated response. Open discussion forums or virtual discussion groups are other methods of soliciting feedback. In many cases, the use of after-action reviews occurs as a method of soliciting feedback and evaluation of the performance of the organization during simulated or actual emergencies.

Feedback is essential when evaluating emergency plans, and provides "valuable information that can allow improvements in training, group performance, skills development, and greater understanding of both the simple and complex issues involved in handling emergency events and trying to manage crises."[84] Through the use of feedback, the conclusions and recommendations concerning the emergency plan can be improved and strengthened. Such conclusions and recommendations must be evaluated regarding their merit, feasibility, and applicability to determine whether they can be included within the emergency plan and implemented during future testing exercises or actual emergencies. The process of crafting and maintaining an emergency plan is cyclical and continuous. Although feedback is the final stage of the formal process, it represents the evaluating and improving of the plan before a new iteration of the planning cycle process begins again.

Hazard Preparedness

The notion of prevention involves "the act of preventing or hindering."[85] Within the context of emergency management, this notion involves not experiencing some undesirable event, if possible. Avoiding the risks of both man-made and natural emergencies is a primary concept of disaster preparedness. Preparing for hazards is a unique endeavor that is specific to individual hazards before an event occurs, and involves "unique actions" as attempts to avert the experiencing of emergency situations.[86]

From a national perspective, the National Response Framework "establishes a comprehensive, national, all-hazards approach" regarding national emergencies to "prepare for and provide a unified national response to disasters and emergencies."[87] Such preparation accommodates "lessons learned" from previous

---

[84]Ibid., 142.

[85]"Prevention," *Merriam-Webster Dictionary*, 2012, http://www.merriam-webster.com/dictionary/prevention (accessed January 5, 2013).

[86]Federal Emergency Management Agency, "Plan, Prepare, and Mitigate," 2013, http://www.fema.gov/plan-prepare-mitigate (accessed January 5, 2013).

[87]U.S. Department of Homeland Security, *Introducing National Response Framework*, 2008, http://www.fema.gov/pdf/emergency/nrf/about_nrf.pdf (accessed January 5, 2013).

emergency situations and the defining of "core principles for managing incidents."[88] It also establishes the architecture for chains-of-command, authorities, roles, and responsibilities regarding the facets of "who, what, and how of emergency preparedness and response."[89] This paradigm also incorporates the primary doctrines of emergency preparedness and response. These considerations are applicable to large catastrophes and small incidents.[90] These doctrines are presented within Figure 10.6.[91]

These tenets of emergency doctrine embellish preparedness both tangibly and intangibly. They facilitate relationships among emergency response organizations and an awareness of their resources and capabilities. Having such knowledge reduces confusion and ensures that the responding organizations are aware of their attributes of their peers. Facilitating such relationships also enhances organizational familiarity and scalability of resources among the responding entities. These doctrines also provide a basis for understanding within chains-of-command and for facilitating communication among organizational leaders.

Response Doctrine: Key Principles

➤ **Engaged Partnership.** Leaders at all levels must communicate and actively support engaged partnerships by developing shared goals and aligning capabilities so that no one is overwhelmed in times of crisis.

➤ **Tiered Response.** Incidents must be managed at the lowest possible jurisdictional level and supported by additional capabilities when needed.

➤ **Scalable, Flexible, and Adaptable Operational Capabilities.** As incidents change in size, scope, and complexity, the response must adapt to meet requirements.

➤ **Unity of Effort Through Unified Command.** Effective unified command is indispensable to response activities and requires a clear understanding of the roles and responsibilities of each participating organization.

➤ **Readiness To Act.** Effective response requires readiness to act balanced with an understanding of risk. From individuals, households, and communities to local, tribal, State, and Federal governments, national response depends on the instinct and ability to act.

Figure 10.6 Emergency doctrines.

---

[88] Ibid., 4.
[89] Ibid.
[90] Ibid., 1.
[91] Ibid., 4.

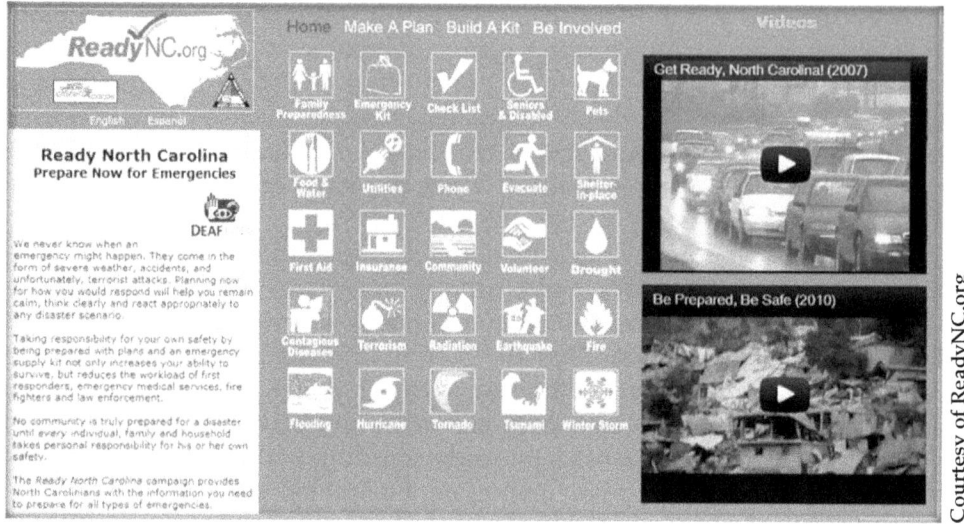

Figure 10.7 North Carolina emergency preparedness site.

Preparedness is not conducted solely from a national or federal perspective. Individual states and localities also facilitate preparedness initiatives. For example, from a state perspective, the state of North Carolina expresses a variety of methods that affect its preparedness attributes. Through its Internet site, it provides numerous materials from which individuals and organizations may craft emergency plans, obtain emergency supplies regarding an array of possible hazards, and become involved with emergency management. Figure 10.7 shows this Internet presence.[92]

Within this resource, advice is given regarding an array of preparedness scenarios. For instance, regarding preparedness for evacuation, the site indicates that one should have stored and made available any necessary medicines, a "first aid kit," and "copies of important documents."[93] Examples of the primary items to embellish preparedness for evacuation include:[94]

- Mortgage insurance
- Birth certificates
- Social security cards
- Wills

---

[92]State of North Carolina, "Ready North Carolina," 2013, http://readync.org/index.cfm?espanol=0&on=Home (accessed January 7, 2013).

[93]State of North Carolina, "Emergency Evacuation Kit Checklist," 2013, http://readync.org/index.cfm?espanol=0&topic=36&on=Checklist (accessed January 7, 2013).

[94]Ibid.

- Tax information
- Bank information
- List of family physicians
- List of important family information, such as style and serial number of medical devices like pacemakers
- Video or picture inventory of household items

The remaining options within the North Carolina site provide lists that are commensurate with the uniqueness of the potential hazard for which preparedness is sought. Through preparedness, the chance of any individual or organization surviving an adverse situation is improved. Preparedness also "reduces the workload of first responders, emergency medical services, fire fighters, and law enforcement."[95] In any case, preparedness should incorporate the following attributes:[96]

- "Ensuring that the resources necessary for responding effectively in the event of a disaster are in place, and that those faced with having to respond know how to use those resources."[97]
- "Developing planning processes to ensure readiness."[98]
- "Formulating disaster plans."[99]
- "Stockpiling resources necessary for effective response."[100]
- "Developing skills and competencies to ensure the effective performance of disaster-related tasks."[101]
- "Actions designed to enhance the ability to undertake emergency actions in order to protect property and contain disaster damage and disruption, as well as the ability to engage in post-disaster restoration and early recovery activities."[102]

---

[95]State of North Carolina, "Family Preparedness," 2013, http://readync.org/index.cfm?espanol=0&on=Make%20A%20Plan (accessed January 7, 2013).

[96]Jeannette Sutton and Kathleen Tierney, "Disaster Preparedness: Concepts, Guidance, and Research," Natural Hazards Center, University of Colorado, 2006, http://www.colorado.edu/hazards/ (accessed January 7, 2013).

[97]Ibid.
[98]Ibid.
[99]Ibid.
[100]Ibid.
[101]Ibid.
[102]Ibid.

Preparedness is an essential aspect of the domain of emergency management. The notion of preparedness conceptually represents the intersecting of pre-emergency and post-emergency periods (i.e., both before and after the impacting of an event).[103] Avoiding some events is possible, but it is impossible to avoid all emergency incidents. Therefore, both individuals and organizations must exhibit levels of preparedness that are commensurate with the expected hazards and vulnerabilities that affect them uniquely and circumstantially.

Hazard Mitigation

Mitigation is defined as causing sometime to "become less harsh or hostile" or to "make less severe or painful."[104] Within the context of emergency management, the notion of mitigation involves the diminishing or avoiding of the harmful effects of an adverse emergency incident.[105] Mitigation initiatives are usually "most effective" when employed before an emergency incident occurs.[106] Given such notions, from a perspective of emergency management, mitigation is viewed as the "pre-impact actions that provide passive protection at the time of disaster impact so there is less need for emergency response actions."[107]

Whether they arise from man-made or natural origins, emergency incidents will occur in due time. Therefore, the concept of mitigation must be mindful throughout the entirety of the emergency management planning process. Emergency plans should accommodate methods through which the negative effects of emergency incidents are diminished. Organizational policies, processes, and procedures should incorporate facets of mitigation that are appropriate for the uniqueness of their individual situations and circumstances that may necessitate emergency management activities.[108]

The outcomes of hazards analysis and vulnerability analysis influence the crafting of mitigation activities. Therefore, mitigation activities are unique among organizations with respect to the different types of hazards they may anticipate experiencing through time. For instance, among coastal regions that experience catastrophic storms, mitigation is involved with the designing of building and structures to improve their resiliency to withstand the effects of natural disasters

---

[103]Ibid.

[104]"Mitigate," *Merriam-Webster Dictionary*, 2013, http://www.merriam-webster.com/dictionary/mitigate (accessed January 7, 2013).

[105]Michael Lindell, Carla Prater, and Ronald Perry, *Introduction to Emergency Management* (Hoboken, NJ: John Wiley and Sons Publishing, 2007).

[106]Ibid., 193.

[107]Ibid.

[108]Daniel Alesch, Lucy Arendt, and William Petak, *Natural Hazard Mitigation Policy: Implementation, Organizational Choice, and Contextual Dynamics* (New York, NY: Springer Publishing, 2012).

(e.g., hurricanes).[109] Incorporating the principle of mitigation improves the survivability not only of the buildings and structures but also of their human occupants.

Mitigation is also found within the medical domain. This context involves the uses of "surveillance programs" to observe the spreading of disease, which are essential regarding the mitigation of the effects of "contagious disease."[110] Through surveillance, the benefit of mitigation is exhibited through the early recognition of an event. Early recognition of significant illnesses facilitates "earlier treatment" and the implementing of "preventive measures" in order to "prevent the spread of illness to healthcare workers and responders, as well as the rest of the community."[111]

Additionally, mitigation is necessary to embellish the activities of "response and recovery" that are implemented during the aftermath of "a large-scale mass casualty event."[112] Organizations should ensure that personnel are trained regarding their roles, responsibilities, and activities regarding mitigation to support the aftermaths of disasters. Therefore, training programs that emphasize "risk reduction" methods should be implemented organizationally.[113] Through accommodating such facets of mitigation, organizations improve the chances of maintaining operational and functional continuity throughout periods of danger and during the aftermath of an emergency incident.[114]

Despite the best efforts of preparedness and mitigation, it is impossible to avoid all possible events that necessitate emergencies that arise from both man-made and natural origins. This notion permeates the entirety of the crafting and maintaining of an emergency plan. The existence of the emergency plan demonstrates thoughtfulness regarding the hazards and vulnerabilities that may be experienced by organizations, and represents a means of mitigating their effects.

Although mitigation may diminish the effects of various emergency situations, it does not prevent the situations from occurring. All organizations and individuals are susceptible to the dangers of both man-made and natural emergencies. Therefore, organizations must consider their unique hazards and vulnerabilities, and craft mitigation activities commensurately that will diminish potentially any negative effects of emergency incidents.

---

[109]David Godschalk, David Brower, and Timothy Beatley, *Catastrophic Coastal Storms: Hazard Mitigation and Development Management* (Durham, NC: Duke University Press, 1989).

[110]Gregory Ciottone, *Disaster Medicine* (Philadelphia, PA: Mosby-Elsevier Publishing, 2006), 143.

[111]Ibid.

[112]Ibid.

[113]I. Sundar and T. Sezhiyan, *Disaster Management* (New Delhi, India: Sarup and Sons Publishing, 2007), 67.

[114]Ken Doughty, *Business Continuity Planning: Protecting Your Organization's Life* (Boca Raton, FL: CRC Press, 2001).

Hazard Contingencies

Neither an emergency plan nor the enacting of its processes and procedures will be absolutely perfect. Contingencies will exist within the emergency domain. A contingency is defined as "an event (as an emergency) that may but is not certain to occur" or as "something liable to happen as an adjunct to or result of something else."[115] Most folks are probably familiar with the basic notion of Murphy's Law: "what can go wrong, will go wrong, and at the worst possible time."[116] Murphy's Law highlights the necessity of planning for contingencies within the emergency management domain.

Contingencies will arise from a variety of origins. Storms may change course over land or sea; communications devices mail fail; vehicles may not start; emergency personnel may become ill necessitating hospitalization; and weather delays may impact the transporting of personnel and resources. Certainly, many other contingencies may be identified and itemized. Regardless, the effects of potential contingencies must be considered within the context of the emergency planning process. All efforts should be made to identify possible contingencies, and derive appropriate alternative courses of actions.

In some cases, various alternative courses of actions may be identified and crafted that accommodate such contingencies that may necessitate the pursuit of different courses of action. Accommodating the potential of contingencies improves the ability of an organization to adapt to the dynamics of "uncertainty and change."[117] If a decision to pursue an alternative course of action is rendered, then it must be announced among the emergency responders to ensure that the appropriate contingency measures are implemented as efficiently and effectively as possible.

Contingencies may or may not be insurmountable. It is impossible to imagine scenarios "for every possible contingency," but one may "consider several most probable scenarios."[118] The potential "advantages and disadvantages" of each identified alternative must be contemplated when choosing an alternative course of action.[119] One may also imagine the "worst possible" outcome of a situation, and plan accordingly regardless of the probability of the event actually occurring.[120] In

---

[115] "Contingency," *Merriam-Webster Dictionary*, 2013, http://www.merriam-webster.com/dictionary/contingency (accessed January 8, 2013).

[116] Ben Gerwick, *Construction of Marine and Offshore Structures*, 2nd ed. (Boca Raton, FL: CRC Press, 2000), 568.

[117] Ricky Griffin, *Fundamentals of Management*, 6th ed. (Mason, OH: South-Western Cengage Publishing, 2012), 84.

[118] Karel Montor, *Naval Leadership: Voices of Experience* (Annapolis, MD: United States Naval Institute, 1998), 180.

[119] Ibid.

[120] Ibid.

any case, one cannot discount the potential of having to alter the primary course of action to favor some secondary or tertiary alternatives.

Commentary Regarding Emergency Planning: The Case of Fukushima, Japan

In some cases, emergencies may be of almost unimaginable scope and magnitude that surpass the expectations of any emergency plan. On March 11, 2011, the nation of Japan experienced such an incident. The events of this tragic day are given as follows:[121]

> The circumstances which led to the huge damage of the Fukushima nuclear reactors arose from one of the worst natural disasters in recent history. At 14.46 pm on Friday, March 11th, a huge earthquake hit the north eastern coast off the Japanese island of Honshu. The epicenter was off the north east coast of Japan around 140 miles from Tokyo and reached 9.0 on the Beaufort scale, one of the largest earthquakes in the world for over a century. The earthquake destroyed considerable parts of eastern Japan's infrastructure and many buildings, but also triggered a huge tsunami which was up to more than 24 meters high in magnitude. The tsunami devastated large swathes of the north east coast and thousands were killed within a matter of minutes. As of 1st June 2011, 15,281 people had died in these combined incidents and 8,492 people are still missing, presumed dead. There were also 5,363 people injured. A total of 88,873 houses were damaged, 3,970 roads were damaged and 71 bridges damaged or completely destroyed. A 14 meter wave smashed into the Fukushima nuclear reactor, vastly higher than its 5.9 m safety walls. These seriously damaged the outer shell of four of the six reactors and initiated a combination of factors that have led to a major nuclear incident and the necessity of initiating the largest peacetime evacuation in Japan's history.[122]

The seriousness and extensiveness of this emergency necessitated the enacting of emergency plans and responses internationally. The nation of Japan was "woefully unprepared" for such a gigantic calamity.[123] It was also acknowledged that the

---

[121] Manchester City Council, "Fukushima Nuclear Incident and the UK Nuclear Safety Review," *Nuclear Free Local Authorities Briefing*, 2012, http://nuclearpolicy.info/docs/briefings/NFLA_NB_100_Nuclear_emergency_planning.pdf (accessed January 8, 2013), 1.
[122] Ibid.
[123] Associated Press, "Naoto Kan, Ex-Prime Minister: Japan Was Unprepared for Fukushima Nuclear Crises," *The Huffington Post*, 2012, http://www.huffingtonpost.com/2012/03/08/naoto-kan-fukushima_n_1330525.html (accessed January 8, 2013).

Fukushima nuclear facility "should not have been built so close to the ocean on a tsunami-prone coast."[124] Some of the additional failures were systematic and organizational.[125] The obtaining of information concerning the emergency incident was slow, and the information was often "inaccurate" and "unreliable."[126]

These comments allude to poor planning. Preparing for potential hazards and vulnerabilities must be accommodated within the emergency planning process. A historical review of tsunami activity probably would have revealed the frequency of ocean events along the coastline that posed danger to the location of the facility. Acknowledging the potential of danger regarding the physical location of the nuclear facility should have been incorporated during the earliest stages of planning not only of the location of the facility but also within any emergency plans involving the possibility of an emergency involving a hazardous nuclear incident.

Similar notions are expressed within the comments of Naoto Kan, the former Japanese prime minister: "If they had thought about it, they wouldn't have intentionally built it at a place so low . . . The plant was built on the assumption that there was no need to anticipate a major tsunami, and that's the very beginning of the problem. We should have taken more adequate safety steps, and we failed to do so . . . It was a big mistake and I must admit that (the accident) was due to human error."[127]

The aftermath of the event resulted in various complexities that will affect the nation of Japan, other nations, and incident responders for a prolonged period. Despite achieving some stability to control the incident, it is estimated that a period of 40 years is required to "decommission the plant."[128] Environmental concerns also are associated with the contaminating of "everything from fruit and vegetables to fish and water."[129] During 2012, a total of eight U.S. sailors filed a $100 million lawsuit against the Tokyo Electric Power Company, which is responsible for operating the Fukushima facility, regarding allegations "that the company lied about the high level of radiation in the area where they were carrying out a humanitarian mission after the tsunami that triggered the reactor crisis."[130] They were crewmen aboard the U.S.S. Ronald Reagan which was transporting water and food into the affected incident area. Because of their exposure to radiation, they necessitated extensive medical evaluation and may be "at risk for developing cancer and a shorter life

---

[124]Ibid.

[125]Ibid.

[126]Ibid.

[127]Ibid.

[128]Ibid.

[129]Ibid.

[130]Associated Press, "U.S. Sailors Sue Japanese Utility Over Radiation," *USA Today*, 2012, http://www.usatoday.com/story/news/nation/2012/12/28/fukushima-sailors-lawsuit/1797053/ (accessed January 9, 2013).

expectancy, and are undergoing considerable mental anguish as a result." The outcome of the lawsuit is pending at the time of this authorship.

This event represents an emergency of enormous severity. Despite the best efforts to facilitate vulnerability analysis and hazards analysis, envisioning the trio of calamities that befell Japan and the continuously cascading series of resulting events would have been nearly unimaginable given the dynamics of the emergency domain and the immensity of the quantity of variables to be considered when formulating emergency plans. However, the incident serves as a reminder that events of gross magnitude and scope do occur, and have both anticipated and unforeseen contingencies.

These considerations of contingencies were pertinent during the incident. Potential mass evacuations were considered if a worst-case scenario had developed.[131] This consideration of contingency planning is expressed as follows:

> Early on in the crisis, Kan said he did consider the possibility of a worst-case scenario in which all six of the reactors plus spent fuel pools melted down out of control. That probably would have resulted in radioactive fallout over a wide area, requiring evacuation of millions of people, including possibly the population of Tokyo.[132]

Luckily, such mass evacuations involving the entire population of Tokyo were unnecessary. However, the possibility of such mass movements of people was considered as a viable course of action had a worst-case scenario occurred during the incident. This consideration shows the incorporation of contingency planning. During the course of the emergency, it was undeterminable whether the worst case would occur. Therefore, the dynamics of the situation necessitated the realistic potential of mass evacuation in addition to the existing scope and magnitude of the emergency.

Emergency situations often unfold in many unforeseen ways that cannot be imagined. The Fukusima incident is certainly representative of this notion. What was originally an undersea earthquake resulted in calamities and contingencies of such extensive severity that even the continuance of Japan to "function as a state" was questionable.[133] During the period of this authorship, the complexities of the aftermath continue to affect the region and other nations.

The scope and magnitude of this emergency provide many lessons concerning emergency planning. According to the Official Report of the Fukushima Nuclear

---

[131] Associated Press, *The Huffington Post*.

[132] Ibid.

[133] Reuters, "Japanese Urge Farewell to Nuclear Power Six Months After Quake," 2011, http://www.reuters.com/article/2011/09/19/japan-nuclear-idAFL3E7KJ1A020110919 (accessed January 9, 2013).

Accident Independent Investigation Commission, "The Commission has verified that there was a lag in upgrading nuclear emergency preparedness and complex disaster countermeasures, and attributes this to regulators' negative attitudes toward revising and improving existing emergency plans."[134] The preceding discussions regarding accommodating change within the context of crafting and maintaining an emergency plan must not be ignored. The inability and unwillingness to adapt to and accommodate change hampered the timeliness and robustness of the emergency plans regarding the Fukushima incident.

The severity of the Fukushima incident also represents the integrating of both man-made and natural emergencies. This notion is expressed as follows:

> The direct causes of the accident were all foreseeable prior to March 11, 2011. But the Fukushima Daiichi Nuclear Power Plant was incapable of withstanding the earthquake and tsunami that hit on that day. The operator (TEPCO), the regulatory bodies (NISA and NSC) and the government body promoting the nuclear power industry (METI), all failed to correctly develop the most basic safety requirements—such as assessing the probability of damage, preparing for containing collateral damage from such a disaster, and developing evacuation plans for the public in the case of a serious radiation release.[135]

The man-made, "direct causes of the accident" were attributed to "collusion between the government, the regulators and TEPCO, and the lack of governance by said parties."[136] Both the undersea earthquake and the resulting tsunami were certainly of natural origins. However, the combination of both categories of emergencies furthered the complexity of the incident domain. These observations also incorporate an additional lesson for emergency planning: although emergencies arise from man-made or natural origins, they may be integrated through the exhibiting of cascading effects. Therefore, when crafting emergency plans and preparedness measures, organizations must be mindful of the potential of an integrated emergency.

The Fukushima incident will long be viewed as one of the greatest cataclysm to strike the Asian region of the Pacific Ocean. Its effects will long be remembered among numerous nations and peoples. Although tragic, the incident shows the necessity of crafting emergency plans robustly, the importance of updating and maintaining emergency plans, and the unforeseen contingencies that arise during

---

[134] Nuclear Information and Resource Service, "The Official Report of the Fukushima Nuclear Accident Independent Investigation Commission," 2012, http://www.nirs.org/fukushima/naiic_report.pdf (accessed January 9, 2013), 19.

[135] Ibid., 16.

[136] Ibid.

emergency responses involving the exercising of emergency plans. It also shows various and unique complexities of the emergency domain that may span many years after the incident. The Fukushima incident also shows that all attempts should be made to envision each of the hazards and vulnerabilities that affect an organization. Even with the best of emergency plans, failure(s) can occur—as was witnessed during the Fukushima incident. Despite the catastrophic effects of the emergency, one concept is pertinent: lessons can be learned from the Fukushima incident through which other organizations, nations, and peoples can better plan and prepare for emergencies.

Chapter Comments and Summary

This chapter introduced the emergency planning process through which organizational emergency plans are crafted, maintained, and exercised through time. The phases of the emergency planning process are as follows: (1) establish a planning group, (2) analyze hazards and capabilities, (3) design and develop the emergency plan, (4) implement the plan, and (5) feedback activities. This process is iterative and reflexive. It may be repeated as necessary to accommodate new endangerments that exist within the emergency domain.

This chapter introduces the concepts of hazards analysis and vulnerability analysis. These activities occur to identify the potential endangerments that may exist that an organization may experience through time and the resources and capacities that organizations possess through which they may be countered. These endangerments may arise from both man-made and natural origins. Despite the best efforts to identify possible hazards, it is impossible to foresee all possible threats that may impact an organization. Regardless, such planning must be accomplished to improve the chances of organizational survivability and the continuity of operations during emergencies.

All organizations will experience some form of emergency at some point in time. Therefore, an organization must consider how it will mitigate the effects of any emergency that it encounters. Such contemplation is accomplished through the use of mitigation planning. Through mitigation activities, organizations may diminish the effects of adverse situations that they may experience.

Given the outcomes of hazard analysis and vulnerability analysis, organizations must be prepared to endure adverse situations and circumstances. Therefore, preparedness must also be integrated within the emergency planning process. Through preparedness, organizations improve their ability to execute specific aspects of their emergency plans relating to the encountered situation. It is essential that preparedness must be considered throughout the entirety of the emergency planning process.

This chapter serves as a reminder concerning the relevancy of Murphy's Law within the context of emergency management. If something can go wrong, then it might do so at the most inopportune time. Therefore, when crafting emergency plans, one must consider the possibility of having to change the primary course of actions because something happens that necessitates change. During the overall emergency planning process, the potential of contingencies must not be ignored. Alternative courses of actions should be crafted with respect to each of the identified contingencies.

Crafting an emergency plan is both an art and a science, and involves considerations of both the intangible and tangible characteristics of the emergency management domain. It involves imagination to conceive of emergency scenarios and trustfulness among participants. It also involves quantitative observations of historical events and rigorous methodology to generate a viable emergency plan. Although the discussions of herein merely highlight the emergency planning process, the crafting of an emergency plan can be both complex and tedious. In any case, the responsibility of crafting an emergency plan is to be taken seriously and reverently.

An old adage indicates that "failing to plan is planning to fail." Another old adage states that one should "plan the work, and work the plan." These adages should be mindful when crafting any emergency plan. Failure to address even the simplest of issues could become catastrophic when the plan is executed during an actual emergency. Therefore, when conducting test exercises of the emergency plan, every effort should be made to address shortcomings in the contents of the plan, and to rectify them as quickly as is possible. Through such diligence, the qualities of the emergency plan and its associated responses are both improved.

Terminology

After-action review
Budget
Cascading effects
Checklist
Contingency
Contingency plan
Development
Directions
Emergency
Emergency drill
Emergency planning
Emergency planning process
Endangerment
Evaluation

Executive summary
Expertise
Feedback
Financing
Full-scale exercise
Functional exercise
Goals
Hazard
Hazards analysis
Hazard avoidance
Implementation
Iterative
Mission atatement
Mitigation

National Response Framework
Objectives
Phase
Planning impediment
Preparedness
Prevention
Process
Reflexive
Risk
Requirements
Resource
Resource analysis
Scheduling
SWOT
Tabletop exercise
Team
Test suite
Testing
Vulnerability analysis

Thought and Discussion Questions

1. The traditional functions of management encompass controlling, coordinating, leading, organizing, planning. Define each of these terms and their relationship to the concept of managing emergencies. Within your discussion, relate these concepts to the crafting, maintaining, and implementing of an organizational emergency plan.
2. This chapter introduced the notion of contingencies and contingency planning. A few examples of contingencies that necessitate the enacting of possible alterative courses of actions were identified herein. In addition to these examples, do some research and determine what other contingences have affected the primary plans of actual emergencies in or near your locale. Write a brief essay that highlights the decision to pursue the alternative course of action and its consequences within the context of emergency management.
3. This chapter introduces the notion that the emergency planning process is both cyclical and continuous. Do some research regarding the emergency management plans of organizations in or near your locale. Write a brief essay that discusses how these plans have been improved through time throughout repeated iterations of the emergency planning process. Within your discussion, provide some commentary regarding the effectiveness and efficiency of these improvements regarding the anticipated emergency response.
4. This chapter alludes to the importance and relevancy of evaluation within the context of the emergency planning process. Examine the emergency planning process cumulatively, and consider how evaluation separately impacts every phase of the process. Write a brief essay that discusses how evaluation may be incorporated within each phase of the emergency management process to generate improvements in the emergency plan and the anticipated emergency response.

References

1. Alesch, D., L. Arendt, and W. Petak. 2012. *Natural hazard mitigation policy: Implementation, organizational choice, and contextual dynamics.* New York, NY: Springer Publishing.
2. Alexander, D. 2002. *Principles of emergency planning and management,* 42. Hertfordshire, England: Terra Publishing.
3. Arslan, H. 2007. *Cognitive radio, software defined radio, and adaptive wireless systems,* 407. Dordrecht, The Netherlands: Springer Publishing.
4. Associated Press. 2012. Naoto Kan, ex-Prime Minister: Japan was unprepared for Fukushima nuclear crises. *The Huffington Post.* http://www.huffingtonpost.com/2012/03/08/naoto-kan-fukushima_n_1330525.html (accessed January 8, 2013).
5. Associated Press. 2012. U.S. sailors sue Japanese utility over radiation. *USA Today.* http://www.usatoday.com/story/news/nation/2012/12/28/fukushima-sailors-lawsuit/1797053/ (accessed January 9, 2013).
6. Baden-Powell, R. n.d. Emergency quotes. http://www.brainyquote.com/quotes/keywords/emergency.html (accessed January 1, 2013).
7. Berkowitz, E. 2010. *Essentials of health care marketing.* 3rd ed., 453. Sudbury, MA: Jones and Bartlett Publishing.
8. Ciottone, G. 2006. *Disaster medicine,* 143. Philadelphia, PA: Mosby-Elsevier Publishing.
9. Contingency. 2013. *Merriam-Webster Dictionary.* http://www.merriam-webster.com/dictionary/contingency (accessed January 8, 2013).
10. Daft, R. 2007. *Organization theory and design.* 9th ed., 58. Mason, OH: Thomson South-Western Publishing.
11. Decker, R. 1999. *Bomb threat management and policy.* Woburn, MA: Butterworth-Heinemann Publishing.
12. Devereaux, R. 2012. Hurricane Sandy: Two weeks on, Coney Island recovery far from complete. *The Guardian.* http://www.guardian.co.uk/world/2012/nov/12/coney-island-struggles-sandy-recovery (accessed January 3, 2013).
13. Doughty, K. 2001. *Business continuity planning: Protecting your organization's life.* Boca Raton, FL: CRC Press.
14. Ericson, C. 2005. *Hazard analysis techniques for system safety,* 32. Hoboken, NJ: John Wiley and Sons Publishing.
15. Evaluate. 2013. *Merriam-Webster Dictionary.* http://www.merriam-webster.com/dictionary/evaluation (accessed January 3, 2013).
16. Faler, B. 2012. Sandy aid worth $60 billion passes Senate as House balks. *Bloomberg.* http://www.bloomberg.com/news/2012-12-29/senate-approves-60-billion-hurricane-sandy-aid-package.html (accessed January 3, 2013).
17. Federal Emergency Management Agency. 2013. Plan, prepare, and mitigate. http://www.fema.gov/plan-prepare-mitigate (accessed January 5, 2013).

18. Feedback. 2013. *Merriam-Webster Dictionary.* http://www.merriam-webster.com/dictionary/feedback (accessed January 3, 2013).
19. Gerwick, B. 2000. *Construction of marine and offshore structures.* 2nd ed., 568. Boca Raton, FL: CRC Press.
20. Godschalk, D., D. Brower, and T. Beatley. 1989. *Catastrophic coastal storms: Hazard mitigation and development management.* Durham, NC: Duke University Press.
21. Greenhaus, J., G. Callanan, and V. Godshalk. 2010. *Career management.* 4th ed., 292. Thousand Oaks, CA: Sage Publishing.
22. Griffin, R. 2012. *Fundamentals of management.* 6th ed., 84. Mason, OH: South-Western Cengage Publishing.
23. Griffin, R. and G. Moorhead. 2008. *Organizational behavior: Managing people and organizations.* 10th ed., 286. Mason, OH: South-Western Cengage Publishing.
24. Hammer, R., B. Moynihan, and E. Pagliaro. 2013. *Forensic nursing: A handbook for practice.* 2nd ed., 424. Burlington, MA: Jones and Bartlett Publishing.
25. Hiles, A. 2007. *The definitive guide to business continuity management.* 2nd ed., 264. West Sussex, England: John Wiley and Sons Publishing.
26. Karmis, M. 2001. *Mine health and safety management*, 220. Littleton, CO: Society for Mining, Metallurgy, and Exploration.
27. Kim, S. 2012. Hurricane Sandy: Losses estimated at $45 billion. *ABC News.* http://abcnews.go.com/blogs/business/2012/10/hurricane-sandy-losses-estimated-at-45-billion/ (accessed January 3, 2013).
28. Lindell, M., C. Prater, and R. Perry. 2007. *Introduction to emergency management.* Hoboken, NJ: John Wiley and Sons Publishing.
29. Manchester City Council. 2012. Fukushima nuclear incident and the UK nuclear safety review, 1. Nuclear free local authorities briefing. http://nuclearpolicy.info/docs/briefings/NFLA_NB_100_Nuclear_emergency_planning.pdf (accessed January 8, 2013).
30. Margusity, H. 2012. No hurricane warning for what could be the most expensive storm in history. *Accuweather.* http://www.accuweather.com/en/weather-news/national-hurricane-center-no-a/839301 (accessed January 3, 2013).
31. McCreight, R. 2011. *An introduction to emergency exercise design and evaluation.* Plymouth, UK: Government Institutes – Scarecrow Press.
32. Mitigate. 2013. *Merriam-Webster Dictionary.* http://www.merriam-webster.com/dictionary/mitigate (accessed January 7, 2013).
33. Montor, K. 1998. *Naval leadership: Voices of experience*, 180. Annapolis, MD: United States Naval Institute.
34. Moscone Center. 2012. *Moscone center emergency preparedness plan*, 18–19. San Francisco, CA: Moscone Center.
35. Nuclear Information and Resource Service. 2012. The official report of the Fukushima nuclear accident independent investigation commission, 19. http://www.nirs.org/fukushima/naiic_report.pdf (accessed January 9, 2013).

36. Perry, R. and M. Lindell. 2007. *Emergency planning*. Hoboken, NJ: John Wiley and Sons Publishing.
37. Prevention. 2012. *Merriam-Webster Dictionary*. http://www.merriam-webster.com/dictionary/prevention (accessed January 5, 2012).
38. Rao, C. A., B. P. Rao, and K. Sivaramakrishna. 2008. *Strategic management and business policy: Text and cases*, 86. New Delhi, India: Excel Books.
39. Reuters. 2011. Japanese urge farewell to nuclear power six months after quake. *Reuters*. http://www.reuters.com/article/2011/09/19/japan-nuclear-idAFL3E7KJ1A020110919 (accessed January 9, 2013).
40. Schwalbe, K. 2006. *Introduction to project management*. Boston, MA: Course Technology.
41. Shelly, G. and M. Vermaat. 2008. *Discovering computers: Fundamentals*. 5th ed., 479. Boston, MA: Course Technology.
42. State of North Carolina. 2013. Emergency evacuation kit checklist. http://readync.org/index.cfm?espanol=0&topic=36&on=Checklist (accessed January 7, 2013).
43. State of North Carolina. 2013. Family preparedness. http://readync.org/index.cfm?espanol=0&on=Make%20A%20Plan (accessed January 7, 2013).
44. State of North Carolina. 2013. Ready North Carolina. http://readync.org/index.cfm?espanol=0&on=Home (accessed January 7, 2013).
45. Sundar, I. and T. Sezhiyan. 2007. *Disaster management*, 67. New Delhi, India: Sarup and Sons Publishing.
46. Sutton, J. and K. Tierney. 2006. Disaster preparedness: Concepts, guidance, and research. Natural Hazards Center, University of Colorado. http://www.colorado.edu/hazards/ (accessed January 7, 2013).
47. Taylor, A. 2013. Huge California bushfire spreads over 6,500 acres. *Business Insider*. http://www.businessinsider.com/california-wildefire-spreading-2013-5 (accessed June 9, 2013).
48. Urist, J. 2012. Hurricane Sandy exposes faults of modern family preparedness. *The Washington Post*. http://www.washingtonpost.com/blogs/on-parenting/post/hurricane-sandy-exposes-faults-of-modern-family-preparedness/2012/11/01/d488b106-2438-11e2-ac85-e669876c6a24_blog.html (accessed January 3, 2013).
49. U.S. Department of Homeland Security. 2008. *Introducing national response framework*. http://www.fema.gov/pdf/emergency/nrf/about_nrf.pdf (accessed January 5, 2013).
50. U.S. Environmental Protection Agency. 1987. *Technical guidance for hazards analysis: Emergency planning for extremely hazardous substances*, 1–3. Washington, D.C.: U.S. Government.
51. Utah Department of Public Safety. 2011. Vulnerability analysis methodology, 54. https://www.google.com/url?q=http://publicsafety.utah.gov/emergencymanagement/

documents/VulnerabilityAnalysisMethodologyMarch2011.pdf&sa=U&ei=XOHnUKuxMYz-8ASRlIHYBg&ved=0CAcQFjAA&client=internal-uds-cse&usg=AFQjCNGl5Y8vTxNw51h14sr1KpnQekNcsw (accessed January 5, 2013).

52. Wahle, T. and G. Beatty. 2007. *Emergency management guide for business and industry: A step-by-step approach to emergency planning, response, and recovery for companies of all sizes*, 10. http://www.fema.gov/pdf/business/guide/bizindst.pdf (accessed January 4, 2013).

53. Wallace, I. 1995. *Developing effective safety systems*, 85. Warwickshire, UK: Institution of Chemical Engineers.

THE EMERGENCY MANAGEMENT PROCESS: DISASTER RECOVERY

"Every post-disaster recovery manifests tension between speed and deliberation. Speed of recovery is important in order to keep businesses alive, rebuild infrastructure, and provide temporary and permanent housing. If official agencies do not act quickly, many victims will begin to rebuild on their own in ways and at locations that they determine."

Robert Olshansky[1]

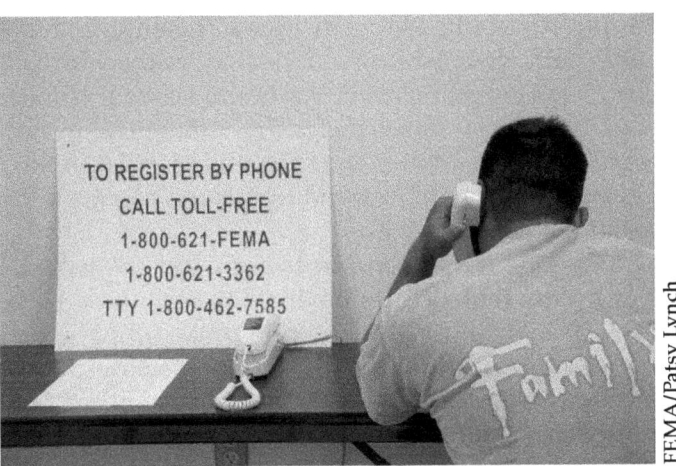

Figure 11.1 Facilitating Disaster Recovery. "La Villa, Texas, August 3, 2008—Texas resident calls to register for disaster assistance from a Disaster Recovery Center. People affected by Hurricane Dolly come into the Disaster Recovery Center (DRC) that have been set up. Photo by Patsy Lynch/FEMA."

---

[1]Robert B. Olshansky, "Planning After Hurricane Katrina." *Journal of the American Planning Association* 72, no. 2 (2006): 148.

OBJECTIVES

The objectives of this chapter are to:
- Introduce the notion of recovery that occurs after calamities;
- Discuss recovery as a component of the disaster management process;
- Discuss recovery from the perspective of regaining a sense of normalcy;
- Discuss the methods through which recovery planning occurs; and
- Discuss recovery planning as a preparedness measure.

Introduction

Every community is susceptible to the effects of man-made and natural disasters. During the aftermath of disasters, various emergency shelters and assistance centers are established to assist refugees from the disaster. Figure 11.1 shows this concept. These types of shelters and assistance centers are often implemented following disasters of varying scopes and magnitudes ranging from tornadoes to major hurricanes.

On May 20, 2013, a major tornado impacted the town of Moore, Oklahoma, killing over 50 individuals, many of whom were children. Figure 11.2 shows the devastation that resulted from the Moore incident. The impact of this storm and the destruction left in its path were extensive. Even before the storm cleared the area, rescue and recovery efforts commenced. As hours passed, resources from other jurisdictions, many from nearby states, including commercial, private, and governmental organizational support, moved into the area to assist in the initial phase of the recovery. These initial efforts to respond to this disaster were critical, and they set the tone for the longer-term recovery.

Communities across the nation, and in fact around the world, are vulnerable to a wide range of events that require emergency response. These events range from localized damage to extensive destruction impacting a wide area. The 2012 hurricane, referred to as Superstorm Sandy, cut a wide path of destruction and is considered the second costliest hurricane to impact the United States.[2] Across the Atlantic basin, at least 147 deaths were directly related to this storm.[3] A total of 72 of these fatalities occurred "in the mid-Atlantic and northeastern United States."[4] This quantity of deaths represents the highest number of direct fatalities, within the United States and outside the South, that were related to a "tropical cyclone" since

---

[2]National Oceanic and Atmospheric Administration, *Tropical Cyclone Report: Hurricane Sandy*, 2012, http://www.nhc.noaa.gov/data/tcr/AL182012_Sandy.pdf (accessed May 24, 2013), 1.
[3]Ibid.
[4]Ibid.

CHAPTER 11 The Emergency Management Process: Disaster Recovery   311

Figure 11.2   Moore, Okla., May 22, 2013—Aerial views of the damage caused by the tornado that touched down in the area on May 20, 2013. FEMA continues to assist disaster survivors and are encouraged to register for assistance. Jocelyn Augustino/FEMA.

the occurrence of 1972's Hurricane Agnes.[5] Financially, the cost of this storm was estimated to be approximately $65 billion.[6] The damage inflicted by the storm will be felt for years. Astoundingly, the United States accounted for and represented approximately 90% of "all the world's insured losses" during the 2012 year.[7]

Major disasters inflict their toll upon communities, causing physical and psychological damage. Some communities prove during recovery to be more resilient than others, but in each case of a disaster, the impacted communities experience their own unique, preparation, response, and recovery initiatives and activities. No two communities are identical. Every disastrous experience is unique, and necessitates its own methods of recovery. All communities must consider the strategic implications of recovering from calamities that may affect them in due time.

---

[5]Ibid.

[6]Doyle Rice, "Hurricane Sandy, Drought Cost U.S. $100 Billion," *USA Today*, 2013, http://www.usatoday.com/story/weather/2013/01/24/global-disaster-report-sandy-drought/1862201/ (accessed May 25, 2013).

[7]Ibid.

The complexities of recovery are diverse. Different events produce myriad of challenges and issues that must be considered when crafting recovery plans. Table 11.1 highlights some of the considerations.

Table 11.1 Relationship Between Magnitude of Event and Type of Recovery Challenges (Source: Phillips and Neal, 2007, p. 212)

Recovery item	Characteristics		
Recovery challenges	Small-scale event (earthquake magnitude 5.0 or less; F2 tornado of limited length)	Normal disaster (F3 tornado of several miles; flash flooding to less than 5 feet)	Catastrophic event (Hurricane Katrina, massive damage to broad region)
Social and psychological needs	Usually short-term effect; opportunity to involve schools and agencies in proactive response	Usually short-term effect, but stress debriefing and counseling should be made available	Potential for significant impact on employees and staff (of government, organizations, businesses, schools) and on families
Housing	Dozens of homes with minor damage; weaker structures badly damaged; up to 2 years for recovery	Thousands affected; most residents able to return home in 2–3 years. Federal, state, and local governments able to handle most housing with help from voluntary sector	Massive and widespread losses; hundreds of thousands of homes destroyed or damaged; 5–10 years anticipated for recovery. All governmental levels overrun and unable to assist all households
Economic sector	Limited impact to economic sector unless a direct hit	Larger businesses return most quickly; smaller businesses challenged to return	Massive disruption to employees and businesses; small businesses unlikely to return; all businesses hit heavily

Environment	Minimal impact; opportunity to improve tree density, address stormwater, and increase public awareness	Opportunity to significantly improve local environmental conditions: increase open space, preserve floodplains, consider density transfers	Massive damage, undermining local ecosystems; disaster often exacerbated by neglect to environment before disaster
Infrastructure and lifelines	Usually a rapid return	Up to years depending on the event; most resources quickly recover	Months to years to repair roads, bridges, hospitals, and other key institutions; some never recover

It is impossible to predict exactly when natural disasters will occur, or the extent to which they will affect a community. However, with careful planning, coordination, and collaboration, public agencies, private sector organizations, and citizens within the community can efficiently respond to the issues that result from natural disasters. Post-disaster recovery planning that takes place before a disaster can help a community more effectively respond to and recover from natural disasters. Establishing recovery strategies prior to the event helps ensure that communities are rebuilt according to the vision that is shared by and benefits all community members.

Emergency Management and Disaster Recovery

Today, emergency management is both an art and a science. Emergency managers and their teams are expected to examine their communities and anticipate the threats most likely to endanger. They have to examine infrastructure and determine vulnerabilities. They have to anticipate response-related needs. They have to determine what resources are available and what resources, within reason, may be required in response to a disaster. They have to draw from the information available and, using their knowledge and insight to work to develop contingency plans for their communities.

Emergency management directors also have to be effective in the political area. Seldom do they control the resources they will need to in mounting a response to a major event. Search and rescue, medical, fire, public utilities, housing, transportation, fuel and emergency food and water, to supply victims and responders have to be planned, coordinated, and sourced prior to an event. Marshalling and exercising these resources necessitates strong processes of strategic management and strategic planning.

Emergency management is often described in terms of phases or components, using terms such as mitigation, preparedness, response, and recovery. Some sources add or substitute different terms, but virtually every source policy documents, plans, manuals, textbooks, journals, and research reports agree that recovery is an essential part of emergency management, but it is also the phase that one desires to avoid.[8] Recovery is the phase in which strategic plans are enacted to restore a sense of normalcy and through which infrastructure is rebuilt during the period succeeding a calamity.

Depending on the nature and/or location of the event, the hazard may be relatively innocuous in nature, or may pose a variety of response issues and concerns that have serious implications. Local government is typically the first line of defense in response to an emergency. The response generated by local government may be adequate to remedy the situation, or the event may overwhelm some or all of the resources at the local level.

This chapter discusses the recovery phase within the emergency management process. Included within this chapter is a consideration regarding the two types of recoveries: short-term recovery and long-term recovery. This dichotomy includes considerations of actions in the immediate post-event environment, including search and rescue, emergency evacuation, temporary housing, and health care. Within these categories, disaster impact involves considerations of: (1) assessment, (2) declaration, and (3) assistance. The dimensions of recovery accommodate the following attributes: (1) household, (2) critical infrastructure, (3) environmental, and (4) economic. Facilitating conditions for recovery involves considerations of: (1) identifying the essential resources needed, (2) identifying the asset assistance stakeholders (government, nongovernment, and private sector), and (3) implementing the asset allocation process.

The process of recovery, as well as the individual, community, nonprofit, and business partners who will likely be involved in recovery operations, is addressed in this chapter. Because of the prominent role of the Robert T. Stafford Disaster Relief and Emergency Assistance Act in providing federal assistance funds for recovery in the event of a disaster or emergency, this legislation should again be revisited within this chapter. The responsibilities of emergency management agencies (EMAs) are reasonably well defined in laws and ordinances, and most of those responsibilities fit into the "short-term" category. The financial assistance for recovery that flows through the EMAs is defined by the Stafford Act and focuses on damage repair and replacement.

---

[8]Malcolm E Baird, The Recovery Phase of Emergency Management, January 2010, http://www.memphis.edu/ifti/pdfs/cait_recovery_phase.pdf (accessed May 20, 2013).

Disaster Categories, Impacts, and Risk

Disaster may be defined in many ways, but for the purpose herein, disaster is viewed as any natural, technological, or civil emergency that causes damage of sufficient severity and magnitude to result in a declaration of a state of emergency by a county, the governor, or the president of the United States. Disasters are identified by the severity of resulting damage in three major categories: catastrophic, major, and minor disasters. A catastrophic disaster requires massive state and federal assistance, including immediate military involvement. This level of disaster, especially in some of the most severe cases, may inflict damage across portions of multiple states.

In the cases of Hurricane Katrina and Hurricane Sandy, the damages destroyed much infrastructure, inflicted billions of dollars in damage, and resulted in thousands of injuries and hundreds of deaths. Economically, many businesses were completely dissolved, never to open again. Although other businesses did reopen, they did not experience the same scope and magnitude of operations and profitability that they enjoyed before disaster occurred. It is estimated that over approximately 40% of businesses that experience a major disaster never reopen.[9]

A major disaster likely exceeds local capabilities and requires a broad range of state and federal assistance. The Moore, Oklahoma, tornado of 2013 is a classic example of this category. This event was classified as a major disaster by presidential declaration.[10] During the time of this authorship, this disaster was known to have resulted in the deaths of "24 people, including seven children."[11] Much of the physical infrastructure within the affected area was complete obliterated.

Major disasters are not uncommon in the United States and represent a variety of hazards. Many such events occur throughout the nation regardless of location. These events contained are varied, and range from flooding to Hurricane Sandy whose effects were felt nationally (i.e., both physically and economically). Derived from Federal Emergency Management Agency (FEMA) data, Table 11.2 shows recent examples of such events that were declared as major disasters during just the first six months of 2013.[12]

---

[9]Federal Emergency Management Agency, "Protecting Your Businesses," 2013, http://www.fema.gov/protecting-your-businesses (accessed May 24, 2013).

[10]WJLA News, "Oklahoma Tornado: Barack Obama Declares Major Disaster After Storm," 2013, http://www.wjla.com/articles/2013/05/oklahoma-tornado-barack-obama-declares-major-disaster-after-storm-89084.html (accessed May 24, 2013).

[11]Ibid.

[12]Federal Emergency Management Agency, "Disaster Declarations," 2013, http://www.fema.gov/disasters?field_state_tid=All&field_disaster_type_term_tid=All&field_disaster_declaration_type_value=DR&items_per_page=10&page=2 (accessed May 24, 2013).

Table 11.2 Recent major disasters

Location	Event	Date on which event was declared a major disaster
Oklahoma	Tornado	5/18/2013
Illinois	Severe storms and flooding	5/10/2013
South Dakota	Severe winter storm and snowstorm	5/10/2013
Iowa	Severe winter storm	5/6/2013
Minnesota	Severe winter storm	4/26/2013
Kansas	Snowstorm	4/26/2013
New York	Severe winter storm and snowstorm	4/23/2013
Massachusetts	Severe winter storm, snowstorm, and flooding	4/19/2013
Oklahoma	Severe winter storm and snowstorm	4/8/2013
Maine	Severe winter storm, snowstorm, and flooding	3/25/2013
Rhode Island	Severe winter storm and snowstorm	3/22/2013
Connecticut	Severe winter storm and snowstorm	3/21/2013
New Hampshire	Severe winter storm and snowstorm	3/19/2013
Navajo Nation	Severe freeze	3/5/2013
Eastern Band of Cherokee Indians	Severe storms, flooding, landslides, and mudslides	3/1/2013
Louisiana	Severe storms and flooding	2/22/2013
Mississippi	Severe storms, tornadoes, and flooding	2/13/2013
Arkansas	Severe winter storm	1/29/2013
Pennsylvania	Hurricane Sandy (occurred in 2012)	1/10/2013
Ohio	Hurricane Sandy (occurred in 2012)	1/3/2013

A minor disaster represents the third category of disaster. Such events are likely within the response capabilities of local government and results in only a minimal need for state or federal assistance. Although the word "minor" may make it appear as though such events are nominal, they can be quite dangerous and damaging. In some instances, even minor disasters have the potential of generating a crisis situation in which necessary resources for "survival and well-being" are scarce or unavailable.[13]

During 2003, as the result of flooding, Governor Jeb Bush declared a minor disaster in Deltona, Florida.[14] Certainly, the flooding was destructive. However, it caused secondary effects that produced "hazardous sanitary conditions" that undeniably posed "an imminent threat to public health."[15] As a method of responding and recovering from this incident, actions of construction and excavation were undertaken to create a drainage system to "lower the water level of Lake Doyle," reduce flooding, and for "alleviating the public health threat."[16]

Disasters of any category have the potential of threatening life, causing much physical damage, and interrupting the flows of goods and services that permeate an economy. Disaster may be associated with climatic conditions annually that affect geographic regions differently throughout the year. For example, hurricane season begins in the month of June and ends in the month of November of each year. Contrastingly, blizzards occur only during the coldest months.

Every community exhibits some chance of experiencing a disaster through time. This concept embodies the basic concept of incident risk within the emergency domain. Risk represents the chance of an event either occurring or not occurring through time. Throughout the nation, different communities manifest different incident risks. A classic example of this notion is simply the fact that folks living in Florida are much more likely to experience a hurricane than a snowstorm or severe blizzard.

All communities should identify the types of threats that may impact them and assess the associated risks. Assessing such risk is often accomplished through threat assessment matrices or other forms of analysis. The succeeding chapters of this textbook provide discussions of these concepts. Regardless, within the context of recovery planning, a simple lesson regarding risk is simply stated: the higher the chance of an event occurring, the higher the risk that may be associated with recovery; the lower the chance of an event occurring, the lower the risk that may be associated with recovery.

Every community faces uncertainty when considering the types of threats that may incite a disaster recovery. No guarantee exists that at some point in time, a future recovery will or will not be necessary. This notion is present when forecasting potential events that may be calamities. For example, regarding the 2013 hurricane season, the National Oceanic and Atmospheric Administration predicted a

---

[13]Michael Zakour and David Gillespie, *Community Disaster Vulnerability: Theory, Research, and Practice* (New York, NY: Springer Publishing, 2013), 9.

[14]State of Florida, "State of Florida, Office of the Governor, Executive Order Number 03-60," 2003, http://edocs.dlis.state.fl.us/fldocs/governor/orders/2003/03-60.pdf (accessed May 25, 2013).

[15]Ibid.

[16]Ibid.

"70% chance of an above-normal season, a 25% chance of a near-normal season, and only a 5% chance of a below-normal season."[17]

These statistics are only speculations regarding what may or may not happen. No one knows the future with exact certainty and can predict future events with full accuracy. No guarantee exists regarding which of these states of existence will occur, or whether these probabilities may be revised through time. Therefore, risk exists. However, for those who may be impacted by hurricane season, they must examine such forecasts with a strong consideration of experiencing an event that may necessitate recovery planning and recovery operations.

The Concept and Phases of Recovery

Examples of recovery activities include providing temporary shelter, restoring power, critical stress debriefing for emergency responders and victims, job assistance, small business loans, debris clearance, and so forth. Many agree that the recovery component of emergency management is more complex than the other components and involves a much larger group of diverse stakeholders with sometimes conflicting objectives. Recovery consists of those activities that continue beyond the emergency period to restore lifelines and a sense of normalcy (in due time). Recovery involves the restoration of services to the public and returning the affected area(s) to pre-emergency conditions. A disaster may strike quickly, leaving the need for recovery operations in its wake, or it can be a prolonged event requiring recovery activities to begin while the response phase is still in full activation.

Recovery actions occur in three general phases. The actions in each phase and the timing vary according to the nature and the severity of the disaster. The first phase overlaps with emergency response and consists of immediate actions taken to reduce life-safety hazards and make short-term repairs to critical lifelines. The second phase provides for ongoing social needs before permanent rebuilding is complete. This phase may continue for weeks or perhaps months. The third phase includes planning for and implementing the rebuilding of damaged buildings and other facilities and infrastructure and the resumption of normal social and economic life in the community. It may include a reconsideration of pre-disaster conditions. This third phase continues for several years.

Strategic Management Aspects of Recovery

Recovery planning and the act of recovering from a calamity necessitate a consideration of strategic management. When examining the threats that may impact a locale, one must be mindful of the fact that some recoveries may require large

---
[17]National Oceanic and Atmospheric Administration, "NOAA Atlantic Hurricane Season Outlook," 2013, http://www.cpc.ncep.noaa.gov/products/outlooks/hurricane.shtml (accessed May 23, 2013).

amounts of time whereas others may require little time. For the events that necessitate long recovery periods that transcend several years, attributes of strategic management must be present when crafting recovery plans. This notion is expressed within a recovery plan through the use of strategy.

Within the context of recovery planning, a common method of exercising strategic management to craft a recovery strategy may be itemized as follows:

1. Determine the mission to be accomplished.
2. Set goals and objectives that are commensurate with the established mission.
3. Identify threats and opportunities that exist within the external environment that exists outside the organization that crafts the recovery plan.
4. Indentify strengths and weaknesses that exist internally within the organization crafting the recovery plan.
5. Determine primary and alternative methods of action and recovery that capitalize upon strengths and opportunities, that rectify weaknesses, and that mitigate threats. These methods must also consider the potential of contingency actions. These methods must conform to the established mission and be commensurate with goals and objectives.
6. Implement the strategic recovery plan.
7. Adapt and adjust the plan according to any contingencies that arise.

Although this methodology may seem simple, it can be quite complex. Any organization that crafts a strategic recovery plan must have a vision regarding exactly what end-state is desirable in due time through the implementing of the recovery plan. From the conceptualizing of a future state of existence is derived a mission. Essentially, the mission succinctly expresses what is to be accomplished toward the achieving of the conceptualized, visionary end-state of existence and being. Strategy represents the overall plan through which this conceptualized, visionary end-state of existence and being is pursued through time.

Further, although strategy may be crafted, it is a dynamic entity. Therefore, it must be capable of changing with respect to the complexities and attributes of the emergency domain representing the calamity. Hence, it must be capable of being altered to facilitate a wide range of contingencies. Basically, a contingency represents the acknowledging of Murphy's Law—anything that can go wrong, will go wrong. If something unforeseen occurs during the exercising of the recovery plan, which impedes or endangers its successful completion, then the overall strategy must be dynamic enough to overcome this obstacle. Figure 11.3 shows various aspects of contingencies that affect strategy.

The mission must incorporate both goals and objectives. A goal is something that an organization desires to accomplish, but is difficult to measure. Conversely, an objective is something an organization desires to accomplish and is measureable. Therefore, goals are qualitative aspects of strategy whereas objectives are

Figure 11.3 Contingency considerations.

quantitative aspects of strategy.[18] Further, goals and objectives must be complements, both of which contribute positively to the exercising of the strategic plan.

Both goals and objectives must be realistic and achievable. For example, a recovery plan may incorporate the goal of providing for the "efficient and safe movement of people and goods."[19] Similarly, the same recovery plan may incorporate the objective of improving "intersections in the downtown so they function no worse than Level of Service E, at or near the capacity of the roadway, during morning and afternoon peak hours."[20] The former consideration, representing goals, is both realistic and achievable, but it is not quantifiably measurable. The latter consideration, representing objectives, is also realistic and achievable and is also quantifiably measurable.

For example, the Florida Division of Emergency Management (FDEM) maintains a strategic business plan that governs the functioning of the organization over long-term periods.[21] The contents of the FDEM plan range from the developing of pertinent economic impact models for disasters to researching and recommending disaster transportation options.[22] An excerpt of this document is given within Figure 11.4.

---

[18] Milind Phadtare, *Strategic Management: Concepts and Cases* (New Delhi, India: PHI Learning Private Ltd., 2011), 31.

[19] Municipal Research and Services Center of Washington, *Local Government Policy-Making Process*, 1999, http://www.mrsc.org/publications/policyprocess.pdf (accessed May 25, 2013), 22.

[20] Ibid.

[21] Florida Division of Emergency Management, *Florida Division of Emergency Management Business Plan*, 2013, http://www.floridadisaster.org/documents/2012–2013%20FDEM%20Business%20Plan.pdf (accessed May 25, 2013).

[22] Ibid.

CHAPTER 11 The Emergency Management Process: Disaster Recovery

Goal: RE S48					
Description of Goal:	Establishment of Logistical Staging Area's MOU's				
Date of Goal Completion:	August 30, 2012	Lead Staff Assigned to Goal:	Connie Nunn		
Gubernatorial Transition Team Report Reference:		Strategic Planning Reference: Goal 6.1: Maintain national accreditation Goal 12.1: Stabilize impacted jurisdictions, 12.3, 12.6			
Tasks Anticipated to Complete Goal	Date of Task Completion	Outcome	Intended Metric	Staff Assigned to Task	
1. Review of each State LSA site survey to determine of it will be a lease or an MOA.	8/31/11	Determinations if a lease or MOA is required for the site based on if we must pay a fee or not to use the site during an emergency.		Connie Nunn	
1. Begin process of generating appropriate document and route for signatures	06-30-12	Process		Connie Nunn, Legal and Logistics Planner	
2. Routing of documents and completion of MOA's	07-30-12	Process		Legal and Logistics Planner	
3. Completion of project	08-30-12	Written agreements with each facility	28 completed State LSA sites	Connie Nunn and Logistics Planner	
4. Mail out of the MOA's or Leases to facility owners.	08-30-12			Logistics Planner	

Courtesy of the Florida Department of Emergency Management Business Plan

Figure 11.4 Excerpt of FDEM strategic plan.

This excerpt shows a goal of establishing memorandums of understanding (MOU) regarding staging areas. It also contains a metric that may be viewed as an objective which identifies the completing of the goal. This metric is the completion of 28 state LSA sites. This quantity is complementary to the stated goal. Further, both the goal and the metric were realistic and achievable, as is indicated by the completion of the goal on August 30, 2012.

The use of strategic management during planning processes is not uncommon. A variety of entities maintain strategic plans which they revise, as necessary, through time according to the dynamics of a changing threat environment. Periodically, strategic recovery plans must be updated for a variety of reasons ranging from changes in threat perceptions to the accommodating of new or improved technologies that may be useful during calamities. Regardless, in any case, the use of strategy is a tool through which long-term recovery plan may be crafted, and that definitively expresses the attributes of vision, mission, goals, and objectives.

The aforementioned concepts represent the notions of strategy that are incorporated within recovery plans. The second half of the phrase—management—must also be considered within the strategic context. Traditionally, management is defined as the "attainment of organizational goals in an effective and efficient manner through planning, organizing, leading, and controlling organizational resources."[23] The acts of planning, organizing, leading, and controlling are deemed to be the four primary functions of all management. In some cases, a fifth category, controlling, may also be included. These items may be considered strategically within the context of recovery planning.

Any recovery plan must accommodate the long-term use of strategic management resources that are necessary to recover from a disaster. Probably the most understandable and most important resource that must be considered is simple and straightforward: people. Recovery plans cannot exist or be implemented without people. People provide leadership and must be led in order to accomplish goals and objectives to fulfill mission in the pursuit of vision. People exercise control over the implementing of the recovery plan involving both intangible (i.e., processes, software, etc.) assets and tangible (i.e., food, shelter, vehicles, etc.) assets. By far, humans are the most important aspect of crafting, revising, and exercising strategic recovery plans.

Post-Disaster Recovery Planning

The immediate post-disaster period is obviously one with immense potential for confusion, or at least for many of those involved to take actions that serve opposite or divergent purposes. Decisions must be made quickly, with little time for reconsideration before new problems urgently demand resolution. Thus, an essential purpose of the plan for post-disaster recovery and reconstruction is to provide some vision that serves as a beacon for decision makers and some framework within which decisions will be taken.

Post-disaster recovery planning is defined as developing a set of strategies to assist a community in rebuilding after a disaster occurs. Recovery planning can also be thought of as building the blueprint for reconstruction of the community after a disaster. There are a number of goals and activities that communities can consider when addressing post-disaster recovery and its associated planning.[24]

---

[23]Richard Daft and Dorothy Marcic, *Understanding Management*, 8th ed. (Mason, OH: South-Western Publishing, 2013), 9.
[24]University of Oregon Community Service Center, *Post-Disaster Recovery Planning Forum: How-To-Guide. Partnership for Disaster Resilience*, 2007, http://nthmp.tsunami.gov/Minutes/oct-nov07/post-disaster_recovery_planning_forum_uo-csc-2.pdf (accessed April 1, 2013).

The goals of post-disaster recovery planning are to: (1) identify and prioritize key issues, (2) establish partnerships, (3) develop a recovery strategy, (4) effectively direct internal and external resources, and (5) identify pre-disaster mitigation projects and enhance response and preparedness capabilities.[25]

True disaster recovery occurs over time. Communities must have time to heal from the impact of the disaster, physically, economically, socially, politically, culturally, and psychologically. For the emergency management professional, recovery planning must include anticipated time lines for recovery operations. To provide a framework for planning, the majority of the activities expected to be conducted during the period of initial disaster response should occur during the first one to seven days.

Figure 11.5 shows the signage for a disaster recovery center. During the initial response, focus is placed on emergency assistance for those injured and displaced: establishment of basic services to include temporary housing, security, sanitation, emergency medical assistance, food services, and identification of survivors and those lost. Initial and emergency debris removal and cleanup include opening avenues of access (roads); emergency, short-term repair of lifeline utilities; emergency, short-term repair of transportation systems and provision of interim transit services; initial building safety and casualty inspections; and the coordination of state/federal damage assessments.

Figure 11.5 FEMA opens disaster recovery center in devastated area in the aftermath of Hurricane Sandy on November 15, 2012, in Breezy Point, NY.

---

[25]Ibid.

Certainly, these items are not the only items and activities that must be considered during planning stages. Other items and activities consist of the following:

- Re-occupancy of buildings
- Mid-term planning (7–30 days)
- Provision of interim housing
- Restoration of lifeline utilities (power, water, sewers)
- Restoration of social and health services
- Restoration of normal city services
- Establishment of new ordinances governing location and nature of rebuilding
- Examination of building standards
- Economic recovery measures, including interim sites for business restoration
- Long-term reconstruction (several years)
- Rebuilding
- Restoration of transportation systems
- Hazard mitigation
- Reconstruction of permanent housing
- Reconstruction of commercial facilities
- Development and implementations of long-term economic recovery targeting impacted and critical industries

A post-disaster redevelopment plan (PDRP) is useful for all levels of disaster—minor, major, or catastrophic. In general, however, the scale of long-term recovery and redevelopment is proportional to the severity of the disaster. Therefore, the plan will be most valuable in the event of a major or catastrophic disaster affecting a large segment of the community or region. Particular components of the plan and certain actions, such as acquisition of damaged properties, could also occur in a minor or localized disaster. A minor disaster may also be an excellent time to exercise the plan and practice implementation of post-disaster actions. If the plan is implemented in such a fashion, it must be evaluated to determine how it can be improved in due time.

The objective of the PDRP is to guide the redevelopment decision-making process following a disaster in a manner consistent with local comprehensive plans (especially the future land use and coastal management elements, where applicable), the local mitigation strategy, the comprehensive emergency management plan, and other relevant plans or codes such as the long-range transportation plan, land development regulations, and economic development and redevelopment plans. Each of these plans, and potentially others, has pre-existing policies or procedures that affect post-disaster redevelopment.

For instance, the comprehensive plan has many policies that determine where and to what extent redevelopment can occur. Ultimately, the PDRP acts as a guide for utilizing the policies and procedures found in other documents when making post-disaster redevelopment decisions. The planning process provides an opportunity to examine how local plans and codes will impact redevelopment and to recommend changes that could result in a faster and more sustainable recovery.

The plan identifies policies, operational strategies, and roles and responsibilities for implementation that will guide decisions that affect long-term recovery and redevelopment of the community after a disaster. It emphasizes seizing opportunities for hazard mitigation and community improvement consistent with the goals of the local comprehensive plan and with full participation of the citizens.

Recovery topics addressed include sustainable land use, housing repair and reconstruction, business resumption and economic redevelopment, infrastructure restoration and mitigation, long-term health and social services support, environmental restoration, financial considerations, and short-term recovery actions that affect long-term redevelopment as well as other long-term recovery issues identified by the community. The following issues must be considered when crafting recovery plans:

1. *Faster and more efficient recovery without a comprehensive, long-term recovery plan, ad hoc efforts in the aftermath of a significant disaster will delay the return of community stability.* Creating a process to make smart post-disaster decisions and prepare for long-term recovery requirements enables a community to do more than react, prompting post-disaster action rather than time-consuming debate. By identifying appropriate planning mechanisms, financial assistance, and agency roles and responsibilities beforehand, a community begins the road to recovery more quickly. Being able to show efficient and effective use of taxpayer dollars after a disaster is incredibly important for the public's perception of the recovery.[26]

2. *Opportunity to build back better.* A disaster, while tragic, can also create opportunities to fix past mistakes or leap forward with plans for community improvements. In the immediate aftermath of a disaster, local officials are under significant pressure to restore the community to its pre-disaster condition. Without a guiding vision, short-term decisions may inadvertently restrict long-term, sustainable redevelopment and overlook opportunities to surpass the status quo. A PDRP strengthens the recovery process, and communities benefit from assessing their risk levels and crafting a long-term redevelopment plan under "blue skies." Local officials and the public can thoughtfully analyze and debate issues, linking

---

[26]*Post-Disaster Redevelopment Planning: A Guide for Florida Communities,* Florida Department of Emergency Management, http://www.floridadisaster.org/Recovery/documents/Post%20Disaster%20 Redevelopment%20Planning%20Guidebook%20Lo.pdf (accessed May 1, 2013).

redevelopment goals with other important community plans. Careful thought and planning achieves a more sustainable and resilient outcome than decisions made under emergency circumstances, compromised budgets, and political pressures.

3. *Local control of recovery.* Developing a PDRP provides local government officials, residents, and businesses the opportunity to determine long-term redevelopment goals and develop policies and procedures that will guide redevelopment before well-intended outside agencies and nongovernment organizations rush to aid the community. While outside resources are needed and welcomed in a major or catastrophic disaster, a locally developed plan will best channel those resources to effectively meet the community's specific needs and goals.

4. *Post disaster.* The redevelopment plan shows outside agencies and donors that the community is prepared to play an active role in the recovery process and promote its capabilities to wisely use donated and loaned resources. There will always be rules and, occasionally, strings attached to external sources of funding, but a community that has researched the allowable uses of federal and state assistance can better work within their boundaries in an effort to fund projects that further local redevelopment goals.

All response activities that contribute toward recovery are sensitive to the passing of time.[27] Therefore, recovery may be viewed chronologically regarding the immediacy of activities that are necessary to achieve some sense of normalcy. A short-term perspective involves a consideration of activities and initiatives that occur immediately following an incident and those that may begin before the incident is completed. Any recovery planning endeavor must accommodate short-term endeavors. Examples of short-term resources include medical care and medicines, foodstuffs, sheltering, and so forth. These endeavors and activities provide a foundation for the successfulness of long-term recovery activities and plans.[28]

Similarly, long-term activities are also within the scope of recovery planning. Depending on the severity of an incident, long-term endeavors may necessitate quite long periods, spanning many years. Regardless of the amount of time required, a major emphasis of long-term recovery is the instantiating of societal normalcy through time. Examples of long-term endeavors include rebuilding of roadways and buildings, economic restoration, and so forth.

All recovery plans must incorporate both short-term and long-term perspectives of disaster recovery. Table 11.3 highlights some typical attributes that characterize both the perspectives of short-term and long-term recovery planning.

---

[27]Federal Emergency Management Agency, *Guide for All-Hazard Emergency Operations Planning* (Washington, DC: U.S. Government, 1996), 1–5.
[28]Ibid.

Table 11.3 Short-term vs. long-term perspectives of disaster recovery characteristics and outcomes

Short-term perspective characteristics	Long-term perspective characteristics
• Ad hoc recovery • Issuing building permits without adequate review of reconstruction implications • Limited public participation • Rebuilding to pre-disaster conditions • Over reliance on state and federal recovery funding	• Developing a recovery plan • Establishing a temporary building moratorium • Conducting an in-depth damage assessment • Integrating hazard mitigation techniques into reconstruction • Identifying local resources • Involving the public • Identifying sustainable recovery objectives • Linking recovery objectives with existing community goals
Outcomes	**Outcomes**
• Reduced economic viability • Increased hazard vulnerability • State or federal paternalism • Out-migration of residents • Declining tax base • Declining sense of place	• Greater economic viability • Reduced hazard vulnerability • Greater environmental well-being • Enhanced public health • Enhanced community self-reliance • Increased tax base • Enhanced sense of place

Source: Adapted from Gavin Smith, Holistic Disaster Recovery: Creating a More Sustainable Future, Session 3: Dimensions of Recovery, FEMA (Smith 2004)

Short-Term Recovery Planning and Activities

Preparedness for short-term recovery is part of the all-hazards emergency management cycle. During the preparedness phase, efforts focus on creating emergency plans and procedures, ensuring that agencies are aware of their roles and responsibilities, that responders have sufficient training to perform duties and tasks, and that exercises are conducted to test these plans and procedures.

The initial goal of short-term recovery activities is to rapidly return vital life support systems to minimum operating standards and to take the initial steps to long-term recovery. Some of these activities include the initial cleanup, establishing temporary housing, performing emergency demolition, commencing debris management and initial infrastructure restoration, and commencing disaster assistance.[29]

The role of the plan during the short-term recovery phase is to begin organizing for long-term redevelopment activities and guide short-term recovery decisions that may have long-term implications (e.g., placement of temporary housing or debris sites). Short-term recovery operations are addressed in the CEMP, but the PDRP can provide direction for transitioning to long-term redevelopment during this phase. The short-term recovery phase begins as the emergency response phase is winding down and will continue until critical services are restored. The duration of the short-term recovery phase depends on the severity of the disaster and the level of community preparedness; it could range from several weeks to one year to complete this phase.

Long-Term Recovery Planning and Activities

Long-term recovery after a disaster is always challenging, even if a community has planned for a worst-case scenario; however, by proactively creating a process to make smart post-disaster decisions and prepare for recovery needs, the community can do more than simply react. Long-term recovery activities may continue for a number of years after a disaster. Their purpose is to return life to normal or improved levels. Some of those activities include long-term reconstruction, hazard source control and area protection, infrastructure resilience, land-use practices, historic preservation, reviewing and initiating new building construction practices, coordinating environmental recovery, public and mental health recovery, redevelopment loans, legal assistance, disaster memorialization, and community planning.[30]

Long-term recovery involves more than debris removal and restoring power, which are considered short-term recovery actions. According to FEMA, long-term recovery refers to the "need to re-establish a healthy, functioning community that will sustain itself over time."[31] In its *Long-Term Community Recovery Planning Process: A Self-Help Guide*, FEMA outlines a recovery approach that emphasizes a community-driven process with significant public involvement and local control. The process also emphasizes a project-oriented focus on actions that will have the greatest impact on community recovery.

---

[29]National Governors' Association Center for Policy Research, *Comprehensive Emergency Management: A Governor's Guide* (Washington, DC: National Governors' Association, 1979).
[30]Ibid.
[31]Federal Emergency Management Agency, 2013, http://www.fema.gov/txt/rebuild/ltrc/selfhelp.txt (accessed May 25, 2013).

The plan is used most during this phase. Long-term recovery and redevelopment include efforts to reconstruct and enhance the built environment as well as recover the economy, environment, and social systems. This phase begins as short-term recovery activities are accomplished and can last from a couple of years for a minor disaster to five or more years for a major or catastrophic disaster. Given these amounts of time, long-term recovery and planning are strategic endeavors.

Project Management

Long-term planning often incorporates the emphasis and perspectives of project management regarding recovery endeavors. Project management involves a variety of characteristics that must be acknowledged when considering recovery planning. During the planning period, a variety of considerations must be given to everything from physical assets and their logistics to time and training regimens. For example, using no certain order, incorporating such perspectives within recovery planning accommodates consideration of the following items:

- selecting projects and prioritizing them;
- responsibilities and capabilities of the organization;
- organizational structuring and culture;
- project charters, MOUs, and contracts;
- communication;
- stakeholder analysis;
- project risk assessment and evaluation;
- project resources and their financial and budgetary requirements;
- logistics and supply chains;
- project leadership and management;
- project scheduling chronologically;
- project implementation according to expressed plans and schedules;
- project progression, tracking, quality control, and quality assurance;
- establishing of milestones; and
- collection of metrics for evaluation and analysis.

These items are only subsets of the project management attributes that permeate recovery planning. Many more can be listed. Regardless, every recovery initiative can be viewed from its solitary existence as an entity that must be managed through time whereby project management characteristics are necessary for its effective and efficient implementation and overall level of successfulness. This notion must be accommodated when crafting disaster recovery plans.

Numerous managed projects occur during the aftermaths of cataclysms that involve a variety of issues that affect victims, refugees, organizations, and governments. Commonly, a primary goal of such projects is to "employ victims" throughout various "small-scale, labor-intensive projects" for the purposes of providing income and employment.[32] One example of these concepts is the Emergency Tsunami Reconstruction Project that was initiated during the aftermath of the 2004 tsunami that affected the Asian regions bordering the Indian Ocean.[33] This project was approved in 2005 and closed in 2011.[34] Its scope involved an array of recovery initiatives that spanned everything from social services to housing construction.[35] It also accommodated sanitation issues as well as flood protection.[36] Its themes were also diverse, and encompassed areas of civic engagement as well as natural disaster management.[37]

Monetarily, this project cost approximately US $682.80 million.[38] Money was allocated for the restoration of livelihoods (US$ 36.4 million), housing reconstruction (US$ 596.8 million), public buildings and public works (US$ 19.5 million), implementation support (US$ 19.0 million), and technical assistance and training (US$ 11.1 million).[39]

Certainly, this project represents a large-scale endeavor that is associated with foreign tragedy. Within the U.S., various projects were necessitated in conjunction with emergency scenarios. In no certain order, below are itemized several historical instantiations of project management that were necessitated in conjunction with emergency conditions:[40]

- Hurricane Katrina
- Space Shuttle *Challenger* explosion
- Space Shuttle *Columbia* re-entry disaster

---

[32]John Nicholas and Herman Steyn, *Project Management for Engineering, Business, and Technology*, 4th ed. (New York, NY: Routledge, 2012), 38.

[33]Ibid.

[34]"India: Emergency Tsunami Reconstruction Project," World Bank, 2013, http://www.worldbank.org/projects/P094513/india-emergency-tsunami-reconstruction-project?lang=en (accessed May 27, 2013).

[35]Ibid.

[36]Ibid.

[37]Ibid.

[38]Ibid.

[39]"Emergency Tsunami Reconstruction Project in India," Devex, 2013, https://www.devex.com/en/projects/emergency-tsunami-reconstruction-project-in-india/secure?mem=ua&src=tender (accessed May 21, 2013).

[40]Harold Kerzner, *Project Management: Case Studies* (Hoboken, NJ: Wiley Publishing, 2013).

- Tylenol poisonings of the 1980s
- Controversy associated with Nestle's infant formula
- Oil spill associated with the Exxon Valdez disaster
- Nuclear event associated with Three Mile Island

Although planning for catastrophic events is a valid consideration of disaster recovery, it must be noted that such plans are not impervious to the effects of uncertainty. Despite the best efforts to craft recovery plans, unknown factors within the event domain can nullify or impede an intended recovery plan or project, thereby rendering the recovery plan either completely useless or marginal. Because of such unknown factors and uncertainties, established plans may often become useless during the aftermath of disasters, and responding agencies must be capable of acting without a fully expressed and crafted plan.[41] Therefore, many different decisions must be rendered immediately without any plan contents to influence them. During such conditions, contingency plans may be crafted to accommodate the characteristics of the changed event domain whereby recovery efforts may be guided and implemented.

Another consideration of disaster project management involves the training, education, and credentials of planners. The events of Hurricane Katrina and the Indian Ocean tsunami were the catalysts for developing a project management

"I need a list of specific unknown problems that we'll encounter."

Figure 11.6 Unknown variables affect all domains.

---

[41]Aaron Shenhar and Dov Dvir, *Reinventing Project Management: The Diamond Approach to Successful Growth and Innovation* (Boston, MA: Harvard Business School Press, 2007).

certification for individuals that work with aid agencies.[42] This training regimen and its certification, referred to as the Project Management in Development Professional, were designed especially for individuals "working in charities, humanitarian agencies, and not for profit organizations."[43] Because many personnel involved with disaster projects exhibited "limited training in professional project management concepts common in the commercial sector," it was surmised that the provision of project management training "could have significant impact of the effectiveness of aid" that is associated with catastrophic events.[44]

Project management is an essential aspect of planning for disaster recovery. It represents the "application of knowledge, skills, tools, and techniques to project activities to meet project requirements."[45] It also represents a "process of managing, allocating, and timing resources to achieve a given goal in an efficient and expeditious manner."[46] Although various uncertainties and unknown factors certainly tend to impact the processes and methods through which recovery occurs, recovery plans must incorporate these facets of project management as a means of embellishing both the efficiency and effectiveness of recovery efforts.

Recovery Financing

Financing long-term disaster recovery poses significant and often frustrating challenges for local leaders who must rely on the state and federal government as major sources of disaster recovery funds. Those challenges are exacerbated in the heat of a crisis when funding is urgent, not optional or negotiable. This is the reason why incorporating a framework for financing long-term recovery improves the odds of success when a disaster strikes.[47]

Immediately following a disaster is the optimal time to remove or retrofit damaged buildings, since disaster assistance and FEMA mitigation funds are most

---

[42]APMNG International, "Tsunami and Hurricane Katrina Prompt New Certifications for Project Managers Working in Aid Agencies," 2013, http://www.apmg-international.com/en/news-events/14Sept09PMDPro.aspx (accessed May 23, 2013).

[43]Ibid.

[44]Ibid.

[45]Kathy Schwalbe, *Information Technology Project Management* (Boston, MA: Course Technology, 2010), 35.

[46]Adedeji Badiru and Samuel Osisanya, *Project Management for the Oil and Gas Industry: A World System Approach* (Boca Raton, FL: CRC Press, 2013), xvii.

[47]Christine Becker, "Disaster Recovery: A Local Government Responsibility," *PM Magazine*, vol. 91, no. 2, March 2009, http://webapps.icma.org/pm/9102/public/cover.cfm?author=christine+becker&title=disaster+recovery:++a+local+government+responsibility (accessed May 2, 2013).

available, as is willingness to adopt new technology and accept smarter reconstruction and repair options. Working collaboratively before a disaster to identify appropriate buildings to target for removal or retrofit will make this easier.

Funding is available from a wide variety of sources. Below are some examples of common sources through which funds are procured to support disaster recovery initiatives. Certainly, many other funding sources exist.

Public assistance—Public assistance is aid to state or local governments administered by FEMA to pay part of the costs of rebuilding a community's damaged infrastructure. Generally, public assistance programs pay for 75% of the approved project costs. Public assistance may include debris removal, emergency protective measures and public services, repair of damaged public property, loans needed by communities for essential government functions, and grants for public schools. Public assistance also allows for improved and alternate projects which can provide opportunities for mitigation or other redevelopment needs.

Hazard Mitigation Grant Program (HMGP)—The HMGP provides grants to state and local governments to implement long-term hazard mitigation measures after a major disaster declaration. Authorized under Section 404 of the Stafford Act and administered by FEMA, HMGP was created to reduce the loss of life and property due to natural disasters. The program enables mitigation measures to be implemented during the immediate recovery from a disaster. A state or local match of 25% is required.

Community Development Block Grants—In response to disasters, congress may appropriate additional funding for the Community Development Block Grant (CDBG) and HOME programs as disaster recovery grants to rebuild the affected areas and provide crucial seed money to start the recovery process. Since CDBG disaster recovery assistance may fund a broad range of recovery activities, the U.S. Department of Housing and Urban Development can help communities and neighborhoods that otherwise might not recover due to limited resources.

Community Disaster Loan Program—FEMA also provides loans to jurisdictions in a designated disaster area that has suffered a substantial loss of tax and other revenue and has a demonstrated need for financial assistance to perform its governmental functions. The loans cannot exceed 25% of the local government's annual operating budget, with a maximum of $5 million.

The allocating and expending of these monies are not sporadic events. Instead, they involve considerations of capital budgeting techniques, advanced financial methods, and prioritization whereby funding associated with emergency management may be increased.[48] Examples of these financial techniques consist of the following methods:[49]

- Payback-time method
- Net present value (NPV)
- Internal rate of return (IRR)
- Modified internal rate of return (MIRR)
- Profitability index method (PIM)
- Cost-benefit (i.e., benefit-cost) ratio

The payback-time method represents the notion that a project is deemed to be acceptable provided that "its payback is less than the maximum cost recovery time" that is established by the organization.[50] Basically, this method represents a perspective of time—how long will it take to recover money invested among recovery projects?

The NPV method represents the discounting of the future cash flows, with respect to an established and accepted discount rate, in which the costs of the initial financial investment are subtracted.[51] This method represents a perspective of equating recovery effort costs in terms of current dollars. Once the equating of costs is defined in terms of present dollars, initiatives may then be compared and prioritized according to the needs of the organization. Only positive NPV financial outcomes are generally considered to be acceptable.[52]

Capital budgeting also considers the rates of return that may be associated with the financial attributes of recovery efforts. The techniques involving considerations of return rates represent the "discount rate at which a project's NPV becomes zero."[53] As an example, such methods were used within the Argentina Flood Rehabilitation Project.[54]

---

[48]Naim Kapucu and Alpaslan Ozerdem, *Managing Emergencies and Crises* (Burlington, MA: Jones and Bartlett Publishing, 2013).

[49]Daniel Doss, William Sumrall, and Don Jones, *Strategic Finance for Criminal Justice Organizations* (Boca Raton, FL: CRC Press, 2012).

[50]Scott Besley and Brigham Eugene, *Essentials of Managerial Finance*, 14th ed. (Mason, OH: Thomson South-Western Publishing, 2008), 372.

[51]P. K. Jain, *Financial Management: Text Problems and Cases*, 5th ed. (New Delhi, India: Tata McGraw-Hill Publishing, 2007).

[52]Ibid.

[53]Sriyanie Miththapala, *Integrating Environmental Safeguards into Disaster Management*, vol. 3 (Sri Lanka: International Union for Conservation of Nature and Natural Resources, 2009), 49.

[54]Stefan Hochrainer, *Macroeconomic Risk Management Against Natural Disasters* (Wiesbaden, Germany: Springer Publishing, 2006).

The PIM technique represents a method of examining whether a considered initiative will result in profitability over time.[55] Specifically, the PIM represents the "present value of all future cash flows divided by the absolute value of the initial cash outlay."[56] This method is often used among health-care organizations as a method of examining capital rationing among multiple endeavors through which it becomes a basis of deciding which option to choose for implementation.[57]

The analysis of benefits versus costs is probably the most easily recognized method of examining recovery financing. This method is often used to examine the financial effects and impacts of "natural disasters on the economy and the costs and benefits of undertaking risk management."[58] The use of this method provides a foundation for exploring the prioritization and structuring of effective programs during recovery.[59] This technique was used significantly following Hurricane Katrina.

These capital budgeting techniques provide a basis for understanding the scope, magnitude, and potency that financial investments may have within the context of recovery efforts. They occur within both the government and private sectors. Their use provides a basis for crafting insurance coverage, estimating resource requirements, determining the ordering of recovery efforts, exploring financial amounts that are expended during recovery programs, and so forth. In any case, such techniques are methods through which the financial aspects of recovery initiatives may be judged, and through which human recovery decisions are influenced.

Chapter Comments and Summary

At some point in time, somewhere, a disaster shall occur and recovery shall be required to achieve a sense of societal normalcy. All locales have some amount of risk when considering the potential of experiencing a calamity. No locale is insusceptible to some type of event that destroys infrastructure and that may result in human death or suffering (or both). Therefore, all locales must consider how they could recover from such an incident. All locales must seriously consider recovery planning.

---

[55]Doss, Sumrall, and Jones, *Strategic Finance for Criminal Justice Organizations*.

[56]Robert McLean, *Financial Management in Health Care Organizations*, 2nd ed. (Clifton Park, NY: Delmar Thomson Publishing, 1997), 197.

[57]William Cleverley, Paula Song, and James Cleverley, *Essentials of Health Care Finance*, 7th ed. (Sudbury, MA: Jones and Bartlett Publishing, 2010).

[58]Reinhard Mechler, *Natural Disaster Risk Management and Financing: Disaster Losses in Developing Countries* (Karlsruhe, Germany: Verlag, 2004), 163.

[59]Margaret Arnold, *Natural Disaster Hotspots: Case Studies* (Washington, DC: World Bank Publications, 2006), 160.

Post-disaster recovery planning can be thought of as providing a blueprint for the restoration of a community after a disaster occurs. This can be done through long- and short-term strategies. Specific strategies might include planning, policy changes, programs, projects, and other activities such as business continuity planning. Post-disaster recovery planning is a shared responsibility between individuals, private businesses and industries, state and local governments, and the federal government.

Post-disaster recovery planning outlines and defines a community's vision of how it would like to rebuild in the aftermath of a disaster. If a community engages in post-disaster recovery planning prior to the event, it can more effectively direct outside redevelopment resources from federal, state, or other regional authorities once the disaster occurs. As a result, community redevelopment and recovery can take place in a manner that is consistent with community values.

As a method of becoming more disaster-resilient, local governments should plan for what must happen after rescue and recovery operations are completed in order to return the community to normal or perhaps rebuild an even better community. Through a PDRP, local governments can collaboratively create a long-term recovery and redevelopment strategy in pursuit of a sustainable community.

Disasters may be costly in terms of both human life and money. Recovering from a calamity is not free; instead, costs are involved that range from human resources to the reconstructing of infrastructure. Various recovery programs exist through which affected regions may obtain funding to supplement recovery initiatives. Determining prioritizing which endeavors to pursue, from a financial perspective, is often influenced through the use of capital budgeting techniques.

The dynamics of the recovery phase of managing disasters is dynamic, and often exhibits unforeseen events that can impede or nullify the implementing of established recovery plans. Therefore, emergency management personnel must incorporate some facets of adapting to contingencies when crafting and implementing response plans. Although it is impossible to predict with complete accuracy and certainty the events that may transpire which impact negatively the recovery effort, planning is an essential activity that must be neither ignored nor discounted.

Terminology

Achievable
Asset
Budget
Capital budgeting
Catastrophic
Complexity
Contingency

Controlling
Coordinating
Disaster
Finance
Goal
Leading
Long-term

Measurable
Mission
Murphy's Law
Objective
Organizing
Plan
Planning
Project management
Qualitative
Quantitative
Post-disaster

Prioritization
Realistic
Recovery
Recovery plan
Resource
Short-term
Strategic management
Strategic plan
Strategy
Vision

Thought and Discussion Questions

1. This chapter introduces the concept of disaster recovery. It is highly advisable that all communities have and maintain some type of recovery plans to accommodate a wide range of potential hazards. Spend some time researching whether your locale maintains a recovery plan regarding its anticipated endangerments. Review this plan, and provide a critical analysis of its contents. Write a brief essay that summarizes your findings.
2. This chapter introduces capital budgeting as a method of exploring financially the monetary costs that are associated with recovery initiatives. Perform some research concerning each of the capital budgeting techniques presented herein, and determine how they have been used in a variety of disasters to facilitate recovery. Write a brief essay that highlights your findings.
3. This chapter introduces the notion that approximately 40% of organizations that experience a disaster do not reopen again. Perform some research concerning this situation, and speculate why you believe these businesses do not reopen. Write a brief essay that highlights your findings and opinions.
4. The outcomes of recovery plans often experience varying amounts of success and failure. Perform some research regarding recovery initiatives that were undertaken following both man-made and natural disasters. Locate some examples of both successful and failed recoveries. Write an essay that critically analyzes why you believe these successes and failures occurred.

References

1. APMNG International. 2013. Tsunami and Hurricane Katrina prompt new certifications for project managers working in aid agencies. http://www.apmg-international.com/en/news-events/14Sept09PMDPro.aspx (accessed May 23, 2013).

2. Arnold, Margaret. 2006. *Natural disaster hotspots: Case studies.* Washington, DC: World Bank Publications.
3. Badiru, Adedeji, and Samuel Osisanya. 2013. *Project management for the oil and gas industry: A world system approach.* Boca Raton, FL: CRC Press.
4. Baird, Malcolm E. January 2010. The recovery phase of emergency management. http://www.memphis.edu/ifti/pdfs/cait_recovery_phase.pdf (accessed May 20, 2013).
5. Becker, Christine. 2009. Disaster recovery: A local government responsibility. *PM Magazine,* vol. 91, no. 2, March. http://webapps.icma.org/pm/9102/public/cover.cfm?author=christine+becker&title=disaster+recovery:++a+local+government+responsibility (accessed May 2, 2013).
6. Besley, Scott and Brigham Eugene. 2008. *Essentials of managerial finance.* 14th ed. Mason, OH: Thomson South-Western Publishing.
7. Cleverley, William, Paula Song, and James Cleverley. 2010. *Essentials of health care finance.* 7th ed. Sudbury, MA: Jones and Bartlett Publishing.
8. Daft, Richard and Dorothy Marcic. 2013. *Understanding management.* 8th ed. Mason, OH: South-Western Publishing.
9. Doss, Daniel, William Sumrall, and Don Jones. 2012. *Strategic finance for criminal justice organizations.* Boca Raton, FL: CRC Press.
10. Emergency Tsunami Reconstruction Project in India. 2013. *Devex.* https://www.devex.com/en/projects/emergency-tsunami-reconstruction-project-in-india/secure?mem=ua&src=tender (accessed May 21, 2013).
11. Federal Emergency Management Agency. 1996. *Guide for all-hazard emergency operations planning.* Washington, DC: U.S. Government.
12. Federal Emergency Management Agency. 2013a. http://www.fema.gov/txt/rebuild/ltrc/selfhelp.txt (accessed May 25, 2013).
13. Federal Emergency Management Agency. 2013b. Disaster declarations. http://www.fema.gov/disasters?field_state_tid=All&field_disaster_type_term_tid=All&field_disaster_declaration_type_value=DR&items_per_page=10&page=2 (accessed May 24, 2013).
14. Federal Emergency Management Agency. 2013c. Protecting your businesses. http://www.fema.gov/protecting-your-businesses (accessed May 24, 2013).
15. Florida Division of Emergency Management. 2013. Florida Division of Emergency Management business plan. http://www.floridadisaster.org/documents/2012-2013%20FDEM%20Business%20Plan.pdf (accessed May 25, 2013).
16. Hochrainer, Stefan. 2006. *Macroeconomic risk management against natural disasters.* Wiesbaden, Germany: Springer Publishing.
17. India: Emergency Tsunami Reconstruction Project. 2013. *World Bank.* http://www.worldbank.org/projects/P094513/india-emergency-tsunami-reconstruction-project?lang=en (accessed May 27, 2013).

18. Jain, P. K. 2007. *Financial management: Text problems and cases*. 5th ed. New Delhi, India: Tata McGraw-Hill Publishing.
19. Kapucu, Naim and Alpaslan Ozerdem. 2013. *Managing emergencies and crises*. Burlington, MA: Jones and Bartlett Publishing.
20. Kerzner, Harold. 2013. *Project management: Case studies*. Hoboken, NJ: Wiley Publishing.
21. McLean, Robert. 1997. *Financial management in health care organizations*. 2nd ed. Clifton Park, NY: Delmar Thomson Publishing.
22. Mechler, Reinhard. 2004. *Natural disaster risk management and financing: Disaster losses in developing countries*. Karlsruhe, Germany: Verlag.
23. Miththapala, Sriyanie. 2009. *Integrating environmental safeguards into disaster management*. Vol. 3. Sri Lanka: International Union for Conservation of Nature and Natural Resources.
24. Municipal Research and Services Center of Washington. 1999. Local government policy-making process. http://www.mrsc.org/publications/policyprocess.pdf (accessed May 25, 2013).
25. National Governors' Association Center for Policy Research. 1979. *Comprehensive emergency management: A governor's guide*. Washington: National Governors' Association.
26. National Oceanic and Atmospheric Administration. 2012. *Tropical cyclone report: Hurricane sandy*. http://www.nhc.noaa.gov/data/tcr/AL182012_Sandy.pdf (accessed May 24, 2013).
27. National Oceanic and Atmospheric Administration. 2013. NOAA Atlantic Hurricane season outlook. http://www.cpc.ncep.noaa.gov/products/outlooks/hurricane.shtml (accessed May 23, 2013).
28. Nicholas, John and Herman Steyn. 2012. *Project management for engineering, business, and technology*. 4th ed. New York, NY: Routledge.
29. Olshansky, R. 2006. Planning after Hurricane Katrina. *Journal of the American Planning Association* 72(2): 148.
30. Phadtare, Milind. 2011. *Strategic management: Concepts and cases*. New Delhi, India: PHI Learning Private Ltd.
31. Phillips, Brenda D., and David M. Neal. 2007. Recovery. In *Emergency management: Principles and practice for local government*. 2nd ed., chap. 11.
32. Rice, Doyle. 2013. Hurricane sandy, drought cost U.S. $100 billion. *USA Today*. http://www.usatoday.com/story/weather/2013/01/24/global-disaster-report-sandy-drought/1862201/ (accessed May 25, 2013).
33. Schwalbe, Kathy. 2010. *Information technology project management*. Boston, MA: Course Technology.
34. Shenhar, Aaron and Dov Dvir. 2007. *Reinventing project management: The diamond approach to successful growth and innovation*. Boston, MA: Harvard Business School Press.

35. WJLA News. 2013. Oklahoma Tornado: Barack Obama declares major disaster after storm. http://www.wjla.com/articles/2013/05/oklahoma-tornado-barack-obama-declares-major-disaster-after-storm-89084.html (accessed May 24, 2013).
36. Zakour, Michael and David Gillespie. 2013. *Community disaster vulnerability: Theory, research, and practice.* New York, NY: Springer Publishing.

PROFESSIONALIZATION OF EMERGENCY MANAGEMENT

*"Do not go where the path may lead;
go instead where there is no path and leave a trail"*

—Ralph Waldo Emerson

Figure 12.1 Professionalism and emergency management.

OBJECTIVES

The objectives of this chapter are to:
- Discuss the professionalization of emergency management,
- Identify professional standards of emergency management,
- Discuss ethical codes of emergency management,
- Discuss professionalism standards of emergency management,
- Discuss leadership aspects of emergency management, and
- Discuss legal liability of emergency management.

Introduction

Emergency management remains one of the most challenging professions of any line of work. This is primarily due to the increased interest in emergency management programs, support for best practices, and, of course, as a result of major disasters. The profession of emergency management is characterized by a continually expanding landscape that requires an ever-demanding array of experience, education, and training. The progressive steps of emergency management will always include a bend in the road, which is not the end of the road, but it is about how the curve is navigated. That *curve* being the proverbial learning moment whereby emergency management must truly adapt, improvise, and overcome. Thus, the variances of both the foreseen and the unknown are dynamic aspects that challenge both the administrative and operative functions of emergency management.

Such a vast continuum of subject matter expertise and considerable skill sets range from organization, planning, and supervision to current technology, financial management, public relations, hazardous waste, multiagency coordination, and the political savvy necessary to negotiate, resolve conflicts, and facilitate the phases of emergency management. While an emergency manager may wish to be, or is expected to be, in complete control of circumstances, the reality remains that the conditions of a disaster will command complete collaboration, communication, and coordination.

Thus, this chapter will discuss the professionalization of emergency management, including an understanding of the credential/certification standards of operability required by FEMA and state officials; the interdisciplinary model of education, training, and experience for emergency managers; budget and fiscal accountability; legal liability, professional and public accountability, and emergency manager position descriptions.

Professionalization of Emergency Management

Over the years, there has seemingly been a struggle with respect to proclaiming exactly what terminology to use that practitioners, politicians, and the public would acknowledge and respect as a true and legitimate mode of operation regarding emergency management. For example, in times past, local governments empowered either the fire chief or police chief, or both, as having the responsibility for managing an emergency or disaster event while using such terms as disaster management, emergency practices, or as generic as emergency response.

Invariably, as a result of many devastating disasters, it was realized that there needed to be a designated individual and department that was not primarily a first responder, but rather, a position that was exclusively designated to coordinate the multifaceted operations required during an emergency or disaster event. Somewhat analogous to a symphony conductor, who may have no idea as to how to play each or any of the instruments but can sufficiently lead the practitioners to play in harmony, is the responsibility of an emergency manager.

An individual serving as an emergency manager may not be a subject matter expert specific to the duties of a first responder, but does maintain a level of competency as an organizer, administrator, and communicator to effectively conduct business during an emergency situation. Thus, in an effort to demonstrate the tremendous progress that the field of emergency management has made over the past decade, it is important to recognize the progressive process of emergency management, which has evolved into a legitimate, a viable, and a necessary profession.

The profession of emergency management is an unusual field, in that it has varied definitions. Likewise, the term professionalism is perhaps one of the most commonly overused, misused, and abused words in all facets of society. Prior to discussing the profession of emergency management, a brief summary of the definition, concepts, and characteristics of professionalism will be explained. Then, as a preface to recognizing notable definitions of emergency management, characteristics of what the profession of emergency management is and is not will be described.

Principles of Professionalism

Basically, the word profession can be measured against certain key characteristics associated with a particular occupation. Generally speaking, the definition of profession consists of an employment-based position that requires an advanced knowledge base in a specialized field that is conditioned by standards of training and definitive skills necessary to develop an individual as a subject matter expert.

The root word of profession is the term to profess, which simply means to candidly assert or acknowledge one's ability to perform the required job functions of a specific position. Depending upon the occupation, performing the essential functions requires an individual to either currently possess, or have the ability, to acquire the knowledge and skills to perform the work responsibilities pertinent to a specific position.

Unfortunately, more often than not, many people who work in a particular field make the claim of being a professional with respect to a specific occupation. Yet, such an incongruous claim is often done with an entitlement attitude of being part of an agency, organization, or occupation without truly having the complete compliment of experience, education, and training. As simplistic as possible, the term professional describes one who has become competent in a particular field or occupation as evidenced by a proven record of success, which is often contingent upon proactive communication, discretionary decisions, the ability to overcome adversity, and the attitude to persevere.

The characteristics of professionalism should include objectives that are foundationally integral to the safety and welfare of the public such as the following:

- Is the occupation a "service" to others?
- Does the occupation assess the needs of the public?
- Does the occupation identify how the public can best be served?
- Does the occupation promote continuing education and training?
- Does the occupation exemplify community support and partnership?
- Are there professional associations that support and recognize the occupation?
- Does the occupation maintain standards for hiring, best practices, and accountability?
- Does the occupation require a theoretical body of knowledge obtained through research?

The concept of professionalism is about someone who professes to know more than others about the nature of their specialty. Specific to this subject matter, that specialty may involve such public safety services as law enforcement, fire and emergency medical personnel, communications, and emergency management. The quality of professionalism will vary among individuals based upon how compatible, competent, and committed a person is to the expectations and standards distinctive to that particular occupation.

What Emergency Management is and is Not

First, emergency management is the profession of organizing, planning, and administering processes, personnel, and provisions in an effort to prevent, prepare for, respond to, and recover from harmful or catastrophic consequences, which may befall communities, states, or the entire country. Effective emergency management consists of integrating emergency strategies at all levels of government (local, state, and federal), along with nongovernment organizations (NGOs).

Emergency management is not the exclusive responsibility of governing agencies or NGOs. Rather, the public must also be diligent to protect their lives and property through garnering personal resources (safety kits), having a game plan to either evacuate or shelter-in-place, being involved in a local emergency management committee (LEPC), availing themselves of the broad network of community resources, and maintaining the wherewithal to make right decisions regarding their safety and welfare.

Contexts of Emergency Management

Over the years, emergency management has been defined in numerous ways using different verbiage to mean basically the same thing, or that may center exclusively on certain phases. For example, under the leadership of Michael Selves, former president of the International Association of Emergency Managers (IAEM), emergency management was defined as "The managerial function charged with creating the framework within which communities reduce vulnerability to hazards and cope with disasters."[1]

Next, Waugh states, "In simplest terms, emergency management is the management of risk so that societies can live with environmental and technical hazards and deal with the disasters that they cause" (Waugh, 2000). According to Hoetmer, emergency management is, "The discipline and profession of applying science, technology, planning, and management to deal with extreme events that can injure or kill large numbers of people, do extensive damage to property, and disrupt community life."[2]

These are different perspectives about the emergency management profession, which for some continues to be a challenge regarding a succinct comprehension of exactly what the emergency management profession consists of. Experientially, it is also interesting to note that occasionally there is a lack of consensus and fluctuation among practitioners, academia, and the public as to exactly what emergency management is, which often occurs after a major disaster that no one expected, or that was not well managed.

Consequently, such confusion may adversely influence policy makers, a.k.a. politicians, who then seemingly struggle with decisions regarding funding and other support facets of the emergency management profession. This is a challenge

---

[1] W. Waugh, *Local Emergency Management in the Post-9/11 World*, in *Emergency Management Principles and Practice for Local Government*, eds. W. Waugh and K. Tierney (Washington, D.C.: International City/County Management Association, 2007).

[2] T. E. Drabek and G. J Hoetmer, *Emergency Management: Principles and Practice for Local Government* (ICMA Press: Washington, D.C., 1991), 17–34.

Figure 12.2 Tornado damage in Joplin, Missouri, from the May 2011 tornado that struck the area, killing 159 people and injuring more than 1,000 others. At that time, the tornado was deemed the single deadliest in U.S. history.

that emergency managers themselves must overcome as they continue to educate key stakeholders and navigate through the vigorous political sandbox.

Thus, it is important to clarify exactly what the profession of emergency management is about in a practical and user-friendly manner that expands beyond a sentence or two, by examining the foundations of emergency management, the hazards that must be dealt with by emergency managers, the phases of emergency management, and the impact upon stakeholders.

Pathway of Professionalism

The evolution of the emergency management profession has been a work-in-progress for many years. Most experts agree that the first formal and comprehensive dynamics of emergency management officially developed with the advent of the Cold War in the 1950s. The threat of nuclear attack and subsequent fallout had emergency management centered on how to prepare and recover from this threat alone. This prompted President Harry S. Truman in December of 1950, to issue Executive Order 10186—the Federal Civil Defense Act, which created a new civil defense organization known as the Federal Civil Defense Administration (FCDA).

CHAPTER 12 Professionalization of Emergency Management 347

The purpose of the FCDA was to transition the responsibilities of the National Security Resources Board (NSRB), which had been responsible for all emergency planning since its inception in September of 1939, when the first official Office of Emergency Management was created under the leadership of President Franklin D. Roosevelt.[3]

Thereafter, in 1951 President Truman established the first in a series of alert and warning systems that began with the implementation of the Control of Electromagnetic Radiation (CONELRAD) program (NHC, http://www.nhcgov.com). Throughout Truman's tenure, the planning and preparedness process was primarily of a civil defense flavor, and the creation of shelters and the introduction of duck and cover drills for school kids became the norm.

As the 1960s progressed, the United States was affected by a series of major natural disasters that included several destructive Hurricanes, including Donna (1960) and Carla (1961), along with The Ash Wednesday Storm in 1962, which devastated over 620 miles of shoreline on the East Coast, and resulted in over $300 million in damage.[4]

As a result, in 1961 President John F. Kennedy determined that it was necessary to change the federal approach to disasters and develop a professional method for managing major disasters and emergencies. During the Kennedy Administration, the Office of Emergency Planning (OEP) was created for the purpose of managing natural disasters exclusive from the civil defense issues related to the nuclear threats of the cold war that had occurred during the Truman years.[5]

Subsequent to Kennedy's actions, the country continued to experience additional devastating events during the 1960s that included the infamous 1964 earthquake in Alaska's Prince William Sound, which measured 9.2 on the Richter scale, and produced tsunamis along the West Coast as far down to California that resulted in the deaths of 123 people. Two other Hurricanes Betsy (1965) and Camille (1969) significantly impacted the Gulf Coast resulting in the deaths and injury of hundreds of people and causing of millions of dollars in damage.[6] Such catastrophic disasters led to another pivotal professionalization period of emergency management under the presidency of Lyndon Johnson.

In February of 1968, President Johnson's Commission on Law Enforcement and Administration of Justice issued a report that recommended every police department in the United States have a single emergency number that could be used nationwide by the public. The recommendation was based on input from the

---

[3]White House Executive Order (WHEO) 10186, Federal Civil Defense Act of 1950, *President Truman Establishes Federal Civil Defense Administration* (December 1, 1950), http://www.historycommons.org (retrieved March 7, 2013).
[4]FEMA, http://training.fema.gov.
[5]Ibid.
[6]Ibid.

Commission's Task Force on Science and Technology, and led to the developing of the 911 call number.[7]

Thereafter, in October 1968, the professionalization of emergency management took another leap forward when President Johnson renamed the Office of Emergency Planning to the Office of Emergency Preparedness. In doing so, Johnson's efforts included the reorganization of numerous federal agencies in an effort to develop stronger response continuity, and reaffirmed the federal government's commitment to the American people to be as prepared as possible for all major disasters.[8]

It was not until 1978, however, that the United States initiated the most extensive measures in preparing for natural disasters when President Jimmy Carter submitted Reorganization Plan Number 3 to Congress, which created the Federal Emergency Management Agency (FEMA). For most practitioners, the advent of FEMA is generally considered to be the period of a comprehensive professionalization of emergency management. Yet, the evolution was a slow matriculation due to FEMA being led to focus more on civil defense during the height of the cold war in the 1980s, rather than disaster preparedness, which some attribute to the U.S.-held hostages in Iran, the Marine barracks bombing in Lebanon, and numerous deadly airline crashes. Consequently, the Unites States was simply unprepared for a series of major natural disasters that included numerous destructive hurricanes (Alicia, Gilbert, Joan, and Hugo), the eruption of Mount St. Helens, and the devastating drought of 1988.[9]

Eventually, it was not until the presidency of Bill Clinton that the United States began to determine, develop, and deploy a significant domestic preparedness emergency management plan for natural disasters. In 1993, President Clinton appointed James Lee Witt to be the director of FEMA, and was the first director in its history to have prior professional experience in emergency management. In addition, due to the fact that FEMA was responsible for the oversight of 22 federal agencies, FEMA was also transitioned into the President's Cabinet, which President Clinton believed was necessary in an effort to streamline the bureaucratic decision-making process in the event of a major disaster.

During his tenure as Director of FEMA, Witt initiated a strategic planning process that truly exemplified the professionalization of emergency management. Major operational changes were implemented within the national preparedness plan, which emphasized promoting and maintaining the safety and welfare of the public. The proactive efforts of Witt's leadership included the professional development of

---

[7]Emergency Notifications Services (ENS), *The Origins of 911*, http://www.911broadcast.com (retrieved March 7, 2013).

[8]New York Times, *Office of Emergency Planning Is Renamed Office of Emergency Preparedness*, 1968, http://www.historycommons.org (retrieved March 7, 2013).

[9]George Haddow and Jane Bullock, *Introduction to Emergency Management* (Elsevier Butterworth-Heinemann: Burlington, MA, 2011).

the basic functions of emergency management: mitigation, preparedness, response and recovery. To this day, the progressive changes to FEMA under the direction of Witt remain the foundational epoch of professionalizing emergency management.[10]

Thereafter, the Federal Communications Commission (FCC) created the Emergency Alert System in 1994 and for the first time began enforcing requirements that mandated all broadcast stations to have EAS equipment installed by 1997. The Emergency Alert System (EAS) is a national public warning system that requires broadcasters, cable television systems, wireless cable systems, satellite digital audio radio service (SDARS) providers to provide the communications capability to the president to address the American public during a national emergency. The efforts of the FCC greatly complemented the new FEMA initiatives as well, and continued to be yet another significant period of professionalization within the field of emergency management.[11]

The events of September 11, 2001 brought about considerable change in the United States, which shifted the paradigm back to civil defense with the focus on terrorism now in the forefront and the mitigation and preparation for natural disasters on the back burner. Under the direction of President George W. Bush, the Homeland Security Act of 2002 was passed, which established the Department of Homeland Security (DHS) as an executive department of the United States. The passage of this Act brought about the largest reorganization of the federal government since the Truman administration.[12]

Essentially, the Homeland Security Act of 2002 remains another significant period of professionalization for emergency management by integrating many emergency or security-related agencies into DHS. The newly created DHS included such agencies as the FEMA, which was moved from the President's Cabinet, as well as the U.S. Customs and Border Protection (CBP), the U.S. Immigration and Customs Enforcement (ICE), the U.S. Coast Guard, the Transportation Security Administration, and the U.S. Secret Service Border Patrol, just to name a few.

The basic purpose of DHS was to manage the threat of terrorism. As stated in Title 1—Department of Homeland Security, of the Homeland Security Act of 2002 the DHS's primary mission is to:

- prevent terrorist attacks within the United States,
- reduce the vulnerability of the United States to terrorism, and
- minimize the damage, and assist in the recovery, from terrorist attacks that do occur within the United States.

---

[10]R. Steven Daniels *Transforming Government: The Renewal and Revitalization of the Federal Emergency Management Agency*, 2000, http://www.fema.gov/pdf/library/danielsreport.pdf (retrieved March 5, 2013).

[11]GS, http://www.globalsecurity.org.

[12]George Haddow and Jane Bullock, *Introduction to Homeland Security* (Elsevier Butterworth-Heinemann: Burlington, MA, 2006).

The first assignment of DHS was to produce a National Strategy for Homeland Security. The National Strategy for Homeland Security involved 8 months of consultation throughout the United States. The framers of this initiative consulted with political figures, citizens, foreign leaders, public safety professionals, and research scientists. The purpose was to get an overview of the needs and concerns of the professions most affected by terrorism in the now newly evolved post-9/11 world.[13] Eventually, the National Strategy for Homeland Security created a comprehensive plan based on cooperation and partnership amongst federal, state, and local resources to reduce the country's vulnerability of terrorist attacks. The National Strategy for Homeland Security sought to answer four basic questions:[14]

- What is "homeland security" and what missions does it entail?
- What do we seek to accomplish, and what are the most important goals of homeland security?
- What is the federal executive branch doing now to accomplish these goals and what should it do in the future?
- What should nonfederal government organizations, the private sector, and citizens do to help secure the nation's homeland?

Further professionalization of emergency management continued to evolve in February of 2003, when President George W. Bush executed Homeland Security Presidential Directive-5 (HSPD-5) which established the National Incident Management System. The purpose of HSPD-5 was to create a comprehensive, systematic approach to incident management that combines a set of preparedness concepts and principles for all hazards.

The most notable aspects of HSPD-5 was that it instructed all federal agencies to collaborate with the six components of the National Incident Management System (NIMS), and for the first time ever included a requirement that all levels of government were required to implement and exercise the use of the Incident Command System for all domestic response, as a part of the command and management component.[15]

The other five components to NIMS are preparedness, resource management, communication and information management, supporting technologies for emergency response, and ongoing management and maintenance. The benefits of NIMS contributed significantly to the professional development of emergency management by providing standards for organizational structures, processes, and

---

[13]Ibid.

[14]U.S. Department of Homeland Security (DHS), *The National Strategy for Homeland Security,* October 2007, http://www.dhs.gov/.

[15]Waugh, *Local Emergency Management,* 2007.

planning, as well as the necessary personnel qualifications, certification standards, and communication systems that promoted stronger interoperability to perform the functions and responsibilities to manage an emergency or disaster.[16]

Another significant step to professionalize emergency management began in April 2007, when FEMA established the Integrated Public Alert and Warning System (IPAWS), which initiated a modernization and integration of the nation's alert and warning infrastructure that would authorize emergency management officials to provide the public with life-saving information during an emergency, imminent threat, or pending disaster.[17]

The purpose of IPAWS was to allow all federal, state, tribal, and local emergency managers the ability to integrate local warning systems utilizing the IPAWS infrastructure. By doing so, emergency managers would have an effective option to alert and warn the public during an emergency by using a single operable source that includes the Emergency Alert System, the Commercial Mobile Alert System, NOAA Weather Radio, and other public alerting systems.

During the development period of IPAWS, for the first time ever, on November 9, 2011, the FEMA, in coordination with the Federal Communications Center (FCC), conducted a nationwide test of the Emergency Alert System (EAS). Then, in March of 2012, IPAWS became officially operable and fully capable of delivering digitally based alert and warning messages to radio and television stations, personal computers, cell phones, and other consumer wireless devices. The successful implementation of IPAWS became another profound improvement within the professionalization process of emergency management.[18]

Professional Standards of Emergency Management

Many professions are defined as occupations that require specialized knowledge, the mastery of specific skills, and certification and/or licensing, and they use professional organizations to advance proficiency and provide peer accountability. The most significant components of professionalism include ethics, education, training, leadership, and experience.

Principles of Professionalism

Essential to any profession, principles should exist that readily define the purpose of an occupation, demonstrate the objectives, provide the degree of competency required, and establish the methods and practices necessary to ensure the

---

[16]Perry, 2007.
[17]FEMA, http://www.fema.gov/.
[18]DHS, http://www.fema.gov.

proper delivery of a service. For example, the C.A.T. principle of Communication + Accountability = Trust is a demonstrable approach to promoting and maintaining qualities of competency and commitment necessary to ensure that those responsibilities entrusted to others by the public are being practiced and provided within that profession.

Another significant step in the professionalization of emergency management occurred in March of 2007. Under the direction of Dr. Cortez Lawrence, Superintendent of FEMA's Emergency Management Institute, a working group was formed consisting of twelve stakeholders that included practitioners and academicians, which became known as the Principles Roundtable. Led by Dr. B. Wayne Blanchard, the purpose of the EMI project was to discuss, determine, and develop specific principles that would serve as the prolific benchmark for the emergency management profession. Before this EMI project, there was no clear consensus as to what should be the fundamental principles of emergency management that could truly serve as the foundational guide for practitioners, politicians, and the public.

As a result of several months of laborious work, the EMI stakeholders successfully developed and endorsed the Principles of Emergency Management, which consisted of eight tenets considered to be the "doctrine of emergency management." The adoption and implementation of the principles developed now serve as the basis for developing the necessary administrative and operational functions to reduce vulnerability within communities and manage disasters. Throughout the United States, agencies, organizations and jurisdictions now recognize these principles as the foundation of the emergency management profession. These principles are given as follows.[19]

Principles of Emergency Management

1. Comprehensive—emergency managers consider and take into account all hazards, all phases, all stakeholders, and all impacts relevant to disasters.
2. Progressive—emergency managers anticipate future disasters and take preventive and preparatory measures to build disaster-resistant and disaster-resilient communities.
3. Risk-driven—emergency managers use sound risk management principles (hazard identification, risk analysis, and impact analysis) in assigning priorities and resources.
4. Integrated—emergency managers ensure unity of effort among all levels of government and all elements of a community.
5. Collaborative—emergency managers create and sustain broad and sincere relationships among individuals and organizations to encourage trust, advocate a team atmosphere, build consensus, and facilitate communication.

---

[19]FEMA, http://training.fema.gov.

6. Coordinated—emergency managers synchronize the activities of all relevant stakeholders to achieve a common purpose.
7. Flexible—emergency managers use creative and innovative approaches in solving disaster challenges.
8. Professional—emergency managers value a science and knowledge-based approach based on education, training, experience, ethical practice, public stewardship, and continuous improvement.

Bottom of Form

The Professional Emergency Manager

Professionalism is an integral aspect of emergency services organizations and their respective personnel. A variety of recommendations exist through which professionalism may be embellished among emergency services personnel.

Ethical Codes—Although no consensus exists regarding a solitary code of ethics among emergency services organizations and personnel, the Code of Ethics of the International Association of Emergency Managers is generally accepted among emergency managers. This code emphasizes the characteristics of respect, commitment, and professionalism.

Professional Organizations—Professional organizations, such as the National Emergency Manager's Association (NEMA) and the International Association of Emergency Managers (IAEM), exist as resources through which emergency services personnel gain insight regarding professionalism. These organizations provide opportunities for networking, crafting ethical codes, and continuing education. They may also have national, regional, state, and local bodies.

Certifications—Certifications indicate that emergency services personnel have satisfied some minimum qualifications that are deemed necessary for pursuing both vocational and professional careers in emergency management. After the initial certification is earned, additional levels of certification and achievement may exist through which personnel demonstrate continuing education and the maintaining of skills and knowledge.

Specialized body of knowledge—Three primary areas exist regarding the basic level of knowledge that is expected of emergency managers: (1) historical disasters (including those of the locality); (2) a familiarity with literature regarding disasters, homeland security, emergency management, and so forth; and (3) emergency management practices, standards, and guidelines.

Standards and best practices—The primary standards within emergency management are the Emergency Management Accreditation Program (EMAP)

Standard and the NFPA 1600 standard. Standards provide a foundation of congruency within the context of emergency management organizations and personnel. Through the use of standards and best practices, emergency services organizations may determine their strengths and weaknesses. They may also benchmark themselves against similar organizations.

The aforementioned characteristics that describe a professional emergency manager exemplify significant qualities necessary to fulfill the responsibilities, challenges, adversities, and opportunities to successfully serve in this role. Incorporating these attributes organizationally provides a foundation for bolstering organizational integrity and for embellishing public faith and confidence in the emergency services organization.

Ethics

Essentially, the position of emergency manager within a community or state and federal level is commonly known as a public employee, which means that the position is exclusively funded by taxpayers. As such, such a position requires a high degree of expectations with respect to work performance, leadership, and stewardship, which involves compliance of local, state, or federal laws and standard operating procedures. For example, the following ethical expectations for federal emergency managers that are also applicable to the state and local agencies:[20]

- Compliance with laws and regulations regarding providing equal opportunity to all persons, regardless of race, color, religion, sex, heritage, age, parental status, sexual orientation, disability, or genetic information.
- Give an honest effort in the performance of duties.
- Act impartially to all groups, persons, and organizations.
- Do not use public office for private gain.

The International Association of Emergency Managers (IAEM) is one of the most notable organizations dedicated to promoting the goals of saving lives and protecting property by mitigating, preparing for, responding to, and recovering from disasters/emergencies. In an effort to promote and maintain the professionalism of the IAEM certification process afforded to emergency managers, a code of ethics and professional conduct was developed. The IAEM Code of Ethics and Professional Conduct is an expectation that emergency managers promise to uphold, adhere to, and proficiently demonstrate during their professional career.

---

[20]FEMA, http://training.fema.gov.

All members of the International Association of Emergency Managers (IAEM) are expected to adhere to the highest standards of ethical and professional conduct. The IAEM Code of Professional Conduct for IAEM-certified emergency managers exemplifies the character and proper conduct prescribed by the moral beliefs of society regarding efforts to promote and maintain the well-being of the public during a disaster. See Appendix B for the entire IAEM Code of Ethics and Professional Conduct.

Education

Formal education is evidence of an individual's achievement of a diploma or academic degree from an accredited higher education institution. Ideally, the minimum recommendation of those who are just entering the field of emergency management would include the completion of a four-year degree that exclusively consists of an integrated academic curriculum within emergency management. Because of the highly complex nature of the emergency management profession, such a degree tract within the discipline of emergency management serves to familiarize students and practitioners with basic concepts, principles, and practices that will be advantageous to a person's career aspiration.

The educational progression of the emergency management profession is most notably attributed to FEMA's development of the Higher Education Program, which gives current and aspiring emergency management practitioners an opportunity to achieve an associate, bachelor, master, or doctoral degree in emergency management or related studies. The Higher Education Project website offers a compilation of resources for colleges, agencies, and individuals interested in emergency management. Some states now require new local Emergency Management directors to have a college degree. For example, local emergency management directors in Alabama are required by state law to have at least a two-year college degree.[21]

The FEMA Higher Education Program has helped foster and facilitate development of emergency management certificates and degrees at colleges and universities by providing a plethora of resources including sample syllabi to teach college courses. The FEMA Higher Education Program web page offers hundreds of articles, papers, presentations, books, and links to over 173 higher education institutions that can be used as reference sources in support of a person's educational endeavors.[22]

The FEMA Higher Education Program continues to be a significant resource in promoting the development of emergency management higher education programs.

---

[21]Code of Alabama—Title 31: Military Affairs and Civil Defense, Section 31-9-61, Certified Local Emergency Management Director, http://www.legislature.state.al, 2007.
[22]U.S. Department of Homeland Security, Federal Emergency Management Agency-Emergency Management Institute (FEMA-EMI), http://training.fema.gov.

Likewise, higher education programs in turn provide the body of knowledge that can assist practitioner graduates with not only the opportunity to further their professional career aspirations but also equip them with the ability create a research agenda to further improve the functional practices of emergency management.

The development and commitment to emergency management higher education has proven to be of critical importance with respect to the progressive efforts to professionalize the role of emergency managers. The contribution of higher education to an individual will also build confidence in one's ability to perform the essential functions of an emergency management position, and solidify confidence and trust within the community environment in which he or she serves.

Training

Another essential aspect regarding the progressive evolution of the emergency management profession involves training. Overall, practitioners within emergency management acknowledge that the training component complements one's experiences, prepares for future hands-on practices, and assists with career aspiration promotions. An additional benefit from training is that practitioners also become more empowered, which further develops confidence in one's ability to perform the essential job functions as an emergency manager.

The defining era for emergency management training is aptly described by Waugh who refers to the 1990s as the "golden age" of emergency management (Waugh, 2007). During this period, the following four training and accreditation programs were developed: (1) the Certified Emergency Manager (CEM) program; (2) the National Fire Protection Association (NFPA) 1600 standard; (3) the FEMA Higher Education Project; and (4) the Emergency Management Accreditation Program (EMAP) in 1997.[23]

The CEM program was developed by the International Association of Emergency Managers (IAEM) to raise and maintain professional emergency management standards. A person wanting to be recognized as a CEM must go through a rigorous, peer-reviewed process which demonstrates the candidate's experience, references, education, training, and contributions to the profession. In addition, the candidate must submit an essay and take a 100 question multiple-choice exam. The CEM program is considered to be the most highly valued training certification program in the profession of emergency management.[24]

The National Fire Protection Association helped to define the role of emergency management in the United States by releasing NFPA 1600. The purpose of NFPA 1600 is to "provide the fundamental criteria to develop, implement, assess, and

---

[23]Waugh, *Local Emergency Management*, 2007.
[24]Ibid.

maintain the program for prevention, mitigation, preparedness, response, continuity, and recovery."[25] This standard has been recommended or endorsed by the 9/11 Commission, NEMA, and others.[26]

The IAEM, FEMA, and NEMA collaborated to create EMAP in 1997, for state and local governments. The NFPA 1600 standard is used as a measure for state and local emergency management programs to achieve EMAP accreditation.[27] As of April 2013, 29 states, the District of Columbia, and 11 local jurisdictions have achieved EMAP accreditation.[28]

Thereafter, in 2007 the IAEM began a review of the emergency management profession. Representatives from IAEM, EMAP, and NEMA, the NFPA 1600 committee, the private sector, academia, and other stakeholder groups conducted a roundtable discussion to define emergency management, its mission, and identify its core principles. The work resulting from this discussion continued to clarify the training components necessary to properly prepare an emergency manager for performing the essential functions of their position, and what is needed to achieve successful outcomes in accomplishing the mission of emergency management.

One can never be too educated or possess enough knowledge when serving in the capacity of an emergency manager. The leading agency for training resources is the Federal Emergency Management Agency (FEMA), which offers courses through their Emergency Management Institute and Independent Study Programs. Typically, most position descriptions of an emergency manager require the completion of FEMA's Professional Development Series, as well as the completion of several advanced courses as an introduction to the profession. In particular, FEMA's Professional Development Series, which consists of the following courses, is desirable:

- IS230—Principles of Emergency Management
- IS235—Emergency Planning
- IS242—Effective Communication
- IS241—Decision-Making and Problem Solving
- IS240—Leadership and Influence
- IS244—Developing and Managing Volunteers
- IS139—Exercise Design

---

[25]National Fire Protection Association, NFPA 1600: Standard on Disaster/Emergency Management and Business Continuity (Quincy, MA: NFPA, 2010).

[26]D. McEntire, *Disaster Response and Recovery*, (Hoboken, NJ: Wiley & Sons, Inc., 2007).

[27]F. Edwards and D. Goodrich, *Organizing for Emergency Management*, in *Emergency Management Principles and Practice for Local Government*, eds. W. Waugh and K. Tierney (Washington, D.C.: International City/County Management Association, 2007).

[28]EMAP, http://www.emaponline.org/.

Additional training for emergency managers can be obtained through the National Fire Academy through their direct association with FEMA. Other professional certifications are also available that offer an emergency manager the opportunity to further develop one's knowledge base in an effort to fulfill required responsibilities, and be prepared for future unforeseen and pending disasters and emergencies. The International Association of Emergency Managers (IAEM) offers a Certified Emergency Manager (CEM) program in order to receive professional recognition and to establish enhanced credentials.[29]

Further training resources can also be found at many state levels. For instance, the State of Texas' Department of Homeland Security Governor's Division of Emergency Management offers training in emergency management exercises, the four phases (mitigation, preparedness, response, and recovery), debris management, hazardous materials management, and terrorism and weapons of mass destruction.[30]

Thus, it is necessary for an emergency manager to maintain a proactive and progressive mindset within the profession by availing one's self of training opportunities. A good rule to follow is the more you know the better you can be able to prepare, respond to, and recover from an incident. Preparing for the worst and hoping for the best does not include resting one's success; rather, continued education and training is essential for being as prepared as possible.

Characteristics of a Professional Emergency Manager

Emergency managers face many challenges but the individuals entering this career field seem to enjoy their work even though the hours are long, the stress is intense, and tempers may flare from all sides. Moreover, emergency managers often realize well into their career that the bang is not worth the buck, that is, the pay is poor and support for their program lacking. Depending upon which phase of the cycle is in action, emergency managers may often be misunderstood or blamed within their community, by the media, or aggrieved by the landmines of politics. Therefore, an emergency manager's ability to endure must come from within. Likewise, prior to even engaging in the profession, an individual needs to truly size up the positional responsibilities and determine if one's character and qualifications are aligned with the expectations of the profession. Unfortunately, more often than not, people find themselves in over their heads and then start becoming discouraged, angry, bitter, and eventually whining out loud for relief.

---

[29]IAEM, *What is a Certified Emergency Manager,* 2006, http://www.iaem.com.
[30]Texas Department of Homeland Security (TX DHS), Governors Division of Emergency Management http://www.txdps.state.tx.us (retrieved March 7, 2013).

For many emergency managers, the position and responsibilities may, at times, be overwhelming. Exploring the pressure to professionalize; facing the world of risk, trust, and power; witnessing not only evolutionary change but also revolutionary change; and sensitizing the entire community to the social risks which are present around them are only four components of the environment in which emergency managers operate. Thus, an emergency manager is expected to act in a professional manner, i.e., an individual with specialized knowledge equipped with special abilities and the power to act.[31]

Often with few resources and supporters, the emergency manager is commonly faced with the daunting responsibility of developing an efficient and effective organization and comprehensive management plan. The plan must include actions in mitigation, prevention, hazard identification, and the assessment of risk. It must also address what preparations must be made for events that cannot be prevented. This includes overseeing the equipping and training of others who may render assistance as well as conducting drills and exercises to focus on coordination and communications among responders.

Generally, once a significant event occurs, an emergency manager immediately becomes a coach, a conductor, a subject matter expert, an interpreter, and resource provider, all within an around-the-clock environment. The demanding responsibilities of an emergency manager require character, courage, and competency necessary to fulfill the administrative and operational duties. Such tasks will involve, complying with local, state, and federal guidelines, serving as a liaison between agencies, securing facilities, completing the paperwork process, acting as a principal player in the return to normalcy and restoration of community business continuity of business, operations, and quality of life activities. Consequently, over and beyond the prescribed job description, an emergency manager must be an idealist first, followed quickly by the metamorphosis into a tireless perfectionist!

One crucial characteristic of the successful emergency manager is the necessity of collaboration and leadership outside their own organization. Emergency management entities rarely own the resources and influences needed to mitigate, prevent, protect, respond, or recover from the threats and hazards they face. Emergency managers lead communities of effort, or networks of participants. Thus, the leadership skills required for emergency managers are not the traditional ones expected for more hierarchical professions. Emergency managers are leaders, facilitators, coordinators, and collaborators all at the same time.

Given the nature of the emergency management profession, it is important to identify the necessary qualities an emergency manager should possess to be successful. The following, although not meant to be comprehensive, is a list created by

---

[31]T. Grovier, *Social Trust and Human Communities* (Montreal, Canada: McGill-Queen's University Press, 1997).

IAEM consisting of the knowledge, skills, and abilities that an emergency manager should have to be efficient and compatible with the profession:[32]

- Considerable knowledge of the principles of management, organization, and administration.
- Considerable knowledge of emergency management procedures, practices, and regulations.
- Considerable knowledge of the practices in the fields of local government personnel management, organization, administration, budgeting, and accounting.
- Knowledge of personal computers and effective procedures for their use.
- Ability to plan, direct, and coordinate the work of subordinates when required by work assignments.
- Ability to express ideas effectively, both orally and in writing.
- Ability to serve the public and fellow employees with honesty and integrity.
- Ability to establish and maintain effective working relationships with the general public, coworkers, elected and appointed officials, and members of diverse cultural and linguistic backgrounds.

In addition, FEMA's independent study program IS-1, entitled "Emergency Manager: An Orientation to the Position," states that there are three broad characteristics that apply to an emergency manager: (1) professionalism; (2) individual qualities; and (3) emergency management activities. Being professional is all-encompassing. Whether it is job knowledge, work ethic, or morality, being professional means you are striving to stand above others.

Specific individual qualities that make for success include the ability to: provide distinctive leadership; communicate effectively; make sound discretionary decisions; a proficient steward of resources; work well under stressful conditions; commitment to professional development; care and compassion. Often overlooked until the proverbial 'A' word (accountability) hits the fan, comprehensive emergency management not only consists of more than just knowing what to do but also requires the noble qualities essential to successfully manifest doing what needs to be done.

Leadership

The principles herein serve as a model for successful leadership. This notion is not meant to be a "one size fits all" or ideal standard, but rather a paradigm to challenge one's perspective regarding being a leader with the hope of influencing

---

[32]International Association for Emergency Managers (IAEM), *Emergency Program Manager: Knowledge, Skills and Abilities*, 2005, http://training.fema.gov/emiweb/edu/EmergProgMgr.doc.

individuals who serve the public and desire to do so with a positive, proactive, and productive outcome. The Five A's of Leadership are principles that involve authority, accountability, adversity, attitude, and achievable. Under each of the Five A's are action points to guide and mentor individuals to be successful.

With respect to leadership, although there is a variety of models and theories that can be considered; in reality, leadership is the interpretative application of the perspectives of life. For the purpose of this discussion, leadership is defined as the process of transforming a vision into reality. Leadership involves the ability to influence human behavior toward organizational goals. There is no organization without the people who need it or the people who comprise it.

The process of influencing encompasses the action, behavior, and decision-making a leader might use. For instance, during a tactical situation a leader's behavior is likely to be highly directive; the leader influences by giving orders. At other times, when employees are involved in completing tasks outside the direct observations of the leader, indirect influence may come in the form of prior training, mentoring, or necessary corrective action. The ability of a leader to exhibit respect, instill confidence, and build trust will result in empowering employees.

Leadership is also about human behavior. Basically, there are two underlying concepts about human beings. First, one cannot always predict what a human being is going to do. Certainly, you can make accurate generalizations. But, you cannot know with any degree of certainty how every human will react or act. There are too many variables, too many unknowns. Second, one cannot control people; you can only hope to contain them. What's between a person's ears is known as their free will, which can be under the influence, overly influenced, or adversely influenced. A person's free will is triggered by a smorgasbord of intoxicants such as feelings, emotions, drugs, food, sex, money, music, entertainment, religion, relationships, sports, crime, injury, birth, marriage, death, victimization, cruelty, lies, and so forth.

Lastly, one cannot change people. After a certain age people do not change—a generalization meant for another study. But the reality is that most people over time become set in their ways or exhibit behavior and make decisions based upon life experience. Now, there may be a pilot light inside of them waiting for a spark. But, if there is no gas running to the pilot light all the fire in the world is useless. A leader must accept that all they can do is modify a human being's behavior. The key concept is that leadership involves working with and influencing human behavior, simply due to the fact that human beings remain the center of this definition.

Leader vs. Manager

A leader is defined as having the ability to influence human behavior. Essentially, leadership is management in action. Table 12.1 provides a more definitive distinction between the qualities of a manager and a leader.

Table 12.1 Leaders vs. Managers

Characteristic	Leader	Manager
Approach to situations	Leads and sets direction of organization	Plans according to limitations and constraints
Role attribute	Dynamic; large change	Static; changes little.
Decisions	Facilitates decisions; strategic mindset	Makes decisions tactically and operationally
Concept of future	Strategic; long-term view	Operational; short-term view
Control	Ranges from charismatic to motivational	Formal methods
Culture and policy	Crafts and influences	Enforces and endorses
Actions	Proactive	Situational; may be proactive or reactive.
Risk management	Embraces risk	Minimizes risk
Values	Results and achievements	Results and achievements
Emphasis	Embraces vision; forward-thinking	Exercises management functions toward vision

Regardless of one's position, either leadership or management, communication within the established chain-of-command is essential organizationally. Both leaders and managers must possess excellent communication skills to ensure the flow of information efficiently and effectively within the organization. Leaders must be able to clearly explain vision and strategy in terms that are understandable by

"What if, and I know this sounds kooky, we communicated with the employees."

Figure 12.3 Communication is essential.

subordinate managers. Certainly, managers must communicate effectively and efficiently to ensure that subordinate workers and higher echelons of leadership and organizational administration are informed to facilitate decisions. Thus, all leaders and managers must refine their communication skills.

Authority

A leader needs to define and discern the boundaries of authority relative to the essential job functions of their position, which is often entrusted to an individual per an oath of delegation on behalf of the public, fiscal funding by the public, and accountability to the public. Such authority of one's position is also complemented by qualifications such as education, training, required certification, as stipulated per the specific regulations or legislation of a governing body.

Other questions a leader should be cognizant of include: What expectations do you believe subordinates have with respect to the authority ascribed to your position? What support and resources are needed for you to administer, direct, assist, and supervise others? What aspects of the authority of your position are essential in an effort to implement, coordinate, and respond to the needs of the department, to ensure functional and successful service deliverables to the public whom you serve, and who depend on you for maintaining their safety and welfare?

Accountability

What expectations do you have for being held accountable by your supervisor, subordinates, and the public for performing the essential job functions of your position? What responsibility do you believe is required of you in the position you

"There isn't enough blame to go around, there's only enough for you."

© Cartoonresource, 2014. Used under license from Shutterstock, Inc.

Figure 12.4 Accountability involves accepting responsibility for one's actions.

hold regarding the obligation to report, explain, and be answerable for the resulting consequences of your decisions, actions, and behavior? What does your work ethic consist of? What level of professionalism do you subscribe to? Understand the premise of colleagues, coworkers, and friends that are with you and those who are for you. Those who are with you often take you down the wrong path or road to perdition, and are never around when things go bad—they will only use you for their benefit. Whereas, those who are for you are there to get in your face and remind you of the reality of life, and the responsibilities of the position in which you serve because they care.

Adversity

Adversity is an aspect of all emergencies. In many cases, the proverbial instances of Murphy's Law are present—whatever can go wrong, will go wrong. What adversity are you experiencing in life? How do you manage adversity? Do you isolate yourself? Blame others or have a pity party? Remember, the only person at a pity party is you. Do you substitute reality for relief by engaging in alcohol, drugs, or other destructive behavior? Do you engage in confrontations to justify wrong decisions? It is easier to criticize and attempt to bring someone else down rather than work toward self-improvement.

Do you have a game plan to overcome or work through those circumstances? Assess your situation, ask for advice, be quick to hear, slow to speak, and avoid emotional confrontations. Persevere, be patient, remain focused, and do not be afraid to request assistance from others. Stay on task within the daily boundaries of your work environment. During your decision-making process—discover, determine, and discern the reality of your circumstances by measuring the facts rather than feeding off of perception or assumptions.

Attitude

One should maintain properly placed confidence or what is referred to as humility (an often overlooked and much maligned virtue). Remember, power is misplaced confidence and corruption is displaced confidence. Maintain an attitude of gratitude—be grateful and thankful for what one has or does not acquire. Presently, what is specific to the job that affects one's attitude? How do others perceive you? Do you have a cynical "us" versus "them" perspective? Has the position you serve in become your personality? Self-assess and determine if there is a behavior you are exhibiting that may be adversely affecting others. If so, what aspects of your attitude need corrective action and refinement? Develop and sustain a discretionary continuum as part of your decision-making process, including two profound questions—"can I or should I?" Take time for self-examination regarding the Five C's of Character that affects one's attitude relative to one's personal and professional well-being: commitment, competency, courage, caring, and communication.

Achievable

In an effort to avoid complacency and apathy, and to challenge your skills and abilities, develop goals that are specific, measurable, attainable, relevant, and time-based (SMART). Begin to craft a work-improvement plan that will better equip you to perform your responsibilities and better serve others. Develop and sustain a visionary mindset regarding your career path in terms of education, training, and experience. Do not give up—impossible is just an excuse for not trying. Finally, being a CHAMPION requires an understanding that there are virtues inherent to success irrespective of what position, organization, agency, or profession one is a part of, which is relevant to those who we serve with, and that will benefit the community we serve. The following acronym includes virtues that are meant to promote, develop, and establish individual and organizational ethics.

- Courtesy   To be respectful and considerate of others at all times.
- Honesty   To be free of deception and remain truthful at all times.
- Adaptable   To possess the ability to be flexible in all situations.
- Mature   To possess the ability to make decisions with discretion.
- Prudence   To exercise good judgment in the use of resources.
- Integrity   To remain invincible to unlawful, unethical, and corrupt practices.
- Open-Minded   To be a good listener, receptive to ideas, and unbiased in attitude.
- Noble   To be trustworthy, reliable, and servant-centered not self-centered.

Position

Being a leader and exercising leadership is a challenging facet of one's position. Remember, the bend in the road is not the end of the road—it is all how you take the curve. That curve represents the proverbial learning moment whereby an individual serving in a leadership position must understand the aspects of authority, be held accountable, experience adversity, maintain a proper attitude, and develop attainable goals and objectives linked to performing the essential functions of one's position and serving the public with honor and integrity. One important motto for practitioners, the public, and politicians to remember: cooperate and operate!

"We'll solve the conflict with a water pistol fight."

Figure 12.5 Personnel must be cooperative.

Legal Issues of Emergency Management

The last important consideration of the emergency management profession involves legal issues. The emergency management profession involves the phases of mitigation, preparedness, response, and recovery activities. Emergency managers can be found in all sectors including businesses, universities, hospitals, public utility companies and municipal, provincial, and federal governments. As in any profession, emergency managers face legal issues during the performance of their duties. While there are governmental immunities to tort law claims, emergency managers and their agencies are not always immune from liability, and legal issues can be extremely complex.

Consequently, emergency managers should be well-informed and educated regarding the legal implications of their actions. Moreover, it is equally important that an emergency manager have a source—a reference person with the legal savvy to provide guidance and counsel concerning pending decisions or post-incident outcomes. Essentially, for any emergency manager, liability is the principal legal issue that occurs during the administrative and operational facets of one's position. This is particularly important given the litigious society of this day and age.

Basically, the term liability is about being held accountable for the responsibilities of one's actions. A person's work performance can be measured based upon acts of omission or commission, or both. Unfortunately, a recent study indicates that some agencies are not building the types of cultures where constructive accountability thrives. For example, during the years of 1997 and 2000, a study was conducted by the U.S. General Accounting Office (GAO) that included a survey of 3,816 full-time, mid-, and upper-level managers regarding their perceptions about performance and management issues.[33]

---

[33] U.S. Government Accountability Office (GAO), *Managing for Results: Federal Managers' Views Show Need for Ensuring Top Leadership Skills,* October 2000, http://www.gao.gov.

The results of the GAO report (October 2000), *Managing for Results: Federal Managers' Views Show Need for Ensuring Top Leadership Skills*, found that while 63% of managers said they were held accountable for the results of their programs, only 36% of them stated they had the authority needed to accomplish strategic goals. The GAO report concluded that "Such an imbalance can inhibit the development of an environment conducive to achieving results."[34]

The GAO report also discovered that only 31% of managers indicated that employees received positive recognition for helping to achieve organizational goals. If managers are going to hold employees accountable for results, then employees need to be recognized for their efforts. Overall, shifting to constructive accountability may require an extensive culture change within an agency, but managers will find the results well worth the effort.[35]

The avoidance of legal issues can be mitigated by developing and implementing principles of accountability. Per the U.S. Office of Performance Management, the following five themes are important to successfully manage disaster efforts within the emergency management profession that may reduce the risk of liability:[36]

- **Clearly defined roles and responsibilities.** Having clearly defined roles and responsibilities for all levels of local, state, and federal government agencies, and NGOs is critical in an effort to develop and successfully implement the responsibilities of the emergency management profession.

- **Developing and Assessing Capabilities.** Developing the capabilities needed for catastrophic disasters should be part of an effort designed to integrate and define what needs to be done, by whom, and a communication process that successfully complements the operational coordination of all partners. Ensuring needed capabilities are ready requires effective planning, collaboration, and realistic training exercises in which problems are identified and subsequently addressed in partnership with all stakeholders.

- **Effective coordination and collaboration among relevant stakeholders.** Response to and recovery from a major disaster is a complex process that involves an extensive group of participants both across the federal government and at the state and local level. Recovery may take years. At least 14 federal departments and agencies are responsible for administering dozens of recovery-related programs, many of which rely heavily on active participation by state and local government for their implementation. Because these parties are dependent on each other to accomplish recovery goals, sustained focus and effective coordination and collaboration are essential.

---

[34]Ibid.
[35]Ibid.
[36]U.S. Office of Performance Management, Performance Management: *Accountability Can Have Positive Results*, http://www.opm.gov (retrieved March 23, 2013).

- **Periodic evaluation of and reporting on these coordinated efforts.** Collaboration between recovery partners can be enhanced by periodically evaluating and reporting on what worked, what can be improved, and what progress is still needed to address long-term recovery goals. This will assist decision makers, clients, and stakeholders to obtain the feedback needed to improve both the policy and operational effectiveness of recovery efforts.
- **Accountability.** The purpose of accountability is to ensure that resources are used appropriately. Following a catastrophic disaster, decision makers face the challenges of rapid response and recovery, which includes assisting victims, and implementing attentive means of appropriate accountability. Accountability means being held answerable for accomplishing a goal or assignment. Unfortunately, the word "accountability" often connotes punishment or negative consequences. Certainly, management should not tolerate poor performance and should take action when it occurs. However, when organizations use accountability only as a big stick for punishing employees, fear and anxiety permeate the work environment. Employees are afraid to try new methods or propose new ideas for fear of failure. On the other hand, if approached correctly, accountability can produce positive, valuable results, and ensure the service deliverables expected by the public come to fruition.

The U.S. Office of Performance Management also highlights positive results of practicing a constructive approach to accountability that include:[37]

- improved performance,
- more employee participation and involvement,
- increased feelings of competency,
- increased employee commitment to the work,
- more creativity and innovation, and
- higher employee morale and satisfaction with the work.

These positive results occur when employees view accountability programs as helpful and progressive methods of assigning and completing work. For example, emergency managers who involve employees in setting goals and expectations find that employees understand expectations better, are more confident that they can achieve those expectations, and perform at a higher level. Doing so also marginalizes the concerns or possibility of legal liability as employees will have performed their duties with a good faith effort. Positive results also occur when employees do not associate accountability with only negative consequences. If employees do not fear failure, if emergency managers recognize employees for their accomplishments,

---

[37]Ibid.

and if support their employees when goals become difficult, employees are more likely to be creative, innovative, and committed to their work.[38]

Arguments for practicing constructive accountability are overwhelming. In his book, *The Accountability Revolution,* Mark Samuel states that "Accountability means people can count on one another to keep performance commitments and communication agreements."[39] According to Samuel, accountability can result in increased synergy, a safe climate for purposeful change and improved solutions because people feel supported and trusted. All of these positive results create higher employee morale and a productive work environment.

Implementing Accountability for Positive Results

Emergency managers can practice accountability for positive results by following good performance management principles. Primarily, emergency managers should use an agency's performance appraisal process to establish employee work performance expectations, and use formal award programs to recognize employees. However, merely following the minimum requirements of formal programs is not enough to create the positive environment necessary for constructive accountability. Emergency managers need to:

- Involve employees in setting clear, challenging yet attainable goals and objectives, and give them the authority to accomplish those goals.
- Mentor and support employees in all aspects of the job, not just when they need help.
- Monitor progress toward goals, and provide feedback that includes credible, useful performance measures.
- Provide the training and resources employees need to do the work.
- Recognize employees for good performance, both formally and informally.

Accountability fuels the engine of performance. It puts a fine edge on execution. It replaces the administrative rituals of performance management with engagement in the operations and commitment to results. It fills the void of performance-focused communication with precise and continuing conversations about accomplishments and opportunities as well as about shortfalls and what needs to be done to overcome them. Accountability puts talent in the spotlight, exposes poor performance, and corrects work performance deficiencies before they become problematic.

---

[38]Ibid.
[39]Mark Samuel, *The Accountability Revolution: Achieve Breakthrough Results in Half the Time Facts,* 2nd ed. (Tempe, AZ: On Demand Press, 2001).

The practice of accountability means that every person, either as the boots on-the-ground front liner, or as an administrator, is expected to provide a periodic accounting to someone, i.e., team leader, emergency manager, city or county council/board of directors, state or federal administrators, the media, and the public, regarding the outcome of work performance.

The implementation of accountability principles within the emergency management profession provides reasonable expectations for employees to perform the essential responsibilities of their position. A well-defined game plan for accomplishing planned activities will place an agency in a better position to produce planned results. Accountability establishes a reasonable and regular exchange of ideas, and supports channels of communication for employees both internally within an agency and externally on behalf of the public. The following are basic elements of accountability that employees should be expected to follow:

- Encourage employees when courage is needed.
- Provide information that will help solve a problem.
- Ask questions to see if something has been overlooked.
- Provide support to others by obtaining needed resources.
- Stop a process or practice that either is or will impede success.
- Share a perspective that will influence additional thoughts about a circumstances.

Accountability principles will not only improve the quality of employees but also nurture the trust of the public. The public's awareness of the professional standards required of emergency management profession is essential for promoting and maintaining that trust. Whether it is through a town hall meeting, the chamber of commerce, a local emergency management committee (LEMC), the media, or secondary schools, an emergency manager needs to partner with the public and market the profession on behalf of the public they serve.

As stated earlier in this chapter, Communication + Accountability = Trust. Thus, as the professionalization of emergency management continues to evolve, the leadership must demonstrably be progressive-minded and results-oriented, which often involves going forward where there is no path while fostering a trail of success.

Chapter Comments and Summary

This chapter introduces the concepts of leadership and management among emergency service organizations. All emergency service organizations involve managing and leading regardless of their size, scope, or mission. Leading and managing are separate concepts despite their presence within the emergency management domain. In any case, the exercising of these concepts among emergency services organizations must be performed professionally.

Professionalization among emergency management personnel is uncompromisingly a fiduciary aspect of their job and service obligations. Numerous professional organizations and societies exist in which the leaders and management personnel of emergency service organization may share information, gain both personal and professional networks, and pursue continuing education. Professionalization also necessitates the maintaining of skills periodically to ensure the currency of knowledge and acumen among emergency management personnel.

Professional societies and organizations also exhibit various codes of ethics that delineate the behavioral and professional expectations that are associated with their memberships and the conduct of personnel among emergency services organizations. Such ethical codes are guides for crafting and implementing policies among emergency service organizations. Adhering to the tenets of these ethical codes is also an aspect of professionalization.

Another dimension of professionalization includes accountability among emergency services personnel. Accountability involves accepting responsibility for decisions both personally and professionally. All emergency services personnel, especially organizational leaders and managers, must be held responsible for their actions and decisions. Through exercising accountability, both faith and confidence in the emergency services organization are embellished and maintained.

Terminology

- Accountability
- Achievable
- Administration
- Adversity
- Attitude
- Authority
- Behavior
- Code of Ethics
- Conduct
- Ethics
- Leadership
- Management
- Professionalism
- Professional organization
- Professional society

Thought and Discussion Questions

1. This chapter introduces educational requirements that are necessary for emergency managers. Visit with the leaders of your local emergency services organizations, and discuss the qualifications and educational requirements that are necessary for entering emergency management positions. Write a brief essay that highlights your findings.
2. This chapter introduces professionalism. Many management disciplines have an associated professional organization in which the members have common traits professionally. Do some research regarding the leadership of your local emergency services organizations, and determine what professional societies or

organizations are affiliated with their respective positions and responsibilities. Write a brief essay that highlights your findings.
3. Continuing education periodically also is an aspect of professional development. Perform some research regarding the continuing education requirements of emergency managers in your state. Write a brief essay that highlights your findings.
4. This chapter introduces SMART attributes regarding emergency managers. Discuss these characteristics with the leaders of your local emergency services organizations. Determine how they implement these characteristics among their respective organizations. Write a brief essay that highlights your findings.

References

1. Americon, District of Columbia Emergency Management Agency. http://www.americon-usa.com/projects/dc-ema/index-dcema.html (accessed March 18, 2013).
2. Code of Alabama—Title 31: Military Affairs and Civil Defense, Section 31-9-61. Certified Local Emergency Management Director. http://www.legislature.state.al.us/CodeofAlabama/1975/31-9-61.htm (accessed March 2, 2013).
3. Cwiak, Carol L. 2013. *The pivotal role of FEMA's higher education program in the emergency management professionalization process*. The Public Entity Risk Institute. http://www.riskinstitute.org (accessed March 4, 2013).
4. Daily HaHa—Your daily dose of laughs. http://www.dailyhaha.com (accessed March 19, 2013).
5. Daniels, R. Steven. 2000. *Transforming government: The renewal and revitalization of the Federal Emergency Management Agency*. http://www.fema.gov/pdf/library/danielsreport.pdf (accessed March 5, 2013).
6. Drabek, T. E., and G. J. Hoetmer. 1991. *Emergency management principles and practice for local government* (Introduction, pp. xvii–xxxiv). ICMA Press: Washington, D.C.
7. Edwards, F., and D. Goodrick. 2007. Organizing for emergency management. In *Emergency management principles and practice for local government*, ed. W. Waugh and K. Tierney. Washington, D.C.: International City/County Management Association.
8. Emergency Management Accreditation Program. 2013. *The Emergency Management Standard*. http://www.emaponline.org (accessed February 21, 2013).
9. Emergency Management Accreditation Program. 2007. Texas A&M University. http://homelandsecurity.tamu.edu (accessed March 4, 2013).
10. Emergency Management Accreditation Program. 2010. *Who is accredited?* http://www.emaponline.org (accessed March 21, 2013).
11. Emergency Management Accreditation Program—*Logo*. http://www.emaponline.org (accessed March 21, 2013).

12. Emergency Notifications Services. 2014. *The origins of 911.* http://www.911broadcast.com (accessed March 7, 2013).
13. Federal Emergency Management Agency. 2011. IS-33.11—FEMA initial ethics orientation 2011. http://training.fema.gov/EMIWeb/IS/is33.11asp (accessed March 6, 2013).
14. Grovier, T. 1997. *Social trust and human communities.* Montreal, Canada: McGill-Queen's University Press.
15. Haddow, George, and Jane Bullock. 2006. *Introduction to homeland security.* Burlington, MA: Elsevier Butterworth-Heinemann.
16. Haddow, George, and Jane Bullock. 2011. *Introduction to emergency management.* Burlington, MA: Elsevier Butterworth-Heinemann.
17. International Association of Emergency Management. 2006. *What is a certified emergency manager.* http://www.iaem.com (accessed February 24, 2013).
18. International Association for Emergency Managers. 2014. *IAEM code of ethics and professional conduct.* http://www.iaem.com (accessed March 21, 2013).
19. International Association for Emergency Managers. 2005. *Emergency program manager: Knowledge, skills and abilities.* http://training.fema.gov (accessed March 6, 2013).
20. International Association of Emergency Managers. 2007. *Homeland security act of 2002.* http://www.iaem.com (accessed March 3, 2013).
21. International Association of Emergency Managers—*Logo.* http://www.iaem.com/home.cfm (accessed March 21, 2013).
22. Japan Tsunami. 2011. http://www.weather.com (accessed February 24, 2013).
23. Lindell, M., and R. Perry. 2007. Planning and preparedness. In *Emergency management principles and practice for local government,* ed. W. Waugh and K. Tierney. Washington, D.C.: International City/County Management Association.
24. Lindell, M., C. Prater, and R. Perry. 2006. *Fundamentals of emergency management.* http://archone.tamu.edu (accessed March 4, 2013).
25. Lindell, M. K., C. Prater, and R. W. Perry. 2007. *Introduction to emergency management.* Hoboken, NJ: Wiley & Sons, Inc.
26. McEntire, D. 2007. *Disaster response and recovery.* Hoboken, NJ: Wiley & Sons, Inc.
27. National Emergency Management Association. 2008. *A selective history of emergency management.* http://www.norman.noaa.gov (accessed March 6, 2013).
28. McEntire, D., and G. Dawson. 2007. The intergovernmental context. In *Emergency management principles and practice for local government,* ed. W. Waugh and K. Tierney. Washington, D.C.: International City/County Management Association.
29. National Emergency Management Association—*Logo.* http://www.nemaweb.org (accessed March 22, 2013).

30. National Fire Protection Association. 2010. *NFPA 1600: Standard on disaster/emergency management and business continuity*. Quincy, MA: NFPA.
31. National Oceanic and Atmospheric Administration (NOAA). 2011. Joplin, Missouri, Tornado (May 22, 2011). Kansas City, MO: U.S. Department of Commerce, National Weather Service, Central Region Headquarters. http://www.nws.noaa.gov/os/assessments/pdfs/Joplin_tornado.pdf (accessed March 6, 2013).
32. New Hanover County, North Carolina. *History of the emergency alert system*. http://www.nhcgov.com/News/Pages/EAS_History.aspx (accessed March 6, 2013).
33. New York Times. 1968. *Office of emergency planning is renamed office of emergency preparedness*. http://www.historycommons.org (accessed March 7, 2013).
34. Nicholson, W. 2007. Major issues in emergency management. In *Emergency management principles and practice for local government*, ed. W. Waugh and K. Tierney. Washington, D.C.: International City/County Management Association.
35. Perry, Ronald W., and Michael K. Lindell. 2007. *Emergency planning*. Hoboken, NJ: John Wiley & Sons Inc.
36. Phillips, B., and D. Neal. 2007. Recovery. In *Emergency management principles and practice for local government*, ed. W. Waugh and K. Tierney. Washington, D.C.: International City/County Management Association.
37. Renteria Group. Emergency management, public safety & homeland security. http://www.renteriagroup.com (accessed March 18, 2013).
38. Rogers Historical Museum. 2006. *The life atomic! Growing up in the shadow of the A-bomb*. http://www.rogersarkansas.com (accessed March 7, 2013).
39. Samuel, Mark. 2001. *The accountability revolution: Achieve breakthrough results in half the time facts*. 2nd ed. Tempe, AZ: Facts on Demand Press.
40. Sharples, T. 2008. A brief history of FEMA. *Time Magazine*, August 28. http://www.time.com/time/magazine/article/0,9171,1837231,00.html (accessed March 3, 2013).
41. Texas Department of Homeland Security, Governors Division of Emergency Management. http://www.txdps.state.tx.us/dem/pages/training.htm (accessed March 7, 2013).
42. United State Office of Government Ethics. 2000. *Executive Order 12674 of April 12, 1989*. http://www.usoge.gov (accessed March 2, 2013).
43. U.S. Department of Homeland Security. 2007. The national strategy for homeland security. http://www.dhs.gov (accessed March 3, 2013).
44. U.S. Department of Homeland Security, Federal Emergency Management Agency (FEMA). 2013. *The historical context of emergency management*. http://training.fema.gov/ (accessed March 17, 2013).
45. U.S. Department of Homeland Security, Federal Emergency Management Agency (FEMA). Emergency Management Institute—*Logo*. http://training.fema.gov/emiweb/edu/ (accessed March 22, 2013).

46. U.S. Department of Homeland Security, Federal Emergency Management Agency (FEMA). *The Emergency Management Institute—Higher education program.* http://training.fema.gov/emiweb/edu/ (accessed March 2, 2013).
47. U.S. Department of Homeland Security, Federal Emergency Management Agency (FEMA). *History of the integrated public alert and warning system (IPAWS).* http://www.fema.gov/history-integrated-public-alert-and-warning-system (accessed March 6, 2013).
48. U.S. Department of Homeland Security, Federal Emergency Management Agency (FEMA). IS-1 emergency manager: An orientation to the position. FEMA independent study course. http://training.fema.gov (accessed March 3, 2013).
49. U.S. Department of Homeland Security, Federal Emergency Management Agency (FEMA). *National response framework.* http://www.fema.gov (accessed March 12, 2013).
50. U.S. Department of Homeland Security, Federal Emergency Management Agency (FEMA). Emergency Management Institute. 2010. *Operations based exercise design and evaluation course.* http://www.fema.gov (Accessed March 17, 2013).
51. U.S. Department of Homeland Security, Federal Emergency Management Agency Higher Education Learning Project. *Principles of emergency management supplement.* Accessed February 6, 2013, from: http://training.fema.gov.
52. U.S. Government Accountability Office. 2000. *Managing for results: Federal managers' views show need for ensuring top leadership skills.* http://www.gao.gov (accessed March 23, 2013).
53. U.S. Office of Performance Management. *Performance management: Accountability can have positive results.* http://www.opm.gov/ (accessed March 23, 2013).
54. Virginia Historical Society. *Duck and cover: Civil defense in Virginia in the 1950s.* http://www.vahistorical.org/research/tacl_civildefense.htm (accessed March 7, 2013).
55. Waugh, W. 2000. *Living with hazards, dealing with disasters: An introduction to emergency management.* Armonk, NY: M.E. Sharpe.
56. Waugh, W. 2007. Local emergency management in the post-9/11 world. In *Emergency management principles and practice for local government,* ed. W. Waugh and K. Tierney. Washington, D.C.: International City/County Management Association.
57. White House Executive Order 10186 (December 1, 1950). *Federal Civil Defense Act of 1950. President Truman establishes Federal Civil Defense Administration.* http://www.historycommons.org (accessed March 7, 2013).
58. Wilson, Jennifer, and Arthur Oyoal-Yemaiel. 2001. *The evolution of emergency management and the advancement towards a profession in the United States.* www.sciencedirect.com (accessed February 24, 2013).

Appendix—A

Principles of Emergency Management Supplement

Table of Contents

Foreword 377
Definition, Vision, Mission, Principles..378
    Definition 378
    Vision 378
    Mission 378
    Principles...378
Principles of Emergency Management 379
    1. Comprehensive..379
    2. Progressive..380
    3. Risk-driven..381
    4. Integrated...382
    5. Collaborative 382
    6. Coordinated 384
    7. Flexible 384
    8. Professional...385

Credit: INTERNATIONAL ASSOCIATION OF EMERGENCY MANAGERS
Principles of Emergency Management September 11, 2007

Principles of Emergency Management September 11, 2007

Foreword

In March of 2007, Dr. Wayne Blanchard of FEMA's Emergency Management Higher Education Project, at the direction of Dr. Cortez Lawrence, Superintendent of FEMA's Emergency Management Institute, convened a working group of emergency management practitioners and academics to consider principles of emergency management. This project was prompted by the realization that while numerous books, articles, and papers referred to "principles of emergency management", nowhere in the vast array of literature on the subject was there an agreed-upon definition of what these principles were.

The group agreed on eight principles that will be used to guide the development of a doctrine of emergency management. This monograph lists these eight principles and provides a brief description of each.

Members of the working group are:

Dr. B. Wayne Blanchard, CEM
Higher Education Project Manager
FEMA Emergency Management Institute

Lucien G. Canton, CEM, CBCP, CPP
Emergency Management Consultant
Director of Emergency Services (retired)
City and County of San Francisco, CA

Carol L. Cwiak, JD
Instructor, Emergency Management Program
North Dakota State University

Kay C. Goss, CEM
President
Foundation of Higher Education Accreditation

Dr. David A McEntire
Associate Professor
Emergency Administration and Planning Program
University of North Texas

Lee Newsome, CEM
Emergency Response Educators and
 Consultants, Inc.

Eric A. Sorchik
Adjunct Professor, School of Administrative Science
Fairleigh-Dickinson University
State Emergency Management Training Officer
New Jersey State Police (retired)

Kim Stenson
Chief, Preparedness and Recovery Division
South Carolina Emergency Management
Representative
National Emergency Managers Association

James E. Turner III
Director
Delaware Emergency Management Agency
Representative
National Emergency Managers Association

Dr. William L Waugh, Jr.
Professor, Public Administration and Urban Studies/Political Science
Georgia State University
Representative
Emergency Management Accreditation

Program
Representative
NFPA 1600 Technical Advisory Committee

Michael D. Selves, CEM, CPM
Emergency Management and Homeland Security Director
Managers
Johnson County, Kansas
President
International Association of Emergency

Dewayne West, CEM, CCFI
Director of Emergency Services (retired)
Johnston County, North Carolina
Past President
International Association of Emergency

EMERGENCY MANAGEMENT

DEFINITION, VISION, MISSION, PRINCIPLES

Definition

Emergency management is the managerial function charged with creating the framework within which communities reduce vulnerability to hazards and cope with disasters.

Vision

Emergency management seeks to promote safer, less vulnerable communities with the capacity to cope with hazards and disasters.

Mission

Emergency management protects communities by coordinating and integrating all activities necessary to build, sustain, and improve the capability to mitigate against, prepare for, respond to, and recover from threatened or actual natural disasters, acts of terrorism, or other man-made disasters.

Principles

Emergency management must be:

1. **Comprehensive**—emergency managers consider and take into account all hazards, all phases, all stakeholders, and all impacts relevant to disasters.
2. **Progressive**—emergency managers anticipate future disasters and take preventive and preparatory measures to build disaster-resistant and disaster-resilient communities.

3. **Risk-driven**—emergency managers use sound risk management principles (hazard identification, risk analysis, and impact analysis) in assigning priorities and resources.
4. **Integrated**—emergency managers ensure unity of effort among all levels of government and all elements of a community.
5. **Collaborative**—emergency managers create and sustain broad and sincere relationships among individuals and organizations to encourage trust, advocate a team atmosphere, build consensus, and facilitate communication.
6. **Coordinated**—emergency managers synchronize the activities of all relevant stakeholders to achieve a common purpose.
7. **Flexible**—emergency managers use creative and innovative approaches in solving disaster challenges.
8. **Professional**—emergency managers value a science and knowledge-based approach based on education, training, experience, ethical practice, public stewardship, and continuous improvement.

PRINCIPLES OF EMERGENCY MANAGEMENT

1. Comprehensive

Emergency managers consider and take into account all hazards, all phases, all impacts, and all stakeholders relevant to disasters.

Comprehensive emergency management can be defined as the preparation for and the carrying out of all emergency functions necessary to mitigate, prepare for, respond to, and recover from emergencies and disasters caused by all hazards, whether natural, technological, or human caused. Comprehensive emergency management consists of four related components: all hazards, all phases, all impacts, and all stakeholders.

All Hazards: All hazards within a jurisdiction must be considered as part of a thorough risk assessment and prioritized on the basis of impact and likelihood of occurrence. Treating all hazards the same in terms of planning resource allocation ultimately leads to failure. There are similarities in how one reacts to all disasters. These event-specific actions form the basis for most emergency plans. However, there are also distinct differences between disaster agents that must be addressed in agent- or hazard-specific plans and these can only be identified through the risk assessment process.

All Phases: The Comprehensive Emergency Management Model[1] on which modern emergency management is based defines four phases of emergency

---

[1] National Governors' Association, *Emergency Preparedness Project: Final Report* (Washington, DC: NGA, 1978), 5.

management: mitigation, preparedness, response, and recovery. *Mitigation* consists of those activities designed to prevent or reduce losses from disaster. It is usually considered the initial phase of emergency management, although it may be a component of other phases. *Preparedness* is focused on the development of plans and capabilities for effective disaster response. *Response* is the immediate reaction to a disaster. It may occur as the disaster is anticipated, as well as soon after it begins. *Recovery* consists of those activities that continue beyond the emergency period to restore critical community functions and manage reconstruction.[2] Detailed planning and execution is required for each phase. Further, phases often overlap as there is often no clearly defined boundary where one phase ends and another begins. Successful emergency management coordinates activities in all four phases.

All Impacts: Emergencies and disasters cut across a broad spectrum in terms of impact on infrastructure, human services, and the economy. Just as all hazards need to be considered in developing plans and protocols, all impacts or predictable consequences relating to those hazards must also be analyzed and addressed.

All Stakeholders: This component is closely related to the emergency management principles of coordination and collaboration. Effective emergency management requires close working relationships among all levels of government, the private sector, and the general public.

2. Progressive

Emergency managers anticipate future disasters and take preventive and preparatory measures to build disaster-resistant and disaster-resilient communities.

Research and data from natural and social scientists indicate that disasters are becoming more frequent, intense, dynamic, and complex. The number of federally declared disasters has risen dramatically over recent decades. Monetary losses are rising at exponential rates because more property is being put at risk. The location of communities and the construction of buildings and infrastructure have not considered potential hazards. Environmental mismanagement and a failure to develop and enforce sound building codes are producing more disasters. There is an increased risk of terrorist attacks using weapons of mass destruction.

Emergency management must give greater attention to prevention and mitigation activities. Traditionally, emergency managers have confined their activities to developing emergency response plans and coordinating the initial response to disasters. Given the escalating risks facing communities, however, emergency managers must become more progressive and strategic in their thinking. The role of the emergency manager can no longer be that of a technician but must evolve to that of a manager

---

[2]William L. Waugh, Jr., *Living with Hazards, Dealing with Disasters: An Introduction to Emergency Management* (Armonk, New York: M.E. Sharpe, 2000).

and senior policy advisor who oversees a community-wide program to address all hazards and all phases of the emergency management cycle. Emergency managers must understand how to assess hazards and reduce vulnerability, seek the support of public officials and support the passage of laws and the enforcement of ordinances that reduce vulnerability. Collaborative efforts between experts and organizations in the public, private, and nonprofit sectors are needed to promote disaster prevention and preparedness. Efforts such as land-use planning, environmental management, building code enforcement, planning, training, and exercises are required and must emphasize vulnerability reduction and capacity building, not just compliance. Emergency management is progressive and not just reactive in orientation.

3. Risk-driven

Emergency managers use sound risk management principles (hazard identification, risk analysis, and impact analysis) in assigning priorities and resources.

Emergency managers are responsible for using available resources effectively and efficiently to manage risk. That means that the setting of policy and programmatic priorities should be based upon measured levels of risk to lives, property, and the environment. NFPA 1600 states that emergency management programs "shall identify hazards, monitor those hazards, the likelihood of their occurrence, and the vulnerability of people, property, the environment, and the entity [program] itself to those hazards."[3] The Emergency Management Accreditation Program (EMAP) Standard echoes this requirement for public sector emergency management programs. Effective risk management is based upon (1) the identification of the natural and man-made hazards that may have significant effect on the community or organization; (2) the analysis of those hazards based on the vulnerability of the community to determine the nature of the risks they pose; and (3) an impact analysis to determine the potential effect they may have on specific communities, organizations, and other entities. Mitigation strategies, emergency operations plans, continuity of operations plans, and pre- and post-disaster recovery plans should be based upon the specific risks identified and resources should be allocated appropriately to address those risks. Communities across the United States have very different risks. It is the responsibility of emergency managers to address the risks specific to their communities. Budgets, human resource management decisions, plans, public education programs, training and exercising, and other efforts necessarily should focus on the hazards that pose the greatest risks first. An all-hazards focus ensures that plans are adaptable to a variety of disaster types and that, by addressing the hazards that pose the greatest risk, the community will be better prepared for lesser risks as well.

---

[3]*NFPA 1600 Standard on Disaster/Emergency Management and Business Continuity Programs,* . Section 5.3, 2007 edition (Quincy, MA: National Fire Protection Association), 6.

4. Integrated

Emergency managers ensure unity of effort among all levels of government and all elements of a community.

In the early 1980s, emergency managers adopted the Integrated Emergency Management System (IEMS), an all-hazards approach to the direction, control, and coordination of disasters regardless of their location, size, and complexity. IEMS integrates *partnerships* that include all stakeholders in the community's decision-making processes. IEMS is intended to create an organizational culture that is critical to achieving unity of effort between government, key community partners, nongovernmental organizations (NGOs) and the private sector. Unity of effort is dependent on both vertical and horizontal integration. This means that at the local level, emergency programs must be integrated with other activities of government. For example, department emergency plans must be synchronized with and support the overall emergency operations plan for the community. In addition, plans at all levels of local government must ultimately be integrated with and support the community's vision and be consistent with its values.

Similarly, private sector continuity plans should take into account the community's emergency operations plan. Businesses are demanding greater interface with government to understand how to react to events that threaten business survival. Additionally, businesses can provide significant resources during disasters and thus may be a critical component of the community's emergency operations plan. In addition, given the high percentage of critical infrastructure owned by the private sector, failure to include businesses in emergency programs could have grave consequences for the community. The local emergency management program must also be synchronized with higher-level plans and programs. This is most noticeable in the dependence of local government on county, state, and federal resources during a disaster. If plans have not been synchronized and integrated, resources may be delayed.

Emergency management must be integrated into daily decisions, not just during times of disasters. While protecting the population is a primary responsibility of government, it cannot be accomplished without building partnerships among disciplines and across all sectors, including the private sector and the media.

5. Collaborative

Emergency managers create and sustain broad and sincere relationships among individuals and organizations to encourage trust, advocate a team atmosphere, build consensus, and facilitate communication.

There is a difference between the terms "collaboration" and "coordination" and current usage often makes it difficult to distinguish between these words. Coordination refers to a process designed to ensure that functions, roles, and responsibilities are identified and tasks accomplished; collaboration must be viewed as an attitude or an organizational culture that characterizes the degree of unity and cooperation that exists within a community. In essence, collaboration creates the environment in which coordination can function effectively. In disaster situations, the one factor that is consistently credited with improving the performance of a community is the degree to which there is an open and cooperative relationship among those individuals and agencies involved. Shortly after Hurricane Katrina, *Governing* magazine correspondent, Jonathan Walters wrote: "Most important to the strength of the inter-governmental chain are solid relationships among those who might be called upon to work together in times of high stress. 'You don't want to meet someone for the first time while you're standing around in the rubble,' says Jarrod Bernstein, a spokesman for the New York Office of Emergency Management."[4] It is this kind of culture and relationship that collaboration is intended to establish. A commitment to collaboration makes other essential roles and functions possible. Comfort and Cahill acknowledge the essential nature of collaboration within the emergency management function: "In environments of high uncertainty, this quality of inter-personal trust is essential for collective action. Building that trust in a multi-organizational operating environment is a complex process, perhaps the most difficult task involved in creating an integrated emergency management system."[5] Thomas Drabeck[6] suggests that collaboration involves three elements:

1. We must commit to ensuring that we have done everything possible to identify all potential players in a disaster event and work to involve them in every aspect of planning and preparedness for a disaster event.
2. Having achieved this broad involvement, we must constantly work to maintain and sustain the real, human, contact necessary to make the system work in a disaster event.
3. Finally, our involvement of all of our "partners" must be based on a sincere desire to listen to and incorporate their concerns and ideas into our planning and preparedness efforts. This element is probably the most critical because it is this sincere interest that engenders trust, cooperation, and understanding and allows us to truly have a "team" approach to protecting our communities in times of disaster.

This principle can perhaps best be encapsulated by remembering: "If we shake hands before a disaster, we won't have to point fingers afterwards."[7]

---

[4]Jonathan Walters, GOVEXEC.com, December 1, 2005.
[5]Louise K. Comfort and Anthony G. Cahill, *Managing Disaster, Strategies and Policy Perspectives* (Durham, NC: Duke University Press, 1988).
[6]Thomas E. Drabek, *Strategies for Coordinating Disaster Responses* (Boulder, CO: Program on Environment and Behavior, Monograph 61, University of Colorado, 2003).
[7]Michael D. Selves, *Oral Testimony Before the United States House Subcommittee on Emergency Management of the Committee on Transportation and Infrastructure,* April 26, 2007.

6. Coordinated

Emergency managers synchronize the activities of all relevant stakeholders to achieve a common purpose.

Emergency managers are seldom in a position to direct the activities of the many agencies and organizations involved in the emergency management program. In most cases, the people in charge of these organizations are senior to the emergency manager, have direct line authority from the senior official, or are autonomous. Each stakeholder brings to the planning process their own authorities, legal mandates, culture, and operating missions. The principle of coordination requires that the emergency manager gain agreement among these disparate agencies as to a common purpose and then ensure that their independent activities help to achieve this common purpose.

In essence, the principle of coordination requires that the emergency manager think strategically, that he or she see the "big picture" and how each stakeholder fits into that mosaic. This type of thinking is the basis for the strategic program plan required under the National Preparedness Standard (NFPA 1600) and the Emergency Management Accreditation Program. In developing the strategic plan, the emergency manager facilitates the identification of agreed-upon goals and then persuades stakeholders to accept responsibility for specific performance objectives. The strategic plan then becomes a mechanism for assessing program progress and accomplishments.

This same process can be used on a smaller scale to develop a specific plan, such as a community recovery plan; it is also an inherent component of tactical and operational response. The principle of coordination is applicable to all four phases of the Comprehensive Emergency Management cycle and is essential for successful planning and operational activities related to the emergency management program. Application of the principle of coordination provides the emergency manager with the management tools that produce the results necessary to achieve a common purpose.

7. Flexible

Emergency managers use creative and innovative approaches in solving disaster challenges.

Due to their diverse and varied responsibilities, emergency managers constitute one of the most flexible organizational elements of government. Laws, policies, and operating procedures that allow little flexibility in the performance of duties drive more traditional branches of government. Emergency managers are instead encouraged to develop creative solutions to solve problems and achieve goals.

A principal role of the emergency manager is the assessment of vulnerability and risk and the development of corresponding strategies that could be used to reduce or eliminate risk. However, there can be more than one potential mitigation strategy for any given risk. The emergency manager must have the flexibility to choose not only the most efficient course of action but the one that would have the most chance of being implemented.

In the preparedness phase, the emergency manager uses many resources to create and maintain a well-organized community response structure. One such resource is the development of a risk-based community emergency operations plan. While most policies and procedures in government are specific and designed to offer little room for interpretation, the emergency operations plan is designed to be flexible and applicable to all community emergency operations. It is based on the consequences of the event, not the promulgating action.

The most dramatic phase of emergency management is response. In this phase the emergency manager coordinates activities to ensure overall objectives are being met. The emergency manager must be flexible enough to suggest variations in tactics or procedures and adapt quickly to a rapidly changing and frequently unclear situation. The emphasis is on creative problem solving based on the event and not on rigid adherence to preexisting plans.

As part of the community team that will determine recovery priorities the emergency manager must be capable of dealing with the political, economic, and social pressures in making these decisions. It is natural to focus on short-term efforts in disaster recovery. However, the emergency manager cannot lose sight of the long-term needs of the community and it is this aspect of recovery that often must be driven by the emergency manager.

Flexibility is a key trait of emergency management and success in the emergency management field is dependent upon it. Being able to provide alternate solutions to stakeholders and then having the flexibility to implement these solutions is a formula for success in emergency management.

8. Professional

Emergency managers value a science and knowledge-based approach based on education, training, experience, ethical practice, public stewardship, and continuous improvement.

Professionalism in the context of the principles of emergency management pertains not to the personal attributes of the emergency manager but to a commitment to emergency management as a profession. A profession, as opposed to a discipline or a vocation, has certain characteristics, among which are:

Code of ethics—while no single code of ethics has yet been agreed upon for the profession, the Code of Ethics of the International Association of Emergency Managers, with its emphasis on respect, commitment, and professionalism, is generally accepted as the standard for emergency managers.

Professional associations—emergency managers seeking to advance the profession of emergency management are members of professional organizations such as the National Emergency Manager's Association (NEMA) and the International Association of Emergency Managers (IAEM). They also participate in appropriate state, local, and professional associations.

Board certification—emergency managers seek to earn professional certification through such programs as the Certified Emergency Manager program of IAEM. Professional certification demonstrates the achievement of a minimum level of expertise and encourages continued professional development through periodic recertification.

Specialized body of knowledge—the knowledge base for emergency managers consists of three principal areas. The first is the study of historical disasters, particularly as it pertains to the community for which the emergency manager is responsible. Secondly, the emergency manager must have a working familiarity with social science literature pertaining to disaster issues. Third, the emergency manager must be well versed in emergency management practices, standards, and guidelines.

Standards and best practices—the principal standards used in emergency management are NFPA 1600 and the Emergency Management Accreditation Program (EMAP) Standard. These two standards provide the overarching context for the use of other standards and best practices.

Appendix—B

IAEM Code of Ethics and Professional Conduct

(IAEM, http://www.iaem.com/page.cfm?p=about/code-of-ethics)

Principles

The members of the Association agree to conduct themselves in accordance with the basic principles of **RESPECT**, **COMMITMENT**, and **PROFESSIONALISM**.

Respect

Respect for supervising officials, colleagues, associates, and most importantly, for the people we serve is the standard for IAEM members. We comply with all laws and regulations applicable to our purpose and position, and responsibly and impartially apply them to all concerned. We respect fiscal resources by evaluating organizational decisions to provide the best service or product at a minimal cost without sacrificing quality.

Commitment

IAEM members commit themselves to promoting decisions that engender trust and those we serve. We commit to continuous improvement by fairly administering the affairs of our positions, by fostering honest and trustworthy relationships, and by striving for impeccable accuracy and clarity in what we say or write. We commit to enhancing stewardship of resources and the caliber of service we deliver while striving to improve the quality of life in the community we serve.

Professionalism

IAEM is an organization that actively promotes professionalism to ensure public confidence in Emergency Management. Our reputations are built on the faithful discharge of our duties. Our professionalism is founded on Education, Safety, and Protection of Life and Property.

Code

1. **Quality**: Members shall aim to maintain high quality work at all times and apply the Principles of Emergency Management in their professional undertakings. Quality may be assessed by audits, monitoring, quality processes, or other appropriate means.

2. **Professional Independence**: IAEM Members, however employed, owe a primary loyalty to the people in the community they serve and the environment they affect. Their practice should be performed according to high standards and ethical principles, maintaining respect for human dignity. Emergency management practitioners shall seek to ensure professional independence in the execution of their functions. The term professional independence relates to the function of the practitioners within the organization in which they practice. Their role may be advisory or executive.
3. **Legal Requirements**: Members must abide by the legal requirements relating to their practice, and practitioners have a duty to make themselves aware of the appropriate legal requirements for the territory in which they practice.
4. **Objectivity**: Members called to give an opinion in their professional capacity shall be honest and, to the best of their ability, objective, and reliable. Objectivity and reliability is based on the best current available knowledge, or in the absence of such knowledge, reference to appropriate emergency planning and management principles.
5. **Competence**: Members shall not undertake responsibilities as emergency management practitioners if they do not believe themselves competent to discharge them. Members shall acknowledge any limitations in their own competence. In pursuit of this members shall take all reasonable steps to obtain, maintain, and develop their professional competence by attention to new developments and shall encourage others working under their supervision to do so. Competence is defined as "the possession of sufficient knowledge, experience and skill to enable a person to know what he or she is doing and to be able to carry out a task in the way in which a person competent in the activity would expect it to be done and to have an appreciation of one's own limitations." Competence is maintained by undertaking continuing professional development and certification (CEM®) and may be supplemented at appropriate levels by membership of other specialist bodies.
6. **Abuse of Membership**: Members shall not improperly use their membership of IAEM for commercial or personal gain.
7. **Conflict of Interest**: Members shall avoid their professional judgment being influenced by any conflict of interest and shall inform their employer, or client, of any conflict between their own personal interest and service to the relevant party. For example, a consultant may be aware that his/her recommendations are not being implemented, but continues to advise the organization in order to avoid losing the revenue associated with the contract. The members, officers, and agents of the Association shall act in the best interest of the Association at all times and shall avoid activities resulting in actual or implied personal gain in keeping with the highest standards of ethics and professionalism.
8. **Confidentiality**: Members shall not improperly disclose any information which may reasonably be considered to be prejudicial to the business of any present or past employer or client.

9. **Professional Responsibility**: Members shall accept professional responsibility for all their work and shall take all reasonable steps to ensure that persons working under their authority or supervision are competent to carry out the tasks assigned to them; are treated with fairness and equal opportunity; and accept responsibility for the work done on the authority delegated by them. Where members have good reason to believe that their professional advice is not being followed, they shall take all reasonable steps to ensure that persons overruling or neglecting their advice are made aware of the potential adverse consequences which may result. In such instances it is advisable that such actions are recorded in writing.
10. **Upholding the Aims and Objectives**: Members shall have regard to the reputation and good standing of The Society, other members' professional practice and standards, and shall not knowingly bring them into disrepute. Disrepute amounts to the loss of a previously good reputation. It may arise from the conduct of a member who by act or omission lowers the professional reputation of The Association and its members in the view of right thinking members of The Association generally. At all times members shall seek to uphold the bylaws and APPs of the IAEM.
11. **Professional Reputation**: Members shall not in the course of their practice recklessly or maliciously injure, or attempt to injure, whether directly or indirectly, the professional reputation, prospects, or business of another.
12. **Members Relations Inter Se**: Members shall at all times treat other members of the Association with the utmost respect and fairness, and at no time undermine their integrity and dignity. Members will at all times seek to work in a cooperative and productive way with each other. Elected representatives have a particular responsibility in this regard; to ensure that all members and their views are heard, valued, and respected.
13. **Financial Propriety**: Members shall maintain financial propriety in all their professional dealings with employers and clients. Any inducements which may be seen as prejudicial to professional independence or in breach of contractual or moral obligations should be discouraged.

From IAEM-Global Administrative Policies & Procedures
Approved Jan. 31, 2011
(IAEM, http://www.iaem.com/page.cfm?p=about/code-of-ethics)

INTERNATIONAL EMERGENCY MANAGEMENT

"The nuclear meltdown at Chernobyl 20 years ago this month, even more than my launch of perestroika, was perhaps the real cause of the collapse of the Soviet Union five years later."

—Mikhail Gorbachev

Courtesy of the Federal Aviation Administration

Figure 13.1 Destruction of the Chernobyl Nuclear Power Plant. This 1986 disaster occurred because of "a flawed reactor design that was operated with inadequately trained personnel."[1] The incident "released at least 5% of the radioactive reactor core into the atmosphere and downwind."[2] As a result, "two Chernobyl plant workers died on the night of the accident, and a further 28 people died within a few weeks as a result of acute radiation poisoning."[3] Following the incident, the city was abandoned. The city of Chernobyl remains unoccupied, uninhabitable, and unlivable to this day.

---

[1] World Nuclear Association, "Chernobyl Accident 1986," 2013, http://www.world-nuclear.org/info/Safety-and-Security/Safety-of-Plants/Chernobyl-Accident/#.Ub6S6thTpyE (accessed June 16, 2013).
[2] Ibid.
[3] Ibid.

OBJECTIVES

The objectives of this chapter are to:
- Identify some of the major emergency management events that have occurred on the international scene,
- Discuss some of the major challenges facing international emergency management,
- Discuss the role of NGOs in international disaster relief operations,
- Discuss common attributes shared among international emergencies, and
- Discuss the contributions of the United States to international emergencies.

Introduction

International emergency management is as diverse as nations themselves. Typically, the more affluent nations, depending on the threats they feel they face, have proven more capable to respond to major events. Natural disasters, accidents, and man-made events that significantly impact a population occur around the globe. Some of these events, especially those which are more severe, often overwhelm local and even regional government response efforts. As one would expect, impoverished nations typically have a less developed disaster response capability than nations which are more affluent.

When the topic of international disaster management is broached, what typically comes to mind are the frail emergency management systems found in the developing world, rife with insufficient funding, poor training, corruption, and other obstacles. However, there are a great number of highly successful emergency management systems found in the many industrialized nations of the world, and a handful in the developing world. [4]

Emergency management experts typically draw a distinction between humanitarian aid (man-made) and disaster relief (natural) operations. The difference usually lies in the degree of preparedness and response time involved. Humanitarian crises (such as the violence in Kosovo, the former Yugoslavia, East Timor, and Rwanda) rarely happen at a moment's notice or overnight, and are usually monitored by the aid agencies in an attempt to give themselves time to prepare and alert the remainder of the international community when a catastrophe is about to happen. Natural disasters (such as in Mozambique, Ethiopia, Bangladesh, and Turkey), while slowly becoming more predictable, can still strike with little warning

---

[4]Damon Coppola, "The Importance of International Disaster Management in the Field of Emergency Management," http://training.fema.gov (retrieved on May 1, 2013).

and rely more on the training, education, and preparedness of those in the actual disaster zone to hang on until the relevant concerned organizations and agencies can mobilize their resources. While this inevitably takes time, military forces are seen as a pool of prepared, disciplined, and available source of assistance while the international aid community gears itself for action.

International Emergency Management: An Overview

Disasters occur globally. The last several decades have witnessed in Haiti, China, and New Zealand, earthquakes and tsunamis in Japan, the Bhopal chemical accident in India, or the Chernobyl nuclear accident in Russia. Every country has its own hazard profile and faces challenges to its population and infrastructure. As in the United States, all disasters are local in nature; that is, disasters have a primary point of impact, an area most affected, and a population that is most in need. In the international setting, the threats facing nations and regions are as diverse as the nations themselves. More impoverished nations often face starvation and disease, especially in times of civil unrest or major disaster. In these more impoverished nations, resources available for disaster relief support operations are typically lacking.

The international community has a stake in emergency management. Specifically, these concerns involve vulnerability fluctuation and evolution or demise of emergency management systems, as well as unique cultural, economic, and political characteristics.[5] The countries most affected by natural and man-made disasters are usually developing or third world countries. According to the World Health Organization, approximately 85% of all natural disasters occur in Asia. In 2009, Asia accounted for six out the 10 disasters in the world. These countries frequently require international assistance because they lack the necessary resources and training to provide an adequate emergency response. Once a request is made for emergency assistance by the victimized country the international community usually mobilizes their response teams according to the immediate need.

Historical Scope of the Challenge

Disasters have impacted mankind since before recorded history. Some of these events were so severe that they became milestones in world history. In August 79 A.D., the Italian volcano Mount Vesuvius erupted, blanketing the towns and of residents of Pompeii, Stabiae, and Herculaneum. This incident was neither the first nor would it be the last disaster to impact a population, and it is an example of a devastating natural disaster, even 2,000 years ago.

---

[5]Damon Coppola, *Introduction to International Disaster Management* (Burlington, MA: Butterworth-Heinemann, 2007), ISBN 13:978-0-7506-7982-4.

In July 1201, an earthquake struck the Middle East in what is now Egypt and Syria. The earthquake, the deadliest in recorded history, is considered one of the 10 worst natural disasters of all times. This disaster rocked the eastern Mediterranean and killed over 1.1 million people, destroying countless homes. Nearly every major city within the near east felt the effects of this quake.

Such incidents affect the socioeconomic, political, and environmental conditions in which people live. During 1769, in India, a great famine took over ten million people's lives. This was nearly one third of the population of India at the time. It was caused by a shortfall in crops followed by a severe drought. As populations were devastated by the deaths, many areas returned to jungle, further decreasing food supplies. This famine lasted until 1773.

Global and International Scopes of the Challenge

Internationally, the challenge is great. The world is so diverse and the threats that face the world's population so wide ranging that the attempt to describe scope of the challenge must be done in the most general of terms. Natural disasters, war, famine, and accidents place populations at risk. Internationally and globally, disasters share some common attributes. Some of the most important aspects of global and international disasters include:

Emergency relief—Emergency relief consists of disaster management, rapid needs assessments, beneficiary targeting, distribution planning, and management.

Shelter—Sheltering consists of needs assessment and program design implementation and coordination in emergency and nonemergency settings; and planning for and provision of tarps, tents, iron sheeting, and simple construction.

Command and control—Much of the command and control exercised among disaster operations is militaristic.[6] This paradigm emphasizes the centralizing of authority, adhering to "procedure and protocol," and adhering to "specification of plans."[7]

Coordination and logistics—Coordinating resources involves the "set of methods used to manage interdependence between organizations" and establish cooperation among the responding entities and affected nations.[8] Logistics

---

[6]Malcolm Cook, Jan Noyes, and Yvonne Masakowski, *Decision Making in Complex Environments* (Burlington, VT: Ashgate Publishing, 2007).
[7]Ibid.
[8]Sabine Schultz, *Disaster Relief Logistics: Benefits of and Impediments to Cooperation Between Humanitarian Organizations* (Bern, Germany: Haupt-Berne Publishing, 2009), 64.

considers the methods through which personnel and necessary resources are obtained and moved into the affected disaster area.

Corruption—Corruption involves considerations of fraudulent uses of materials, funds, and so forth. Oversight, checks, and balances must exist to ensure that corruption does not impede the relief, response, and recovery initiatives.

Project management—Numerous aspects of project management are relevant for disaster and emergency management. For instance, resources must be obtained and scheduled for implementation in due time; metrics must be collected and analyzed to determine levels of efficiency and effectiveness; planning must occur regarding every aspect of the response and recovery; emergency initiatives must be financed; and so forth.

Natural Disasters

Natural disasters come in many forms. In 1556, the Shaanxi earthquake in China resulted in an estimated loss of over 800,000 people, over 50% of the population. In 1883, the eruption of Krakatoa in what was at the time the Dutch East Indies, resulted in the death of between 36,000 and 120,000.[9] During modern times, natural disasters are just as catastrophic. For instance, during 2004, the Asian coasts experienced a tsunami that resulted in the deaths of approximately 225,000 to 230,210 people.[10]

Accidents

Major industrial accidents are not uncommon. This chapter opened with reference to the Chernobyl incident in the Ukraine. Chernobyl's immediate consequences were severe—the accident destroyed the plant, killed 31 workers and firefighters, seriously injured nearly 300 workers, and released a great amount of radioactive material that severely contaminated a large area around the plant.[11] This incident represents one of the most severe disasters of the twentieth century because of the extensiveness of radiation. The city remains uninhabited to this day. The following figure shows a contemporary view of the ghastly city.

---

[9]Armand Vervaeck, "Krakatau, Indonesia—Volcanic Earthquakes Are Creating Unrest in Coastal Villages," October 6, 2011, *Earthquake Report*, http://earthquake-report.com/2011/10/06/krakatau-indonesia-volcanic-earthquakes-are-creating-ner/ (retrieved on May 5, 2013).

[10]"Indian Ocean Earthquake and Tsunami," 2013, Livescience, http://www.livescience.com/33316-top-10-deadliest-natural-disasters.html (accessed June 17, 2013).

[11]"Post-Chernobyl Global Co-operation: 5 Years Later," *IAEA News briefs*, 6, no. 2 (49), March/April 1991 International Atomic Energy Agency, http://www.iaea.org/newscenter/features/chernobyl-15/cooperation.shtml (retrieved on May 1, 2013).

Figure 13.2 Modern View of Chernobyl. This city remains uninhabitable in modern times following the 1986 nuclear accident. Natural flora and fauna are reclaiming the city.

Pandemics

Pandemics remain a great concern for many, including health-care and emergency management professionals. Medical advances in the last century have limited the potential for a widespread pandemic, but the potential for biological and pathological disasters is continuous. Pandemics occur during conditions involving the emergence of "a novel strain of a virus . . . that causes readily transmissible human illness for which most of the population lacks immunity."[12]

One of the most famous pandemics to strike was the bubonic plague or "Black Death" killed almost 33 percent of the entire population of Europe when it struck between 1347 and 1350. It also affected millions in Asia and North Africa. Scientists believe that the plague was a zoonotic disease caused by *Yersinia pestis* bacterium and spread because of poor hygiene and fleas carried by rats.

At the end of World War I, the world was plagued by what became known as the 1918 pandemic, which lasted until the end of 1920. This was an unusually deadly influenza pandemic, and was the first of the two pandemics involving H1N1 influenza virus, the second occurring in 2009. The 1918 flu pandemic infected an estimated 500 million people and ultimately was responsible for an estimated 50–100 million deaths, making it one of the deadliest natural disasters in human history.

---

[12]Federal Communications Commission, "Pandemics Information," 2013, http://www.fcc.gov/encyclopedia/pandemics-information (accessed June 17, 2013).

CHAPTER 13 International Emergency Management   397

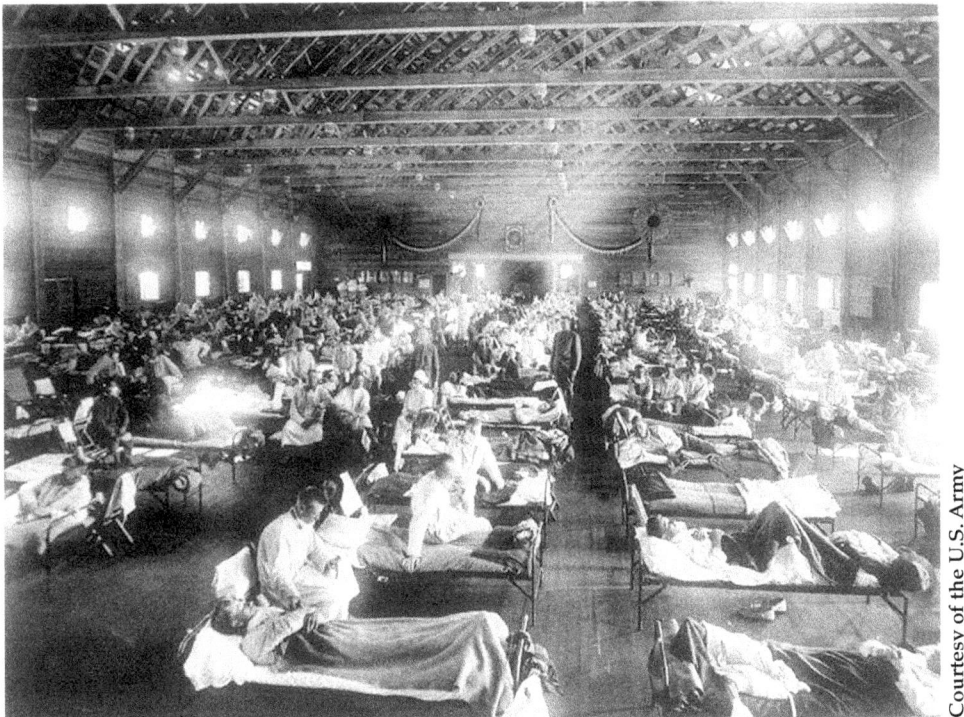

Figure 13.3 Soldiers from Fort Riley, Kansas, ill with Spanish influenza at a hospital ward at Camp Funston were just a few of the millions affected by the 1918 Influenza Pandemic.

While modern medicines have somewhat ameliorated the devastating effects of many pandemics, they still have the potential of being quite catastrophic. Within the United States, it is estimated that approximately "40 percent of the nation's workforce—including personnel supporting our critical communications infrastructure—will be absent during the height of a pandemic."[13] During modern times, pandemics have the capacity to inflict mass fatalities. For instance, the death toll associated with the 2009 H1N1 swine flu was revised recently thereby indicating that approximately "284,400 people actually died in the first year the virus was circulating around the world."[14]

---

[13]Ibid.

[14]Krisha Desai, "Global H1N1 Death Toll may be 15 Times Higher than Previously Reported," *CNN News*, 2012, http://thechart.blogs.cnn.com/2012/06/25/global-h1n1-death-toll-may-be-15-times-higher-than-previously-reported/ (accessed June 17, 2013).

Conflicts and Violence

The twentieth century represented a most violent phase in history. It witnessed two world wars, the Korean War, the Vietnam war, and a myriad of other conflicts and struggles. Conflicts have continued throughout the world such as Kosovo (1999), Liberia (until 2003), Afghanistan (2002), pre-war Iraq (2003), Haiti (2004), and Darfur (since 2003).[15] As a consequence of conflicts, the international community has attempted to provide humanitarian aid to the refugees who have been displaced.

International Assistance

The majority of the developing world's residents lack the physical, economic, and social protections from disaster that wealthier countries enjoy. Earthquake- and hurricane-resistant building codes and land-use restrictions do not always exist or are not well enforced. Property, business, and crop insurance markets are less sophisticated or out of reach to many.

International disaster response occurs when foreign governments, agencies, and organizations assist a society whose government and civil institutions cannot adequately address the humanitarian needs of its disaster-affected populations. Assistance includes immediate and longer-term efforts designed to save lives, alleviate suffering, maintain human dignity, and help people prevent, mitigate, prepare for, and respond to future crises.[16]

International Support Involving Other Nations

International support from other nations may be in the form of a wide range of assistance. Medical aid and personnel, search and rescue, and food supplies are common aspects of international emergencies. One of the hallmarks of international emergency response is *Urban Search and Rescue (SAR)*. According to McEntire,[17] search and rescue is defined as "response activities undertaken to find disaster victims and remove them from danger or confinement." Moreover, Mullins claims the term *"urban search and rescue"* was first used in the United States by the Metro-Dade Fire and Rescue for deployment domestically and internationally for natural or man-made disasters where people had been trapped in collapsed buildings.[18]

---

[15]R. Margesson, "International Crisis and Disasters: U.S. Humanitarian Assistance, Budget Trends, and Issues for Congress," *Congressional Research Service,* order code RL33769 (2006), CRS-3.

[16]International Disaster Response, Social Issue Report, http://rootcause.org/documents/DR-Issue.pdf (retrieved on May 15, 2013).

[17]D. McEntire, *Introduction to Homeland Security Understanding Terrorism with an Emergency Management Perspective* (New York, NY: Wiley, 2009), 233.

[18]G. Mullins, "Urban Search and Rescue-developing Australia's Capability," *The Australian Journal of Emergency Management*, 19, no. 1 (2004).

In the United States, the Federal Emergency Management Agency (FEMA) is the governing body for all national SAR teams operating in the country. The SAR national concept originated in 1991, whereby urban search and rescue became a component of the federal response operation under the Federal Response Plan.[19]

Support Function #9 of the Federal Response Plan of 1999 outlines the organizational structure of an urban search and rescue team as follows:

- More than 5,000 individuals make-up the national system.
- There are 28 task forces sponsored by state and local emergency response organizations.
- Each task force is comprised of 62 individuals organized into 31 positions.
- The positions are organized into five functional teams (management, search, rescue, medical, and technical).

Mullins further explains that SAR teams are divided into three categories: (1) first responders such as fire, police, and emergency medical services (EMS); (2) rescue technicians who are trained in the use of advanced rescue and search equipment, tunneling, shoring, etc.; and (3) task force management, command, and control. SAR is an extremely complex operation that demands a high degree of expertise to help mitigate a disastrous incident.

Following a disaster, SAR operations are primarily engaged in the location and extraction of persons from collapsed buildings and structures to include the medical stabilization of victims in these locations. Moreover, SAR teams are trained to assess to control gas, electrical services, and hazardous materials, and evaluate and stabilize damaged structures.[20]

Structural collapse is the most often cause of victims being trapped. This type of rescue operation requires a high degree of expertise and special equipment to carry out the extraction/rescue mission. The U.S. emergency management system is the most advanced and well-funded in the world; U.S. SAR teams are totally self-sufficient for the first 72 hours of deployment. The FEMA lists the composition of a typical U.S. SAR Team as follows:

1. Each task force consists of two 31-person teams, 4 canines, and a comprehensive equipment cache. For every US&R task force there are 62 positions. To ensure that a full team can respond to an emergency, the task forces have at the ready more than 130 highly trained members.

---

[19]J. Bullock, G. Haddow, D. Coppola, E. Ergin, L. Westerman, and S. Yeletaysi, *Introduction to Homeland Security*, 2nd ed. (Burlington, MA: Butterworth-Heinemann Publishers, 2006).

[20]F. Robertson, "Urban Search and Rescue Teams: What Size or Type Is Needed?" 2007, http://www.usfa.fema.gov/pdf/efop/efo4mo.pdf.

2. A task force is really a partnership between local fire departments, law enforcement agencies, federal and local government agencies, and private companies.
3. A task force is totally self-sufficient for the first 72 hours of deployment.
4. The equipment cache used to support a task force weighs nearly 60,000 pounds and is worth approximately $1.4 million. Add the task force members to the cache, and you can completely fill a military-141 transport or two C-130 cargo aircraft.
5. US&R task force members work in four areas of specialization: search, to find victims trapped after a disaster; rescue, which includes safely digging victims out of tons of collapsed concrete and metal; technical, made up of structural specialists who make rescues safe for the researchers who make rescues safe for the rescuers; and medical, which cares for the victims before and after a rescue.

Currently, in the United States there are 28 Urban Search and Rescue Teams. The two most preeminent are generally thought to be the Fairfax, Virginia Fire Department, and the Los Angeles County Fire Department. These two SAR teams have represented the U.S. in numerous international SAR operations throughout the world. The Fairfax County International Urban Search and Rescue resource is commonly referred to as Virginia Task Force 1 (VA-TF1). This SAR team has been recognized throughout the United States and the world as a premier leader in the provision of training in catastrophic event mitigation, readiness, and response and recovery techniques. When activated, the task force is comprised of 70 members including firefighters and paramedics from the Fairfax County FD (both career and volunteer) and highly trained civilians including physicians, canine handlers, structural engineers, communications experts, and heavy rigging specialists. VATF1 was deployed during the aftermath of the incident.

This incident was not the sole deploying of VATF1. It also served during the 2010 Haiti disaster. These deployments of VATF1 represent the robustness of American urban SAR resources. Such deployments embellish the activities of mitigation, response, and recovery. They cooperate with USAID and other international organizations to support operations associated with a variety of calamities ranging from earthquakes to tsunamis.

California Task Force 2 (CA-TF2) is a specialty-trained and equipped 70 person Urban Search and Rescue Team comprised mostly from the Los Angeles County Fire Department. The organizational structure is very similar to that of the Virginia Task Force 1. The mission of CA-TF2 is to conduct around-the-clock search and rescue operations at domestic and international disasters, both natural and manmade. One of the most notable international SAR responses was the catastrophic earthquake that occurred in Japan in 2011. The CA-TF2 was deployed to Ofunato and Kamaishi Japan for a period of five days.

California Task Force 2 is teamed up with the Virginia Task Force 1 in providing the U.S. Agency for International Development's Office of Foreign Disaster

Assistance with humanitarian assistance in the form of search and rescue operations. Furthermore, both SAR teams are members of the International Search and Rescue Advisory Group (ISARAG) which is made up of Urban Search and Rescue Teams from around the world that work together to coordinate international USAR organizations, management, and standards.

The U.S. government incurs all costs and expenses associated with an international urban search and rescue. For example, the CA-TF2 deployment to Japan cost approximately $1 million. The USAR deployments to Haiti for 16 days and a 12-day mission in New Zealand cost $4.8 million and $2.6 million, respectively.

International Support from Nongovernmental Organizations

As part of the international response, the international community including nongovernment organizations (NGOs) set up refugee camps for those displaced individuals who are attempting to flee violence. This emergency humanitarian event requires aid such as food, shelter, medical supplies, water, and sanitation facilities and security for the refugee camps. When many inexperienced NGOs arrive on disaster scenes, they can complicate relief and recovery efforts by acting unilaterally or by producing poor-quality work.

International Military Assistance

In the post-cold war world, and to an unprecedented extent, forces operating under a United Nations (UN) mandate have become involved in a wide range of humanitarian tasks. These tasks have taken the following forms: Protecting humanitarian relief workers, such as those representing international agencies and NGOs, from attacks by belligerents and generally from the danger of war; directly engaging in humanitarian action, to include delivering relief supplies and reestablishing and maintaining essential services; and establishing designated safety zones where protection is intended for those displaced.[21]

The participation of East Asian militaries (Japan, China, Taiwan, and South Korea) in international disaster relief is a relatively new phenomenon having many implications that will shape the global landscape of the twenty-first century. International disaster relief operations present a challenge to a nation's military. With each disaster are numerous challenges and needs that, in many ways, are specific to the region impacted and the nature of the event that caused the need for response. Military organizations can bring a multitude of assets quickly to an impacted area. Typically self-sustaining, the military adds value while limiting its impact onto what may already be an overly strained region.

---

[21]Adam Roberts, "Use of UN Peacekeeping Forces for Humanitarian Purposes," *Refugee Participation Network*, no. 23 (January–April 1997), http://www.fmreview.org/RPN/23.pdf (retrieved on May 1, 2013).

With regard to the United States, the U.S. State Department is authorized to work with the U.S. Department of Defense (DOD) to identify and direct the use of military assets if deployed for humanitarian disaster response (DR) missions overseas. This usually involves an array of special military assets and expertise in the communications, engineering, water production and purification, and medical support arenas, among other contributions.

The U.S. Military derives its statutory authority to respond to an international emergency for humanitarian assistance under Title 10 United States Code. Title 10 outlines the mission and the responsibilities of usually short-range programs aimed at ending or alleviating human suffering. Humanitarian assistance DOD operations encompass short-range programs aimed at ending or alleviating human suffering and are designed to supplement or complement the efforts of civil authorities that have primary responsibility to provide the relief. Once the DOD reviews the request for humanitarian assistance from the U.S. State Department, the Secretary of Defense may authorize the emergency operation by signing what is called a *"third party waiver"* that authorizes U.S. military resources to be used in a nonmilitary operation to assist a third party.

The Office of Military Affairs (OMA) was established in October 2005, as an operational link to improve USAID's coordination of humanitarian assistance with the U.S. military. In addition, the OMA is responsible for overseeing humanitarian assistance training exercises between the U.S. military and NGOs. According to Margesson (2006), the OMA serves as the point of contact between the U.S. military and NGOs who are responsible for providing emergency assistance following a catastrophic disaster. The rationale for this is that it allows both the U.S. military and NGOs to benefit from the other operational experience with regard to administration and delivery of humanitarian assistance.[22]

On December 26, 2004, an earthquake erupted beneath the Indian Ocean that registered 9.0 on the Richter scale. The epicenter of the earthquake was located near the west coast of the Indonesian island of Sumatra. The earthquake-induced tsunami that resulted from the earthquake caused at least 155,000 fatalities, 500,000 injuries, and damages that exceeded $10 billion. Moreover, it is widely believed that approximately five million people were displaced. Following this catastrophic disaster, the U.S. military responded dispatching ships, planes, and relief supplies to the region:

> A total of 15,000 U.S. soldiers and sailors were deployed as part of the 2004 tsunami response to work alongside OFDA in the affected region. More specifically, the U.S. military provided 26 ships, 82 aircraft, and 51 helicopters to help deliver more than 24.5 tons of relief supplies and enable USAID and other disaster relief agencies to move much-needed aid to inaccessible areas affected by the tsunami.[23]

---

[22]Margesson, "International Crisis and Disasters."

[23]C. Perry and M. Travayiakis, "The U.S. Foreign Disaster Response Process: How It Works and How It Could Work Better," (Cambridge, MA: The Institute for Foreign Policy Analysis, Inc., 2008).

CHAPTER 13 International Emergency Management 403

Figure 13.4 U.S. Air Force Chief Master Sgt. Matt Wickham uses a shovel to extract sidewalk bricks at the fishing port in Misawa, Japan, March 19, 2011. U.S. Service members, civilian employees, and family members from Misawa Air Base joined with Misawa residents to clean up the port following an 8.9-magnitude earthquake and tsunami along Japan's eastern coast. *U.S. Navy photo by Petty Officer 1st Class Matthew Bradle*

The contributions of American forces were not unnoticed by Japan. Gratitude was expressed by Japanese Defense Minister Toshimi Kitazawa thusly: "To all U.S. military members, on behalf of the people of Japan, I sincerely express my deep appreciation for the tremendous support provided by the U.S. military, the U.S. government, and the American people at a time of unprecedented crisis in Japan."[24]

American military involvement among international disasters is a tremendous resource from the humanitarian perspective. During the aftermath of the 2010 Haiti disaster, regarding U.S. Navy aircraft carriers, the following remarks were made to quash complaints regarding the presence of U.S. naval forces in the disaster area:[25]

> . . . have three hospitals on board that can treat several hundred people; they are nuclear powered and can supply emergency electrical power

---
[24]Ibid.

[25]Alicia Colon, "Haitian Tragedy Should Transcend Politics And Naïveté," 2010, http://www.irishex-aminerusa.com/mt/2010/01/19/haitian_tragedy_should_transce.html (accessed June 17, 2013).

to shore facilities; they have three cafeterias with the capacity to feed 3,000 people three meals a day, they can produce several thousand gallons of fresh water from sea water each day, and they carry half a dozen helicopters . . . .[26]

The U.S. military is capable of waging war most decisively and devastatingly with speed, surprise, and violence. However, during periods of international emergencies, it is also capable of providing many necessary resources during mitigation, response, and recovery initiatives. It supplements the efforts of responding organizations to render assistance to those in need, and is capable of providing a variety of critical resources ranging from medical services to transportation.

UN Disaster Relief Support

The UN has proven to be a major participant in international disaster relief operations. The humanitarian and disaster-relief efforts of the UN system are overseen and facilitated by the Office for the Coordination of Humanitarian Affairs (OCHA), led by the UN Emergency Relief Coordinator. Among its many activities, OCHA provides the latest information on emergencies worldwide, and launches international "consolidated appeals" to mobilize financing for the provision of emergency assistance in specific situations.[27]

Through the World Food Programme (WFP) and the Food and Agriculture Organization of the UN (FAO), food is made available to those who might otherwise starve. Thanks to the Office of the UN High Commissioner for Refugees (UNHCR) and the International Organization for Migration (IOM), camps and other facilities are set up and maintained for those who have been forced to leave their homes. UN peacekeepers protect the delivery of that aid—whether provided by members of the UN system or such humanitarian bodies as the International Federation of Red Cross and Red Crescent Societies.

The World Health Organization (WHO) helps protect those displaced by natural and man-made disasters from the ravages of disease. The United Nations Children's Fund (UNICEF), with the aid of such bodies as the International Save the Children Alliance, provides education for children who have been uprooted by calamity. And when it is time to begin rebuilding, the United Nations Development Programme (UNDP) is there to ensure that the recovery process has a firm and stable footing.[28]

---

[26]Ibid.

[27]"Humanitarian and Disaster Relief Assistance," Global Issues, United Nations, http://www.un.org/en/globalissues/humanitarian/index.shtml (retrieved on May 15, 2013).

[28]Ibid.

The U.S. Response to an International Crisis

The U.S. Agency for International Development (USAID) Office of Foreign Disasters Assistance (OFDA) is responsible for leading and coordinating the U.S. Government's response to disasters overseas. The OFDA is an organization located within the U.S. State Department and is divided into three divisions:

1. **The Disaster Response and Mitigation (DRM)** division responsible for coordinating with other organizations for the provision of relief supplies and humanitarian assistance.
2. **The Operations Division (OPS)** develops and manages logistical, operational, and technical support for disaster responses. OPS maintains readiness to respond to emergencies through several mechanisms, including managing several search and rescue teams, the ground operations team, disaster assistance response teams (dart), and the Washington response management teams.
3. **The Program Support (PS)** division provides programmatic and administrative support including budget/financial services procurement planning, contract/grant administration, general administrative support, and communications support for both USAID/OFDA Washington D.C. and its field offices.[29]

The USAID reports that in 2011 the OFDA responded to 67 disasters in 54 countries assisting tens of millions of disaster-affected individuals across the world. The United States has been extremely generous with humanitarian aid to those countries that have been victimized by a disaster. Margesson (2006) contends that the $3.83 billion humanitarian aid budget for fiscal year 2003 was the largest since the late 1970s, and slightly higher than the $3.73 billion funding level for fiscal year 2005 (CRS-3).

U.S. Emergency Response Protocol

After the government of a disaster-stricken nation has requested assistance, the local U.S. Embassy contacts the Operations Center of the U.S. State Department via a disaster declaration cable.[30] However, according to U.S. law, the disaster request cable must fulfill the following three criteria before the U.S. can respond:

1. The disaster must be beyond the ability of the host nation to handle on its own.
2. The host nation must formally request U.S. assistance.
3. Such assistance must be in the strategic interests of the United States.[31]

---

[29]Global Corps, The Office of U.S. Foreign Assistance (OFDA), n.d., http://www.globalcorps.com/ofda.html.
[30]Perry and Travayiakis, The U.S. Foreign Disaster Response Process.
[31]Ibid.

Once the request is set in motion, OFDA dispatches regional and technical experts to assess the situation and identify any/all humanitarian needs. In the event of a large-scale disaster, OFDA can deploy a Disaster Assistance Response Team (DART). DART provides specialists trained in a variety of disaster relief skills that assist U.S. embassies and USAID missions in managing the U.S. Government response to disasters.

Chapter Comments and Summary

International disaster responses require the expertise of many specialized actors, including affected government entities, militaries, intergovernmental organizations (typically UN agencies), international and domestic NGOs, the International Red Cross and Red Crescent Movement, and affected civilian populations. No single actor can undertake all facets of relief and recovery. Addressing survivors' needs, which span health, nutrition, water and sanitation, emergency sheltering, and livelihood reconstruction is of paramount importance during international emergencies.

Any nation is susceptible to both natural and man-made disasters. Natural and man-made incidents may have equivalent consequences with respect to fatalities, economic and political impacts, and physical devastation. Historically, pandemics have resulted in the deaths of millions of people globally. The man-made Chernobyl nuclear incident resulted in the city remaining uninhabitable and unlivable during current times.

Any nation may require the assistance of another nation or multiple nations at some point depending upon the severity of an incident. This assistance may be rendered from government sources or NGOs. An example of government involvement is military support whereas an example of nongovernment involvement is a charitable organization.

International disasters have some commonness regardless of their location. Emergency initiatives must accommodate command and control; emergency relief; sheltering; coordination and logistics; corruption; and project management. Underlying these common characteristics is the notion of cooperation. Nations must cooperate to ensure the efficiency and effectiveness of emergency initiatives.

The United States has a long history of assisting other nations during calamities of various types, scopes, and magnitudes. The U.S. military is a resource that facilitates logistics for emergency materials and resources and provides an array of emergency services. Another facet of U.S. emergency contributions involves USAID and its humanitarian programs. Certainly, the United States provides a variety of urban search and rescue resources.

No one can predict with absolute certainty neither exactly when an international disaster will strike nor its associated consequences. However, one thing is certain—international disasters will occur at some point in time. The international community must remain vigilant regarding preparedness to render assistance when necessary to embellish mitigation, response, and recovery initiatives.

Terminology

Command and control
Coordination and logistics
Corruption
Emergency relief
International assistance
International disaster
International politics
Man-made disaster
Military assistance
Natural disaster
Pandemic
Project management
Sheltering
Urban search and rescue
USAID
U.S. Military

Thought and Discussion Questions

1. This chapter introduced the notion of international assistance and discussed a variety of organizations that are leveraged beneficially during international emergencies. Perform some research, and identify three additional NGOs that may also be used during international emergencies. Write a brief essay that summarizes your findings.
2. This chapter introduced the notion of international assistance and discussed a variety of organizations that are leveraged beneficially during international emergencies. Perform some research, and identify three additional government organizations that may also be used during international emergencies. Write a brief essay that summarizes your findings.
3. This chapter introduced the notion of international assistance and discussed a variety of organizations that are leveraged beneficially during international emergencies. Perform some research, and identify three additional foreign government organizations that may also be used during international emergencies. Write a brief essay that summarizes your findings.
4. This chapter introduced the notion of international assistance and discussed a variety of organizations that are leveraged beneficially during international emergencies. Perform some research, and identify three additional foreign NGOs that may also be used during international emergencies. Write a brief essay that summarizes your findings.

Table 13.1 The Most Devastating Natural Disasters in History

Historically Devastating Disasters			
Location	Year	Type of Disaster	Deaths
Haiti	2010	Earthquake	Unknown
Myanmar	2008	Cyclone	140,000
Pakistan	2005	Earthquake	40,000
New Orleans, La.	2005	Hurricane Katrina	1,800
Indian Ocean	2004	Earthquake/Tsunami	225,000
Florida	1992	Hurricane Andrew	26
Colombia	1985	Volcano	25,000
China	1976	Earthquake	255,000–650,000
China	1931	Flood	1 million
Indonesia	1815	Volcano/Famine	80,000
India	1737	Typhoon	300,000
China	1556	Earthquake	830,000
Europe	1330–51	Plague/Pandemic	75 million
Syria	1138	Earthquake	230,000

Source: David Crossley St. Louis University, Livescience research and reporting. http://www.livescience.com/9794-worst-natural-disasters.html

Table 13.2 Most Notable Fairfax County FD International Emergency Responses

Notable Virginia Task Force 1 Disaster Responses			
Location	Year	Type of Disaster	SAR Type
Guatemala	2012	Earthquake	AST
Japan	2011	Earthquake	Heavy US&R
Haiti	2010	Hurricane	AST
Chile	2010	Earthquake	Medium US&R
Haiti	2010	Earthquake	Heavy US&R
Burma	2008	Cyclone	DART
Honduras	2007	Hurricane	UNDAC
Peru	2007	Earthquake	UNDAC
S. Asia/Pakistan	2005	Earthquake	DART
Indonesia	2005	Tsunami	DART

Iran	2003	Earthquake	Heavy US&R
Turkey	1999	Earthquake	Heavy US&R
Kenya	1998	Bombing	Heavy US&R
Philippines	1990	Earthquake	Heavy US&R
Armenia	1988	Earthquake	Heavy US&R

Source: Virginia Task Force 1 (http//:www.vatif1.org/about.cfm)

References

1. Bullock, J., G. Haddow, D. Coppola, E. Ergin, L. Westerman, and S. Yeletaysi. 2006. *Introduction to homeland security.* 2nd ed. Burlington, MA: Butterworth-Heinemann Publishers.
2. Buzanowski, J. G. 2011. Face of defense: Airman aids Japan recovery effort. http://www.defense.gov/news/newsarticle.aspx?id=63512 (accessed June 17, 2013).
3. Cook, Malcolm, Jan Noyes, and Yvonne Masakowski. 2007. *Decision making in complex environments.* Burlington, VT: Ashgate Publishing.
4. Coppola, Damon. 2006. The importance of international disaster management in the field of emergency management. http://training.fema.gov (accessed May 1, 2013).
5. Coppola, D. 2006. *Introduction to international disaster management.* Burlington, MA: Butterworth Heinemann Publishers.
6. Crossley, D. n.d. The worst natural disasters ever. http://www.livescience.com/9794-worst-natural-disasters.html.
7. Daniel, Lisa. 2011. Japanese defense minister thanks Reagan crew. http://www.defense.gov/news/newsarticle.aspx?id=63422 (accessed June 18, 2013).
8. Engstrom, Jeffrey. 2013. Taking disaster seriously: East Asian military involvement in international disaster relief operations and the implications for force projection. *Asian Security* 9(1), pp. 38–61.
9. Federal Communications Commission. 2013. Pandemics information. http://www.fcc.gov/encyclopedia/pandemics-information (accessed June 17, 2013).
10. Federal Emergency Management Agency. 1999. The federal response plan: Emergency support function #9—Urban search and rescue. http://www.au.af.mil/au/awcgate/frp/frpesf9.htm.
11. Global Corps. n.d. The Office of U.S. Foreign Assistance (OFDA). http://www.globalcorps.com/ofda.html.
12. Globalsecurity.org. n.d. Title ten United States code. http://www.globalsecurity.org/military/library/police/army/fm13-19-40/chio.htm.

13. Humanitarian and Disaster Relief Assistance. 2013. Global issues. United Nations. http://www.un.org/en/globalissues/humanitarian/index.shtml (accessed May 15, 2013).
14. Indian Ocean earthquake and tsunami. 2013. Livescience. http://www.livescience.com/33316-top-10-deadliest-natural-disasters.html (accessed June 17, 2013).
15. International Disaster Response. 2013. Social issue report. http://rootcause.org/documents/DR-Issue.pdf (accessed May 15, 2013).
16. Lien, Y., H. Jang, and T. Tsai. 2009. A MANET based emergency communication and information system for catastrophic natural disasters. In *2009 IEEE International Conference on Distributed Computing Workshops*.
17. Los Angeles County Fire Department. n.d. CA-TF2 International programs. http://www.fire.lacounty.gov/specialops/CA-TF2 International.asp.
18. Los Angeles Fire Department. n.d. CA-TF2 World class responders: California task force 2. http://www.fire.lacounty.gov/specialists/CA-TF2international.asp.
19. Los Angeles Times. 2011. Japanese counsel thanks Los Angeles Fire Department search-and-rescue team for aid. http:latimes.com/lanow/2011/04/Japanese-counsel.
20. Margesson, R. 2006. International crisis and disasters: U.S. humanitarian assistance, budget trends, and issues for congress. *Congressional Research Service*. Order Code RL33769.
21. McEntire, D. 2009. *Introduction to homeland security: Understanding terrorism with an emergency management perspective*. New York, NY: Wiley.
22. Mullins, G. 2004. Urban search and rescue—Developing Australia's capability. *The Australian Journal of Emergency Management* 19, no. 1 (March).
23. National Geographic News. 2005. The deadliest tsunami in history? http://news.nationalgeographic.com/news/pf/65467352.html.
24. Perry, C., and M. Travayiakis. 2008. The U.S. foreign disaster response process: How it works and how it could work better. Cambridge, MA: The Institute for Foreign Policy Analysis, Inc.
25. Post-Chernobyl Global Co-operation: 5 Years Later. 1991. *IAEA Newsbriefs*, vol. 6, no. 2 (49), March/April. International Atomic Energy Agency. http://www.iaea.org/newscenter/features/chernobyl-15/cooperation.shtml (accessed May 1, 2013).
26. Roberts, Adam. 1997. Use of UN peacekeeping forces for humanitarian purposes. Refugee Participation Network, no. 23 (January–April). http://www.fmreview.org/RPN/23.pdf (accessed May 1, 2013).
27. Robertson, F. 2007. Urban search and rescue teams: What size or type is needed? http://www.usfa.fema.gov/pdf/efop/efo4mo.pdf

28. Schultz, Sabine. 2009. *Disaster relief logistics: Benefits of and impediments to cooperation between humanitarian organizations.* Bern, Germany: Haupt-Berne Publishing.
29. United States Agency for International Development. 2011. USAID/OFDA FY 2011 annual report. http://www.usaid.gov/what-we-do/working-crises-and-conflict.
30. U.S. Agency for International Disaster. 2013. Office of U.S. foreign disaster assistance. http://www.usaid.gov/who-we-are/organization/bureaus/bureau.
31. USAID. 2013. What we do. http://www.usaid.gov/what-we-do (accessed June 17, 2013).
32. Vervaeck, Armand. 2011. Krakatau, Indonesia—Volcanic earthquakes are creating unrest in coastal villages. Earthquake report. http://earthquake-report.com/2011/10/06/krakatau-indonesia-volcanic-earthquakes-are-creating-ner/ (accessed May 5, 2013).
33. Virginia Task Force 1—International Search and Rescue. 2013. About VA-TF1. http://www.vatf1.org/about.cfm.

14

FUTURE OF EMERGENCY MANAGEMENT

"First, natural and technological disasters are still more certain than terrorist attacks and the nation has to be prepared better for the next major disaster."[1]

William L. Waugh, Jr.
Georgia State University

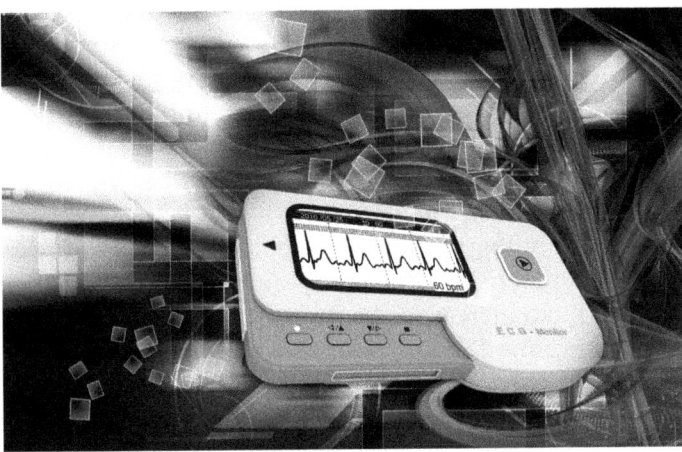

Figure 14.1 Technology may well improve future emergency management endeavors.

---

[1] William Waugh, "The Future of Emergency Management," Georgia State University, 2003, http://search.yahoo.com/r/_ylt=A0oG7thSEKlR3i0AJrFXNyoA;_ylu=X3oDMTEydjVrZDFlBHNlYwNzcgRwb3MDMQRjb2xvA2FjMgR2dGlkA0RGRDVfODg-/SIG=15n7p44fi/EXP=1370063058/**https%3a//training.fema.gov/EMIWeb/edu/docs/emfuture/Future%2520of%2520EM%2520-%2520the%2520Future%2520of%2520Homeland%2520Security%2520and%2520EM%2520-%2520Waug.doc (accessed May 30, 2013).

OBJECTIVES

The objectives of this chapter are to:
- Speculate the economic attributes of emergency management,
- Discuss technologies within the context of emergency management,
- Discuss political aspects of emergency management,
- Discuss both the natural and man-made characteristics of emergency management, and
- Discuss the preceding objectives within the contexts of future paradigms and practices of emergency management.

Introduction

Both at home and abroad, emergency management is in a continual state of evolution. Greater understanding of disasters and disaster response, often drawn from lessons learned, has demonstrated that planning for and responding to an event impacting a community is one of the most challenging tasks facing all levels of government. Despite any increases of knowledge or understanding, the future of emergency management will always be riddled with challenges. No one can predict with full certainty and accuracy the events that will transpire and will affect the emergency management domain.

While advances in technology to include the improvement of response-related equipment, innovations in communications, and the continual maturation of response organizations have increased disaster planning and response, it remains important for both governmental and nongovernmental entities work to ensure disaster response is as synchronized as possible. Today, as in the future, much of response success or failure is dependent upon a realistic risk analysis, effective pre-event planning, the establishment of multiagency relationships and mutual support compacts, including a command and control structure and the identification of resources that can be quickly and efficiently implemented when necessary.

While looking to the future of the discipline, much can be drawn regarding the insights from the lessons learned in the past. The nation has a deep history in responding to events impacting communities. Storms, urban fires, earthquakes, droughts, and even attacks, such as Oklahoma City, New York, Washington, D.C., or Boston are reminders of vulnerabilities to a wide range of threats. For the emergency manager, the challenges of preparation, education, and response will remain a major task. With each event, large and small, costly lessons are learned that improve the ability to management future emergencies.

Future of Domestic Emergency Management

A revolution in emergency management has occurred over the last two decades. The attacks of September 11, 2001 stimulated the nation to examine a wide range of disciplines and functions, to include emergency management and homeland security. In the shadows of the attacks, the public became much more aware of the functions and roles of emergency management. In looking to the future of domestic emergency management, there are many issues to consider. This chapter explores a variety of perspectives regarding the future of emergency management.

Examination of the All-Hazards Approach to Emergency Management Planning and Improved Risk Analysis

While the all-hazards approach to emergency management planning remains valid, disaster anticipation, response planning, and risk analysis must be balanced and based upon a realistic assessment of events most likely to impact the community. From this assessment of likely events, planning can anticipate realistic response and recovery needs and formulate an appropriate strategy which takes into consideration available and anticipated capabilities, time, and resources. All-hazards must be considered as all reasonable hazards, eliminating the improbable. By focusing planning, the emergency management team can realistically prepare for effective event response.

Policy Changes to Refine and Improve the Response Infrastructure

Major disasters will continue to stimulate changes in emergency management-related law, policies, organizations, and processes. In the wake of Hurricane Katrina, all levels of the government assessed the dynamics around event planning, response, resources, roles, and recovery. In the case of Hurricane Katrina, the event was so massive, jurisdictions from Louisiana to Alabama found themselves overwhelmed, confused, and to a great extent paralyzed. Many contingency plans were found to be outdated, inadequate, or unsustainable. As a result, the disaster served as a wake-up call to governments and the emergency management community to conduct ongoing reassessments related to response planning, roles, and policies.

Emphasize Preparedness and Mitigation in Emergency Management in the United States

Federal disaster policy in the United States has been shaped by an ongoing conflict between those who favor federal intervention following a disaster and those who believe disaster response should be the responsibility of state and local governments and charity. Regardless, both factions have a concern for the welfare of the affected incident areas. Both factions advocate the maximizing preparedness and mitigation initiatives within the context of managing emergencies.

Preparedness involves a range of deliberate, critical tasks and activities necessary to build, sustain, and improve the operational capability to prevent, protect against, respond to, and recover from domestic incidents. Preparedness is a continuous process involving efforts at all levels of government and between government and private sector and nongovernmental organizations (NGOs) to identify threats, determine vulnerabilities, and identify required resources. Preparedness must be a consideration of any future endeavors within the emergency management domain.

Mitigation activities are designed to reduce or eliminate risks to persons or property or to lessen the actual or potential effects or consequences of an incident. Mitigation measures may be implemented prior to, during, or after an incident. Mitigation measures are often developed in accordance with lessons learned from prior incidents. Mitigation involves ongoing actions to reduce exposure to, probability of, or potential loss from hazards. Measures may include zoning and building codes, floodplain buyouts, and analysis of hazard-related data to determine where it is safe to build or locate temporary facilities. Mitigation can include efforts to educate governments, businesses, and the public on measures they can take to reduce loss and injury. Mitigation also must be a consideration of future emergency management endeavors.

Continue to Build Community Partnerships Including Government and Nongovernmental Organizations

Community partnerships will remain important. Emergency managers do not control the extensive range of resources that may well be called upon in an emergency. First responders, including search and rescue, firefighters, law enforcement and military-related resources, the National Guard, and state defense forces, are all common to the states. In some cases, among localities, some resources, such as fire departments and police departments, represent grassroots entities that may be activated during emergencies.

Emergency managers must work to educate the public as to potential threats facing their communities and to provide them the knowledge pertaining to the actions they may well need to take in the wake of an event. Using these local and state entities provides a means through which the public may be educated during the future. Such first responder organizations may instigate or complement public awareness campaigns. For example, from a seasonality perspective, they may be used to warn and remind the public of seasonal threats—forest fires, hurricanes, tornadoes, and so forth.

Ultimately, one of the major goals of the emergency manager is the strengthening of the public's trust by collaborating with groups like nongovernmental agencies and volunteer organizations. A variety of such organizations exist through which future cooperation may occur. Public media may be used to solicit critical resources before the occurrence of possible emergency events. For example, American Red Cross blood drives may occur before hurricane season to ensure that blood supplies are increased before a dangerous scenario occurs. Publicizing such events increases the awareness within the public concerning the seasonal need for blood supplies.

For this community, partnerships are not merely standard operating procedures they are essential. In an environment of fiscal constraints and changing government roles and responsibilities, the partnership imperative will have to rise to a new level, involving new associations, broader and deeper interactions, and immense fluidity. Working with communities to understand their needs and where the emergency management community can empower and assist, is a necessary shift in approach. Businesses will continue to serve as a core member of the emergency management team, and they will be crucial to successful service delivery.

Changing Role of the Individual in Disaster Preparedness

Both Hurricanes Katrina and Super Storm Sandy revealed that emergency management involves many challenges, involving response and rescue, among devastated environments. In both of these cases, individuals and their families quickly realized the scope of the disaster was so extensive that they were responsible for much of their sustainment during the early post-event period.

Many believe if the central region of the United States experiences an earthquake along the New Madrid fault at a magnitude similar to the 1811 event, cities from Memphis, Tennessee to St. Louis, Missouri may well receive catastrophic damage, displacing hundreds of thousands and inflicting damages from which it will take decades to recover. Roadways would be damaged thereby impeding the flows of goods and services. Bridges spanning the Mississippi River would be destroyed thereby ensuring the impossibility of physically crossing the river in and possibly near the affected areas.

Increased empowerment of the individual is critical. Humans are the catalysts of all preparedness, response, and recovery initiatives. Individuals represent the smallest unit of emergency response to the highest position of leadership that may be exercised within emergency management chains-of-command. Individuals represent the basest elements of households through which household emergency operations are anticipated, planned, and executed. Individuals represent the constituency that elects political leaders whose responsibilities include emergency management functions.

Such notions are not uncommon within the domain of emergency management. During the 2013 Hurricane Preparedness Week, Texas Governor Rick Perry reminded Texas citizens and residents of the importance of individuals within the context of emergency management. An excerpt of Governor Perry's comments is stated as follows:

> As a new hurricane season begins tomorrow, the most important element of emergency response starts with the individual. Everybody in Texas, particularly those along the coast, needs to have an emergency plan, an emergency supply kit and an evacuation route in place before a hurricane strikes . . . We use these annual exercises to help us keep our system razor-sharp and ready for action at the first sign of trouble. Through practice, simulation and repetition, we've honed one of the most effective and efficient emergency response teams in the country.[2]

These comments reinforce the importance of individuals within the emergency management domain. No organization or organizational initiatives are any better than the qualities and characteristics of their components. In turn, these components are comprised of individual humans. When a disaster occurs, individuals must be prepared to protect not only themselves but also to act on the behalf of others and of societal good.

Throughout the nation, any number of local emergency response resources is comprised of ordinary individuals who contribute their time and skills to emergency services. For example, among rural localities, volunteer fire departments exist as organizations through which any member of the community may participate provided that one completes the requisite training and certification. Such fire departments are comprised of individuals.

When considering the concept of an individual as an emergency responder, one must be mindful of a dichotomy. Individuals must be prepared to save themselves and their households as well as contributing their knowledge, skills, and time toward the maintaining and exercising of emergency response organizations through time. One cannot discount or ignore the relevancy and importance of the individual human as a basic building block of emergency management.

---

[2]State of Texas, "Gov. Perry: Emergency Response Starts with Individual Preparedness," 2013, http://governor.state.tx.us/news/press-release/18598/ (accessed May 31, 2013).

Communities are also comprised of humans. Humans form the basis of all social, economic, political, and other activities that are exhibited among communities. Within the United States, government is derived from the people. Through time, some themes that show a lack of trust in government authorities have emerged. The incidents that occurred during the aftermath of Hurricane Katrina and Hurricane Sandy depict some disgruntlement between the people and their elected political leaders.

Within American society, the people have significant power that is exercised through their votes throughout the political process. Individuals may cast ballots for political candidates that best align with their personal beliefs and philosophies. By doing so, leadership may be elected that influences emergency management activities, resources, and functions. Therefore, the constituency must be mindful of the long-term effects of their votes regarding influences that affect future emergency management endeavors.

Additionally, individuals may also feel a calling to public service. Individuals may become candidates for any public office they desire. By doing so, if elected, they gain some influence regarding the future endeavors and characteristics of emergency management. However, in turn, these people are subject to the will of the populace through the receipt of individual votes from their respective constituencies.

The definition and perception of community is changing from the traditional geographic community to a virtual community. Through the use of electronic technologies, less interaction occurs physically between and among individual humans. Instead, reliance upon technologies is increasingly a common aspect of American society. Therefore, future emergency management domains must consider the methods through which such virtual resources may be leveraged to accommodate community involvement and participation within the emergency management process.

Certainly, such a virtual community is comprised of individual humans. Within the context of emergency management, virtual technologies may be adapted through which individuals participate actively within the emergency management process. For example, individualized emergency messaging may be incorporated among cellular systems through which citizens and residents are warned regarding endangerments that may affect their localities.

Funding Challenges

Budget projections are grim at all levels of government. Balance sheets are still feeling the effects of the economic downturn that began in 2007, and budgets will be squeezed in the long term by health-care and retirement obligations as the Baby Boomers begin to retire. Emergency managers will need to consider how they fund their activities if government budgets continue to be constrained. The solution may well be found through collaborative efforts within and between states, resource sharing, and the leadership needed to effectively identify and fill staffing gaps and resource needs.

Planning will also be impacted by available funding. Over the decade after the attacks of September 11, 2001, funding stimulated the field of emergency management. Unfortunately, current and anticipated reduced funding will impact the emergency management community to make difficult decisions in the establishment of planning and response priorities.

Money is a critical resource within the context of emergency management. It is the resource that pays for training and preparedness. It also pays for mitigation, response, and recovery activities that occur during calamities. When assessing the economic impacts of disasters, these results are often measured and expressed monetarily in terms of dollars. For example, the Northridge earthquake of 1994 caused approximately $67 billion dollars in Californian damages.[3]

Money permeates every facet of individual households as well as the highest echelons of government coffers. Funding is a critical resource that must be considered by individuals within their household budgets. For example, households should expend funds necessary for providing emergency rations and supplies that may become the staples of survival during a calamity. From the government context, funding is necessary to satisfy a wide range of needs and wants within the emergency management domain. For example, the salary of the lowest FEMA employee must be paid as well as paying for the entirety of the costs that are associated with preparedness, mitigation, response, and recovery.

Funding is a limited resource. After all, money does not grow on trees. Instead, governments may obtain funding from a variety of sources ranging from the issuances of municipal bonds and confiscation sales to taxation and municipal licensing fees.[4] Funds obtained through such sources are often expended strategically toward the best interests of societal good and betterment.[5] Therefore, humans must exercise discretion when determining exactly how these funds are expended within the emergency management domain.

Such discretion is not the result of emotion or ad hoc, spontaneous decisions. Instead, it necessitates a business approach to the funding of public endeavors through which the use of limited funds is maximized with respect to generating the highest and best public benefits through time.[6] When considering constrained

---

[3] "Top 10 Costliest Disasters," Bankrate, 2013, http://www.bankrate.com/finance/insurance/top-10-costliest-natural-disasters-4.aspx (accessed May 31, 2013).

[4] Daniel Doss, William Sumrall, and Don Jones, *Strategic Finance for Criminal Justice Organizations* (Boca Raton, FL: CRC Press, 2012).

[5] Ibid.

[6] Daniel Doss, Chen Guo, and Joo Lee, *The Business of Criminal Justice: A Guide for Theory and Practice* (Boca Raton, FL: CRC Press, 2011).

budgets, municipal and government leaders must exercise quantitative methodologies through which human decisions are improved and exhibit greater amounts of objectivity.[7]

When considering the future of emergency management, these quantitative techniques may be used to examine the financial aspects of projects through which public good may be derived through time. Capital budgeting represents a financial methodology through which future benefits of competing projects may be examined to determine which option represents the highest and best use of public monies.[8] Through employing capital budgeting techniques, municipal and government leaders exercise quantitative discretion regarding the future use of limited monies to support the long-term interests of emergency management functions.

Regardless of the quantitative methodologies employed to facilitate the use of funds to support future emergency management initiatives, there will always be various disagreements regarding the use of such funds. Politics, turf wars, and a host of other influences exist regarding the use of monies to support emergency management initiatives. Often, such disputes and challenges affect the long-term expending of funds within the emergency management domain. Therefore, despite the best efforts to quantify objectively the highest and best use of funding, there will always be influences that impact its long-term use.

Aging Critical Infrastructure

The average age of infrastructure in the United States has been rising, meaning that the individual component structures are becoming older and more prone to failure. Beyond that, the very nature of infrastructure could change as America adopts new and different technologies such as alternative energy and enhanced wireless communications networks. Emergency response could be hindered by aging infrastructure, or disruptions to communications networks, and investments in new infrastructure offer opportunities to reduce the impacts of future disasters.

Improved risk assessment incorporated in urban planning and building design is very important, but clearly will not occur without enhanced building codes that take into consideration realistic threats. Most of the existing infrastructure in the United States is not designed to withstand major natural events, such as earthquakes and hurricanes. Additionally, the cost of retrofitting these structures is not considered practical. As a result, the nation's infrastructure remains vulnerable and those vulnerabilities must be a factor in emergency management planning. Aging transportation, communication, energy, and health-care infrastructure may create additional emergency situations.

---

[7]Ibid.
[8]Ibid.

A primary example of aging infrastructure involves the communication and electrical systems that span the nation. Throughout national growth, a vast investment in connecting the nation has occurred to supply both communications and electricity nationally. Although some of this connectivity occurs through underground networks, the vast majority involves the use of physical electrical and communications lines. When many of these resources were developed and erected, they involved the use of wooden poles to support the necessary wiring components.

During the 1950s, the 1960s, and the 1970s, utility organizations developed infrastructure at a "significantly faster pace" than is represented by the replacement rate that is necessary for maintenance and expected systems functioning.[9] Age and condition impact the quality and possible failure rate of wooden poles. Through time, "potential safety and financial risks" become considerations of this type of infrastructure.[10] The following figure represents these concepts involving data collected from two mid-western utilities organizations.

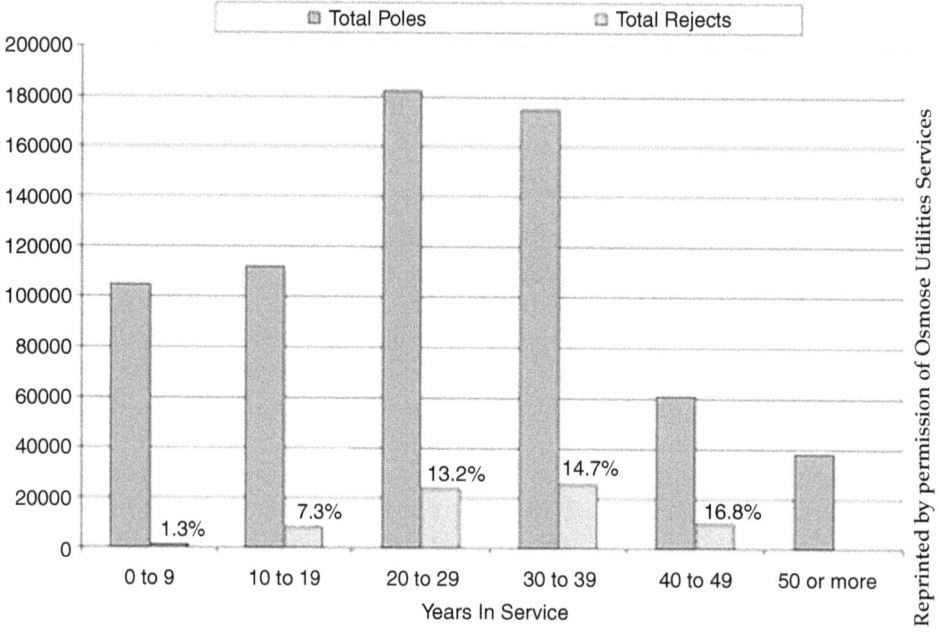

Figure 14.2 Aging infrastructure risk involving wooden utility poles.

---

[9]Osmose Utility Services, "Aging Infrastructure," 2013, http://www.osmoseutilities.com/content/pages/aging-infrastructure (accessed May 31, 2013).
[10]Ibid.

Within this figure, between the ages of 20 and 50 years, a good number of wooden utility poles are shown to be weaker than their intended "design strength" thereby increasing their potential failure risk.[11] The failing of utility poles may occur "during storms or other unusual load situations."[12] Therefore, a strategic approach must be taken when considering the maintaining and improving of wooden pole infrastructures.[13]

Another aspect of aging infrastructure involves a consideration of the national roadway system and its bridges. Within the United States, the average age of roadway bridges is approximately 42 years.[14] Among the approximate quantity of 607,000 bridges that comprise this infrastructure, some 66,749 are structurally deficient whereas approximately 87,748 are functionally obsolete.[15]

During 2007, in Minneapolis, Minnesota, an aging bridge that crossed the Mississippi River collapsed.[16] This disaster resulted initially in the deaths of nine people, the injuring of 60 people, and the reporting of 20 missing individuals.[17] The collapsed bridge was the Interstate 35W bridge which represented a "major Minneapolis artery."[18] The following image shows the collapsed Minnesota bridge.

During 2013, segments of Interstate Highway 5 collapsed in the state of Washington.[19] This bridge was erected in 1955.[20] It was classified as "fracture critical" during modern times before its collapse.[21] Over time, as the integrities of the physical infrastructures of bridges decline, collapses may occur with little or no warning.

---

[11] Ibid.

[12] Ibid.

[13] Ibid.

[14] Marisol Bello, "Bridge Collapse Shines Light on Aging Infrastructure," *USA Today*, 2013, http://www.usatoday.com/story/news/nation/2013/05/24/washington-bridge-collapse-nations-bridges-deficient/2358419/ (accessed May 31, 2013).

[15] Brad Knickerbocker, "Skagit Bridge Collapse: Not the Only One Waiting to Happen," *The Christian Science Monitor*, 2013, http://www.csmonitor.com/USA/USA-Update/2013/0527/Skagit-River-bridge-collapse-Not-the-only-one-waiting-to-happen-video (accessed May 31, 2013).

[16] "9 Thought Dead as Minneapolis Bridge Collapses," *NBC News*, 2007, http://www.nbcnews.com/id/20079534/ns/us_news-life/t/thought-dead-minneapolis-bridge-collapses/#.Uak14Jy87eY (accessed May 31, 2013).

[17] Ibid.

[18] Ibid.

[19] Marisol Bello, "Bridge Collapse Shines Light on Aging Infrastructure," *USA Today*, 2013, http://www.usatoday.com/story/news/nation/2013/05/24/washington-bridge-collapse-nations-bridges-deficient/2358419/ (accessed May 31, 2013).

[20] Ibid.

[21] Ibid.

Figure 14.3 Collapsed minnesota bridge.

Such collapsed bridges and roadway segments are indicative of a weakening infrastructure that cannot be ignored. Although some measures occur to maintain and strengthen infrastructure, a variety of issues impact maintenance and the replacing of significantly hazardous components of infrastructure. Such issues range from insufficient funding to political leveraging. During the future, emergency management must be mindful of the dangers posed by aging infrastructures throughout the nation.

Evolving Terrorist Threat

The use of terrorism against the United States has proven effective in shocking the nation and it is expected that terrorism will be used in the future to advance various radical agendas or to capture attention. The threats of terrorism are unabated, and will continue to endanger the United States. The availability of technical and scientific knowledge may increase terrorist access to "high consequence" weapons, including weapons of mass destruction.

The seriousness of terrorism must be considered within the context of emergency management. Historically, a terrorism emphasis has existed regarding overseas threats and events. However, modern reality dictates that the mainland is now susceptible to the dangers of terrorism that is perpetrated by both groups and individuals. This notion will not change. The emergency management domain must be cognizant of terrorism and its future endangerments.

Events of terrorism may occur anywhere. Notably, in 2009, Major Nidal Hassan, an Army officer, unleashed an attack at Fort Hood, Texas. This event resulted in the deaths of 13 people, and injured 32 others.[22] Sporting events draw mass audiences and large amounts of attention within the general public. During 2013, the Boston Marathon experienced a bombing that was unprecedented in American sports history. Certainly, the nation will never forget the events of September 11, 2001.

Within the emergency management domain, personnel must be vigilantly aware and mindful of the dangers of terrorism, especially regarding the adverse potentials of domestic terrorist events. Future approaches within emergency management may necessitate a more proactive perspective regarding potential terrorist activities among emergency personnel. Within the United States, local law enforcement agencies represent a significant factor in detecting and deterring terrorist events.[23] Future terrorist events may invariably involve the uses of homemade weapons.[24]

For example, during 2011, in California, the San Diego Police Department was credited with such a discovery. This event is summarized as follows:

> In November, San Diego police arrested George Djura Jakubec, 54, a computer software consultant and naturalized U.S. citizen from Serbia, who was living in a house full of homemade explosives and material, including pentaerythritol tetranitrate and hexamethylene triperoxide diamine, both *al-Qaida* favorites. They also found items 'suggestive of armed robberies' and charged the man with two bank robberies and bomb making.[25]

Within the emergency management domain, future expansion and uses of terrorism liaison officers (TLOs) may act as points of contact through which suspected terrorist activities and personnel may be reported by the general public to public safety officials. Currently TLOs exist as a "principle point of contact for a public safety agency in matters related to terrorism information."[26] Through the expanding of TLOs nationally, a greater capacity will exist through which terrorism may

---

[22]Nick Wing, "Nidal Hassan, Fort Hood Shooting Suspect, Has Received $278,00 in Salary Since Arrest," *The Huffington Post*, 2013, http://www.huffingtonpost.com/2013/05/21/nidal-malik-hasan-salary_n_3313519.html (accessed May 31, 2013).

[23]Jim McKay, "The Changing Face of Terror in the U.S.," *Emergency Management: Strategy and Leadership in Critical Times*, 2011, http://www.emergencymgmt.com/safety/Face-Terror-US-Home-Grown.html?page=1& (accessed May 31, 2013).

[24]Ibid.

[25]Ibid.

[26]Terrorism Liaison Officer Information Network, "What is a Terrorism Liaison Officer?" 2013, http://www.tlo.org/what_is_tlo.html (accessed May 31, 2013).

be detected. For example, within California, approximately 25,000 first responder personnel already have received such TLO training.[27]

Future emergency management training functions must also accommodate the potential of terrorist threats. Training programs should unequivocally indoctrinate first-responders with an awareness of the vulnerabilities of all communities to the dangers of terrorism and its associated activities. Anticipating and preparing for future events and activities of terrorism must be accommodated within the emergency management domain.

Global Interdependency

Emergency managers may take on greater role in emergency management internationally. There are times when calamities necessitate an international response involving a variety of nations globally. For example, during the aftermath of the 2004 Asian tsunami, personnel representing the Centers for Disease Control and Prevention were dispatched to the affected region.[28] Their primary responsibility involved providing "critical assistance to the survivors and to monitor for deadly disease outbreaks."[29]

Globalism has inevitably contributed toward increased economic integration and interaction throughout the world. An increased role globally, involving American emergency managers, could have major resource and capability implications. For example, since its origination, the CDC Foundation has "provided $400 million to support CDC's work, launched more than 600 programs around the world and built a network of individuals and organizations committed to supporting CDC and public health."[30] In such instances, American personnel may entertain the thought of international deployments externally from the mainland United States.

Conversely, there are times when American emergency management personnel must cooperate with international entities that operate within the mainland of the United States. During the aftermath of Hurricane Katrina in 2005, components of the Mexican military were deployed internationally to the United States. This event marked the first operation of Mexican military forces on American soil since

---

[27]Jim McKay, "The Changing Face of Terror in the U.S.," *Emergency Management: Strategy and Leadership in Critical Times*, 2011, http://www.emergencymgmt.com/safety/Face-Terror-US-Home-Grown.html?page=1& (accessed May 31, 2013).

[28]CDC Foundation, "Global Disaster Response Fund," 2013, http://cdcfoundation.org/globaldisaster (accessed May 31, 2013).

[29]Ibid.

[30]Ibid.

1846, and was the first time that Mexico rendered disaster assistance to the United States.[31] This event was described as follows:

> Carrying water treatment plants and mobile kitchens that can feed 7,000 people daily, the convoy bound for San Antonio is the first Mexican military unit to operate on U.S. soil since 1846. The first of 45 vehicles in the convoy crossed the international bridge at Laredo at about 8:15 a.m. Military engineers, doctors, and nurses are among the 200 people headed to San Antonio.[32]
>
> Used with permission of The Associated Press. Copyright © 2005. All rights reserved.

During future cataclysms, emergency management personnel must be mindful of the potential of international and global deployments and cooperativeness. During the preceding decade, American emergency management personnel experienced both external and internal events that necessitated international resources both internationally and domestically. Future emergency management training initiatives must be mindful of such scenarios, and provide training that accommodates aspects of international projects that may be beneficial for performing internationally. Examples include intelligence gathering, international travel, public sector operational support, medical services, and so forth.[33] Certainly, training in foreign languages may be beneficial.

Although the United States has exhibited some amounts of isolationism throughout its history, modern times necessitate that it participate in global society. Such participation exposes it to a large array of endangerments ranging from terrorism to overseas natural disasters that may affect the assets of the United States. Regardless, the United States cannot ignore the potentials of international disasters that necessitate its attention, nor can it assume that it will never necessitate disaster assistance from other nations.

Technological Considerations

Within the United States, the communications, energy, and transportation infrastructures are all heavily dependent on technology. Similarly, the defense infrastructure also involves electronic technologies that must be secure. Such uses of

---

[31]"Mexican Troops Aid Katrina Efforts," *Fox News*, 2005, http://www.foxnews.com/story/0,2933,168778,00.html (accessed May 31. 2013).

[32]Ibid.

[33]Global First Response, "Operations," 2013, http://www.globalfirstresponse.com/Operations.html (accessed May 31, 2013).

Figure 14.4 U.S. navy littoral ship visiting pearl harbor.

Source: http://www.navy.mil/view_image.asp?id=145035

technology create a significant vulnerability to cyber-attack, particularly if this reliance creates single points of failure within and among critical systems.

During 2013, an astounding breach of electronic security occurred that jeopardized American national security. It was revealed that America's "most sensitive weapons systems" were compromised by China.[34] This array included "more than two dozen major weapons systems," including combat ships and aircraft, which are the "backbone of the Pentagon's regional missile defense for Asia, Europe, and the Persian Gulf."[35] Examples of the compromised systems range from the F-35 Joint Strike Fighter aircraft to the Littoral combat ships of the U.S. Navy.[36] The following figure shows a Littoral ship.

The defense industry and national defenses are not the only targets of hacking. The electronic systems that comprise the emergency response framework throughout the nation are also susceptible to hackers and infiltration. A recent example of such exploitation occurred within the state of Montana involving the disseminating

---

[34] Ellen Nakashima, "Confidential Report Lists U.S. Weapons System Designs Compromised by Chinese Cyberspies," *The Washington Post*, 2013, http://articles.washingtonpost.com/2013-05-27/world/39554997_1_u-s-missile-defenses-weapons-combat-aircraft (accessed May 31, 2013).
[35] Ibid.
[36] Ibid.

of a fake emergency message.[37] Later, the television station issued a statement indicating that the message had not originated from its location, and that no emergency existed.[38]

Both the compromising of American national security and the compromising of a local emergency alert system have implications for emergency management. During the future, enemies may exploit stolen intellectual properties to better any chances of overcoming American national defenses in the event of a conflict. They may use stolen intellectual property to improve their own systems and weaponries. Both the future and current leaderships of the American government must remain painfully aware of the lives that are threatened through the compromising of such technologies, and must be vigilant to adjust to such contingencies.

The latter incident, associated with the breaching of a local emergency alert system, necessitates a consideration of information legitimacy. Future considerations of emergency management must define and exercise new approaches to securing electronic environments to avoid the disseminating of false reports among emergency communications systems. Failing to do so may result in mass panic ensuing within the general populace regarding perceptions of danger. Such mass panic, in turn, may create a legitimate emergency.

Another consideration of modern technology involves the maturing and changing domain of social media. Access to real-time information will continue to increase as social media and advances in technology create new patterns of communications throughout the populace. Citizens are becoming both producers and consumers of information through their postings among social media sites and social communications networks. Some of the disseminated information may be verifiable, some may be skewed, and some may be entirely false. In any case, the legitimacy and accuracy of information must constantly be questioned and verified. These scenarios may be problematic for future emergency management personnel because of the capacity of spreading rumors, causing confusion, and inciting mass panic within the populace.

Regardless of its numerous benefits, technology may be exploited to hamper functions within the emergency management domain or to imperil American interests both internationally and domestically. No technology is impervious; all technologies are susceptible to infiltration and abuse. During future times, emergency management personnel must be mindful of these concepts, and must make every effort to secure the technologies that are necessary for supporting emergency management functions.

---

[37]KRTV, "Bogus Emergency Alert Message Transmitted," 2013, http://www.krtv.com/news/bogus-emergency-alert-message-transmitted/ (accessed May 31, 2013).
[38]Ibid.

Border Security Issues

In the areas of both emergency management and homeland security, border security issues remain important. Over the last two decades, the federal government has increased the size of the Border Patrol from fewer than 3,000 agents to more than 21,000, built nearly 700 miles of fencing along the southern border with Mexico, and deployed pilotless drones, sensor cameras, and other expensive technologies aimed at preventing illegal crossings at the land borders.

Border security is an essential aspect of emergency management because it buffers American society from much of the violence that exists among the Mexican drug wars and international crime organizations. Numerous projects and operations have been undertaken through which border security is instantiated. For example, during 2005, the state of Texas commenced Operation Linebacker in which the financial amount of $6 million was expended toward strengthening "manpower, specialized equipment and planning resources to border-area law enforcement."[39]

Maintaining border security also protects American society from threats that originate from overseas. In some cases, such threats are discovered accidentally when attempts are made to penetrate the American border. For example, during 2011, Said Jaziri, a dangerous Muslim cleric who was expelled from the nation of Canada, attempted to enter the United States through the U.S.-Mexican border.[40] His attempt to enter the United States was discovered when firefighters inadvertently witnessed him entering the trunk of a BMW vehicle, and reported their observations to law enforcement authorities.[41]

Other dangerous individuals have attempted to cross the U.S. border. It has been observed that people from Algeria, Iraq, Afghanistan, Lebanon, Libya, Nigeria, Pakistan, Saudi Arabia, Yemen, Cuba, Sudan, Iran, Somalia, and Syria have been detected when attempting to illegally enter the United States.[42] The attempts of both individuals and organizations to penetrate the U.S. borders will continue unabated through time. Some of these individuals may be motivated with intentions of harming American society and inflicting mass damages. Future emergency management practitioners and theoreticians must be mindful that the American borders are permeable, and exhibit a variety of potential endangerments.

---

[39] State of Texas, "Gov. Perry Awards $6 Million to Border Counties for Border Security," 2005, http://governor.state.tx.us/news/press-release/2558/ (accessed May 31. 2013).

[40] Richard Marosi, "Controversial Muslim Cleric is Arrested While Sneaking into the U.S.," *The Los Angeles Times,* 2011, http://articles.latimes.com/2011/jan/27/local/la-me-border-cleric-20110127 (accessed May 31. 2013).

[41] Ibid.

[42] "MURDOCK: U.S. Southern Border Open to Terrorists," *The Gleaner,* Henderson, KY, 2013, http://www.courierpress.com/news/2013/may/07/us-southern-border-open-to-terrorists/ (accessed may 31, 2013).

Figure 14.5 U.S.-Mexican border.

Numerous methods exist through which border security is conducted. They consist of everything from physically inspecting vehicles among border entry points to conducting aerial observations along geographic regions. Regardless of the methods employed, no guarantee exists that threatening agents will always be denied access to American society. Therefore, future emergency management initiatives must be cognizant of potential threats that may enter America through its border. Hence, the perpetuating of border security programs and policies, well into future generations, is of the upmost importance to the safety and well-being of American society.

Emerging Technologies

Technology is a constantly changing entity. The technological marvels that now permeate modern society were the wonderments of science fiction, even just a few decades ago. Certain technologies that now are exercised among security environments were unthinkable before the events of September 11, 2001. A primary example involves the use of x-ray scanning machines within airports. Although they may be considered as exceptionally invasive technologies by some individuals, their use is employed with a preventive mindset with respect to averting disasters among aviation environments.

Certainly, other technologies are now employed that bolster the initiatives American emergency management and whose use may prevent disasters or mitigate their associated effects. Examples range from the use of geographic information systems to advanced communications systems. At some point in history, these modern systems did not exist, and their functions were conducted either manually or through predecessor technologies. However, through much research and development, emerging technologies are produced and adapted whereby existing technologies are either embellished or replaced.

The ability to communicate is an essential necessity of emergency management. Communication must be maintained continuously throughout a disaster to ensure the effectiveness and efficiency of the mitigation and response initiatives. During calamities, cellular phone towers may be either damaged or destroyed thereby impeding the ability to communicate among responding entities. Currently, research and development is being performed to examine the potential of portable communications systems that may be employed in a variety of disaster settings.

For example, Carnegie Mellon University is experimenting with a portable communication system that is not dependent upon cellular towers and associated equipment.[43] This type of system represents a "simple, available technology that people can use that would supplement the work of first responders."[44] A possible platform may involve the use of portable, Wi-Fi networking which "can be used to connect people in a social media, even if the cell towers and internet are down."[45]

Numerous other emerging technologies are being evaluated regarding their potential as emergency management resources. Another emerging technology, Intellistreets, involves using typical streetlamps as emergency services resource during and after calamities.[46] Intellistreets may use streetlamps to monitor the levels of rising flood waters as well as using them to disseminate public warnings and directions to the general public.[47]

The use of unmanned drones also has potential applications within the context of emergency management. They may serve a variety of purposes for both law enforcement and firefighting requirements. They may be used to locate "trapped or

---

[43]Tomas Roman, "New Technologies Developed to Help During Disasters," *ABC News*, 2012, http://abclocal.go.com/kgo/story?section=news/local/south_bay&id=8873344 (accessed May 31, 2013).
[44]Ibid.
[45]Ibid.
[46]Elaine Pittman, "3 Emerging Technologies that will Impact Emergency Management," *Emergency Management: Strategy and Leadership in Critical Times*, 2012, http://www.emergencymgmt.com/disaster/3-Emerging-Technologies-Emergency-Management.html?page=2& (accessed May 31, 2013).
[47]Ibid.

missing people" among areas that are inaccessible by helicopters, may be used to "scout wildfires," identify the concealed "hot spots in structure fires," and embellish public safety during "drug busts or hostage situations."[48]

Although these uses of drone technologies may seem perfectly valid and beneficial, much debate is associated with the use of drones. Many people and some state legislatures decry the use of drones because of privacy and Constitutionality issues, and are demanding constraints regarding their uses and applications. In the state of Idaho, legislation was approved that requires "police to obtain warrants before using surveillance drones over homes, businesses or farm fields."[49] In the state of Virginia, a two-year moratorium exists regarding drone use, except for National Guard training exercises and searches for missing persons.[50] Many of the states are pursuing similar legislation.

Unmanned vehicles and driverless cars may also be appropriate for emergency management regarding automotive applications. During 2012, the state of California drafted a legislative bill that "would establish safety and performance standards for the safe operation of autonomous vehicles on California's roads and highways."[51] Similarly, during 2012, the state of Nevada also "approved regulations" that would permit the testing of automated vehicles among its roadways.[52]

Such emerging technology has implications for emergency management. Humans become fatigued; become ill; require rest, food and sleep; incur time off from work (e.g., vacations, etc.); exhibit varying levels of performance through time; and are constrained by physical and cognitive limitations. Automated machinery, despite requiring maintenance, is not susceptible to such factors. Within the context of emergency management, automated vehicles could be used in caravans to ferry emergency resources (e.g., food, medicines, people, etc.) to affected incident areas. They could be used to monitor roadways and perform acts of reconnaissance. They could remove debris from roadways during the aftermaths of cataclysms. They could be used to facilitate mass evacuations of incident areas. They could enter and

---

[48]Colin Wood, "The Case for Drones," *Emergency Management: Strategy and Leadership in Critical Times*, 2013, http://www.emergencymgmt.com/safety/Case-for-Drones.html (accessed May 31, 2013).

[49]Maggie Clark, "Drone Limits Become Law," *Emergency Management: Strategy and Leadership in Critical Times*, 2013, http://www.emergencymgmt.com/safety/Drone-Limits-Law.html (accessed May 31, 2013).

[50]Ibid.

[51]State of California, "California Legislature Approves Driverless Vehicle Bill–Senator Padilla's Legislation Establishes Performance and Safety Standards," 2012, http://sd20.senate.ca.gov/news/2012-08-29-california-legislature-approves-driverless-vehicle-bill-senator-padilla-s-legislatio (accessed May 31, 2013).

[52]Sarah Rich, "Nevada DMV Approves Regulations for Testing Driverless Vehicles," *Government Technology: Solutions for State and Local Government*, 2012, http://www.govtech.com/transportation/Nevada-DMV-Approves-Testing-Driverless-Vehicles.html (accessed May 31, 2013).

operate in conditions that are inhospitable for humans such as wildfires or chemical spills. With a little imagination, many more uses could be envisioned.

Unmanned automobiles are currently the subject of much research and development. During October, 2012, Nissan unveiled a prototype of a driverless vehicle that could become available in 2015.[53] Similarly, General Motors, Toyota, Ford, and BMW are also experimenting with automated vehicles.[54] Notably, Google has expended much effort toward researching and developing the driverless car. Its experimental prototypes have amassed approximately 300,000 miles of roadway driving without accidents.[55]

Only time will tell the outcome of experiments, research, and development involving automated vehicles. Numerous issues must be considered before the technology is publically available and integrated within the emergency management domain. Examples range from insurance and liability to the algorithmic underpinnings that direct the automated navigation software and collision avoidance routines. Regardless, one cannot discount the potential of such automobiles regarding the future of emergency management.

Technology changes through time. Modern technologies will soon be antiquated, and relegated to museums. However, they provide a basis for advancement and improvement through which succeeding generations of technologies will be crafted in due time. As new technologies emerge and are proven, they become publically available and accepted within society. In due time, the emergency management domain will integrate and implement new technologies.

Demographics

The potentials of increasing populations, ethnic diversity, and larger percentages of senior citizens will create additional planning challenges within the emergency management domain. Larger populations will increase the complexities of mass evacuations, and may increase the chances of mass casualties during cataclysms. Ethnic diversity may interject the necessity of emergency management personnel becoming multilingual in order to communicate with various segments of the population. An increasing quantity of senior citizens will increase the burdens of health care that may be necessitated during emergency situations.

Emergency managers will be faced with complex demographics shifts as the United States' population increases, ages, and becomes more culturally and

---

[53]Brian Dumaine, "The Driverless Revolution Rolls On," *Fortune*, 2012, http://tech.fortune.cnn.com/2012/11/12/self-driving-cars/ (accessed May 31, 2013).
[54]Ibid.
[55]Dan Graziano, "Driverless Cars Expected to Go Mainstream by 2025," 2013, http://bgr.com/2013/04/19/google-driverless-cars-release-date-2025-448246/ (accessed May 31, 2013).

linguistically diverse. These shifts call for new capabilities including multilingual proficiencies and close dialogue with community leaders to understand their needs. As information becomes more widely distributed from numerous sources, emergency managers will need to practice omnidirectional knowledge sharing and use the power and influence of social networks to remain relevant in the complex media environment. Therefore, the flows of information must also accommodate the diverse needs of multiple languages to communicate with the residents of an affected incident area.

In a resource-constrained environment, leveraging volunteer support will be crucial, and engaging schools and youth programs to infuse emergency management principles across the entire educational experience will be important in creating awareness of new and evolving threats. Additionally, building an emergency management culture that embraces forward-thinking mentalities will be beneficial regarding the anticipating of emerging challenges.

The geographic locations of populations also change through time as the result of a variety of factors (e.g., economic, political, etc.). If populations increase along the coastlines, then they may be subjected to the dangers of hurricanes or typhoons. If population shifts occur inland, then they may become more susceptible to earthquakes, blizzards, and other natural hazards. Regardless of location, any community may experience the devastating effects of terrorism.

During the coming decades, mega-regions are anticipated to reflect populations within the United States.[56] Immigration, both legal and illegal, may bring populations who do not trust authorities or who are unable to communicate with them well. Therefore, residents may not heed the warnings of emergency service organizations or may have no understanding of any issued warnings. Therefore, emergency organizations may be faced with cultural problems that impede their services through time.

The changing demographics of populations may affect the financing of municipalities. Government coffers may experience restraints and financial shortcomings that affect future emergency services whereas others may be more lucrative. Regardless, funding will always be a challenge within the emergency services domain. Because these services are essential components of society, many of them may become privatized through time.[57] Nonprofit organizations exist to satisfy societal needs for goods and services that are not provided by the commercial

---

[56] America 2050, "Megaregions," 2013, http://www.america2050.org/megaregions.html (accessed May 31, 2013).

[57] Margaret Steen, "Game-Changing Trends Beg Evolution of Emergency Management Planning," *Emergency Management: Strategy and Leadership in Critical Times*, 2011, http://www.emergencymgmt.com/disaster/Population-Shifts-Aging-Infrastructure-Emergency-Planning-041911.html?page=2& (accessed May 31, 2013).

sector or the government. Therefore, an increase in the quantity of emergency services by nonprofit organizations may occur with any decreases in government funding.

Changing demographics will most certainly alter the face of American society through time. No one can predict with absolute certainty and accuracy the changes that will occur or the effects that will be generated from these changes. In any case, change will come in due time. The emergency management domain must examine future projects of demographics to determine how best to serve localities throughout the nation during the coming years.

A Recovery Focus

Recovery is complex and involves many stakeholders. Throughout recovery efforts, diverse interests compete for typically limited resources. Many of the decisions within the recovery effort will not be in the hands of the emergency management professionals. Instead, the recovery efforts are influenced by many interests, often competing to include local governments and a variety of organizations. In any case, any future recovery efforts must work toward achieving a sense of societal normalcy, in due time, following disastrous events.

Short-term recovery focuses on the immediate tasks of securing the impact area, housing victims, and establishing conditions under which households and businesses can begin the process of recovery. Long-term reconstruction actually implements the reconstruction of the disaster impact area and manages the disaster's psychological, demographic, economic, and political impacts. Finally, recovery management monitors the performance of the disaster assessment, short-term recovery, and long-term reconstruction functions. It also ensures they are coordinated and provides the resources needed to accomplish them.[58]

Future of International Emergency Management

A global community exists across the world. International disaster responses will remain complex. Endeavors require coordination among organizations whose internal cultures, mandates, and procedures are not always aligned and well-coordinated among responding organizations and their sponsoring nations. Thus, the emergency manager must continually work to anticipate the future collaborations that may exist among nations and international relief organizations.

---

[58]Michael K. Lindell, Carla S. Prater, and Ronald W. Perry with contribution by William C. Nicholson, *Fundamentals of Emergency Management Electronic Textbook,* Federal Emergency Management Agency, 2006, http://training.fema.gov/EMIWeb/edu/fem.asp (May 1, 2013).

Figure 14.6 All nations are susceptible to disasters.

Disaster operations may be quite complex within the scope and magnitude of international operations. Numerous complexities exist ranging from the logistics of goods and services to support international relief efforts to the problems of human language incompatibilities. Technological problems may plague relief and recovery efforts internationally. For example, complex integrations of computer systems across national boundaries to support relief efforts may require time and may be impeded by politics. International politics, sovereignty issues, and government regulation may also impact international relief efforts.

Certainly, opportunists will attempt to fraudulently misappropriate or interdict disaster assistance for personal gain. Allegations of this type are nothing new within the international context of managing funds during the aftermath of disasters. During 1983, in the nation of Colombia, President Belisario Betancur denied allegations and charges involving "profiteering in distribution of relief aid to Popayan" during the aftermath of an earthquake that affected the region.[59] Within

---

[59]"Colombia Leader Denies Disaster-Aid Profiteering," *The Christian Science Monitor*, 1983, http://www.csmonitor.com/1983/0406/040624.html (accessed May 31, 2013).

the United States, the 2005 incident of Hurricane Katrina also involved allegations and charges of criminal activities involving disaster assistance funding. The owner of a hotel was charged for the crime of "submitting $232,000 in bills for phantom victims" and approximately "1,100 prison inmates across the Gulf Coast apparently collected more than $10 million in rental and disaster-relief assistance."[60] Humans are fallible creations, and will always be tempted by various forms of profiteering. Human nature is unchanging. Therefore, future international disaster assistance efforts must be mindful of the potentials of human corruption when managing assistance funding.

The affluence of nations ranges greatly. In times of disasters, it is not surprising that events impacting some of the world's poorest nations inflict damage that may well serve to initiate further human catastrophes such as an outbreak of disease. In the case of the earthquakes impacting Haiti in 2010, an estimated three million people were affected by the event. Haiti, being one of the poorest nations in the western Hemisphere, was dependent on the international community for both relief funding and relief operations. With over 200,000 deaths and thousands more injured, destruction of the already fragile national infrastructure made international support of relief operations essential. Thus, the goal of providing immediate life-saving assistance, while concurrently supporting long-term development and policy goals, is a complicated one.

The funding of disaster relief operations, especially in the international area, will remain a challenge in the future. Obtaining funds may occur through a variety of methods. In some cases, funding may be derived from governments and charitable organizations. In other instances, donations may be provided by corporations or private individuals. Regardless of the source of funds, decisions must be made regarding how the funding will be expended toward recovery. Analogous with the notions of capital budgeting that is useful for addressing the financial aspects of domestic events, international capital budgeting techniques may be used to influence decisions regarding international disasters. These methods employ similar mathematics that are involved with the domestic considerations of capital budgeting.

The managing of multinational relief operations will be challenging for a variety of reasons. Aligning the logistics of such operations must accommodate an international perspective of project management. Regardless of the scope and magnitude of international emergency management initiatives, all of them must involve various amounts of controlling, coordinating, leading, organizing, and planning. A failure of any of these areas may significantly impair the progression of any emergency management initiative. These notions are applicable for both

---

[60]Eric Lipton, "'Breathtaking' Waste and Fraud in Hurricane Aid," *The New York Times*, 2006, http://www.nytimes.com/2006/06/27/washington/27katrina.html?pagewanted=all&_r=0 (accessed May 31, 2013).

government and nongovernment entities that are the primary entities through which emergency management functions are exercised.

International disaster responses require the expertise of many specialized actors. Examples include affected government entities, militaries, intergovernmental organizations (typically United Nations agencies), international and domestic NGOs, the Salvation Army, and the International Red Cross. Representatives from the affected civilian populations may also contribute to these endeavors. No single actor can undertake all facets of relief and recovery. Addressing the needs of survivors, which span health, nutrition, water and sanitation, emergency shelter, and livelihood reconstruction, will remain a daunting task throughout any future international emergencies.[61] Both government and nongovernment entities will be required to assist with future international emergency management initiatives.

A variety of such organizations exist that have the capacity to support these future emergency management functions. One such example is the Flagstaff International Relief Effort (FIRE). This organization provides humanitarian assistance in the areas of hepatitis vaccinations, health safety training, and orthopedics.[62] Another example is the Denison International Relief Effort (DIRE). This organization provides "funds for the victims and immediate survivors," reconstruction resources, and education concerning the "culture, political situation, economic realities, and environment of the area in need."[63] The DIRE has conducted operations in the nations of "Afghanistan, India, Turkey, Uganda, Zambia, and Zimbabwe."[64] Many other organizations exist that have similar missions and capacities, and can be located through Internet searches.

Military support for disaster relief operations from the international community will remain important. Major nations, such as the United States can quickly, effectively, and efficiently move a wide range of military assets, which are self-sustained, into an affected area and commence relief operations. Humanitarian missions are often the focus of these operations. During modern times, military involvement among humanitarian efforts is significant. The following statements show the relevancy of military assistance:

> . . . military aircraft have air-dropped food into Ethiopia, helicoptered aid into remote villages in Sudan, rescued flood victims (and often their animals) in Bangladesh, rushed pharmaceutical to earthquake sites,

---

[61]International Disaster Response, Social Issue Report, http://rootcause.org/documents/DR-Issue.pdf (accessed May 15, 2013).
[62]Flagstaff International Relief Effort, "Flagstaff International Relief Effort," 2013, http://www.fire-projects.org/ (accessed May 31, 2013).
[63]Denison Community Association, "Denison International Relief Effort," 2013, http://denison.orgsync.com/org/dca/direcommittee (accessed May 31, 2013).
[64]Ibid.

and delivered medical teams to hundreds of major and minor disasters. Engineers have helped rebuild roads and bridges (in some countries, to the point where there are more Bailey bridges than normal bridges!) and have supervised the construction of major flood control works in some regions.[65]

Throughout this discussion is a common theme that will permeate all future emergency management endeavors: flexibility. Emergency management organizations must adapt to the dynamics of changing environments quickly. They must be capable of effectively and efficiently performing when confronted with a variety of circumstances both internationally and domestically. Their sources of funding must be diverse within the international community to lower the risk associated with having only a solitary benefactor. Their personnel must be capable of deploying to a variety of locations and conditions. Their resources must be highly mobile and capable of functioning within a multitude of different physical environments. They must be prepared to encounter different cultures, customs, and traditions. They must be prepared for different political climates, varying levels of cooperation, and acceptances internationally. In any case, future emergency management entities must be flexible.

Chapter Comments and Summary

A major natural disaster, flood, hurricane, or earthquake impacting not only our nation but also other nations is inevitable. Emergency management systems should focus their efforts on preparing for and responding to those events most likely to impact their jurisdictions. Though events such as the bombing in Boston remind emergency managers of the threat posed by terrorist attacks, natural disasters and major industrial accidents continue pose great threats and must be considered as a primary focus of emergency manager.

The maturation of the emergency management profession will continue to be a work in progress based upon consistent evaluation, constant speculation, and proactive integration of continually transformed principles, preparation, protocols, and performance-based best practices. In hindsight, we may definitively discuss the "what-ifs," as well as speculate on future "what-ifs." Yet, irrespective of such live-changing reality, and post-incident experience, there will always be a degree of uncertainty and sometimes no definitive answers with respect to the outcome of a disaster, emergency, or hazardous event. Emergency management is a discipline that deals with countless uncertainties. One can know the events will occur that will impact communities, but one does not know specifically when or where those events will occur. As a result, emergency management must remain flexible and capable of responding in times of great need.

---

[65]Frederick Cuny, "Use of the Military in Humanitarian Relief," *PBS Frontline,* 1989, http://www.pbs.org/wgbh/pages/frontline/shows/cuny/laptop/humanrelief.html (accessed May 31, 2013).

The United States exhibits a history of change throughout its existence. Change will continue to permeate American society in the future. The demographics of the United States will be affected by a variety of factors ranging from an aging population to the effects of both legal and illegal immigration. In any case, emergency management must accommodate demographic change through time in order to effectively and efficiently serve society.

These notions of change also permeate technologies. Change is a common aspect of any technological maturation. During recent years, much research and development efforts have focused upon communications systems and driverless vehicles. Both entities have the potential of greatly affecting future instantiations of emergency management. The former has the potential of facilitating unimpeded, portable communications whereas the latter may support emergency management logistics. In any case, the emergency management domain must accommodate change and adapt new and emerging technologies as they become publically available.

Both internationally and domestically, emergency management will experience a variety of problems and challenges. Examples are diverse, and range from cultural issues and differences to the unending necessity of funding. In any case, no emergency scenario will be perfect—issues will always exist that demand the attention of emergency personnel and nations.

Terminology

Automation
Border security
Capital budgeting
Coordination
Decision
Demographic
Dependency
Disaster
Disaster assistance
Driverless car
Flexibility
Forecast
Funding

Globalism
Interdependency
International
International cooperation
International disaster
Multinational
Population shift
Project management
Relief organization
Revenues
Technology
Terrorism

Thought and Discussion Questions

1. This chapter introduced the scandalous behaviors of government officials and private citizens that attempted to leverage disaster assistance funding for personal gain. Such opportunists will always plague the emergency management domain. Perform some research, and determine what other examples exist that

demonstrate these concepts. Based on your findings, write a brief essay that details how you believe such criminality may be avoided within the context of the future of emergency management?

2. This chapter introduces the concept of international cooperation regarding cataclysmic events. Given the events described herein, what circumstances must exist in order to necessitate the involvement of multiple nations during the aftermath of cataclysms? Write a brief essay that highlights your opinion.
3. This chapter introduces several international challenges that must be overcome when exercising international disaster initiatives. Perform some research, and determine what other factors may be problematic when conducting such operations. Write a brief essay that highlights your findings.
4. This chapter introduces organizations that have the capacity of contributing to international emergency management initiatives. Perform some research, and investigate other organizations whose missions and capacities are similar. Write a brief essay that highlights your findings and that describes these organizations.

References

1. 9 Thought dead as Minneapolis Bridge collapses. 2007. *NBC News*. http://www.nbcnews.com/id/20079534/ns/us_news-life/t/thought-dead-minneapolis-bridge-collapses/#.Uak14Jy87eY (accessed May 31, 2013).
2. America 2050. 2013. Megaregions. http://www.america2050.org/megaregions.html (accessed May 31, 2013).
3. Bello, Marisol. 2013. Bridge collapse shines light on aging infrastructure. *USA Today*. http://www.usatoday.com/story/news/nation/2013/05/24/washington-bridge-collapse-nations-bridges-deficient/2358419/ (accessed May 31, 2013).
4. CDC Foundation. 2013. Global disaster response fund. http://cdcfoundation.org/globaldisaster (accessed May 31, 2013).
5. Clark, Maggie. 2013. Drone limits become law. *Emergency Management: Strategy and Leadership in Critical Times*. http://www.emergencymgmt.com/safety/Drone-Limits-Law.html (accessed May 31, 2013).
6. Colombia leader denies disaster-aid profiteering. 1983. *The Christian Science Monitor*. http://www.csmonitor.com/1983/0406/040624.html (accessed May 31, 2013).
7. Cuny, Frederick. 1989. Use of the military in humanitarian relief. *PBS Frontline*. http://www.pbs.org/wgbh/pages/frontline/shows/cuny/laptop/humanrelief.html (accessed May 31, 2013).
8. Denison Community Association. 2013. Denison international relief effort. http://denison.orgsync.com/org/dca/direcommittee (accessed May 31, 2013).

9. Doss, Daniel, Chen Guo, and Joo Lee. 2011. *The Business of Criminal Justice: A Guide for Theory and Practice*. Boca Raton, FL: CRC Press.
10. Doss, Daniel, William Sumrall, and Don Jones. 2012. *Strategic finance for criminal justice organizations*. Boca Raton, FL: CRC Press.
11. Dumaine, Brian. 2012. The driverless revolution rolls on. *Fortune*. http://tech.fortune.cnn.com/2012/11/12/self-driving-cars/ (accessed May 31, 2013).
12. Flagstaff International Relief Effort. 2013. Flagstaff international relief effort. http://www.fireprojects.org/ (accessed May 31, 2013).
13. Global First Response. 2013. Operations. http://www.globalfirstresponse.com/Operations.html (accessed May 31, 2013).
14. Graziano, Dan. 2013. Driverless cars expected to go mainstream by 2025. http://bgr.com/2013/04/19/google-driverless-cars-release-date-2025-448246/ (accessed May 31, 2013).
15. Knickerbocker, Brad. 2013. Skagit Bridge collapse: Not the only one waiting to happen. *The Christian Science Monitor*. http://www.csmonitor.com/USA/USA-Update/2013/0527/Skagit-River-bridge-collapse-Not-the-only-one-waiting-to-happen-video (accessed May 31, 2013).
16. Lindell, Michael K., Carla S. Prater, and Ronald W. Perry with contribution by William C. Nicholson. 2006. *Fundamentals of emergency management electronic textbook*. Federal Emergency Management Agency. http://training.fema.gov/EMIWeb/edu/fem.asp (May 1, 2013).
17. Lipton, Eric. 2006. 'Breathtaking' waste and fraud in hurricane aid. *The New York Times*. http://www.nytimes.com/2006/06/27/washington/27katrina.html?pagewanted=all&_r=0 (accessed May 31, 2013).
18. Marosi, Richard. 2011. Controversial Muslim cleric is arrested while sneaking into the U.S. *The Los Angeles Times*. http://articles.latimes.com/2011/jan/27/local/la-me-border-cleric-20110127 (accessed May 31. 2013).
19. McKay, Jim. 2011. The changing face of terror in the U.S. *Emergency Management: Strategy and Leadership in Critical Times*. http://www.emergencymgmt.com/safety/Face-Terror-US-Home-Grown.html?page=1& (accessed May 31, 2013).
20. Mexican troops aid Katrina efforts. 2005. *Fox News*. http://www.foxnews.com/story/0,2933,168778,00.html (accessed May 31. 2013).
21. MURDOCK: U.S. southern border open to terrorists. 2013. *The Gleaner* (Henderson, KY). http://www.courierpress.com/news/2013/may/07/us-southern-border-open-to-terrorists/ (accessed may 31, 2013).
22. Nakashima, Ellen. 2013. Confidential report lists U.S. weapons system designs compromised by Chinese cyberspies. *The Washington Post*. http://articles.washingtonpost.com/2013-05-27/world/39554997_1_u-s-missile-defenses-weapons-combat-aircraft (accessed May 31, 2013).
23. Osmose Utility Services. 2013. Aging infrastructure. http://www.osmoseutilities.com/content/pages/aging-infrastructure (accessed May 31, 2013).

24. Pittman, Elaine. 2012. 3 Emerging technologies that will impact emergency management. *Emergency Management: Strategy and Leadership in Critical Times.* http://www.emergencymgmt.com/disaster/3-Emerging-Technologies-Emergency-Management.html?page=2& (accessed May 31, 2013).
25. Rich, Sarah. 2012. Nevada DMV approves regulations for testing driverless vehicles. *Government Technology: Solutions for State and Local Government.* http://www.govtech.com/transportation/Nevada-DMV-Approves-Testing-Driverless-Vehicles.html (accessed May 31, 2013).
26. Roman, Tomas. 2012. New technologies developed to help during disasters. *ABC News.* http://abclocal.go.com/kgo/story?section=news/local/south_bay&id=8873344 (accessed May 31, 2013).
27. State of California. 2012. California legislature approves driverless vehicle bill—Senator Padilla's legislation establishes performance and safety standards. http://sd20.senate.ca.gov/news/2012-08-29-california-legislature-approves-driverless-vehicle-bill-senator-padilla-s-legislatio (accessed May 31, 2013).
28. State of Texas. 2005. Gov. Perry awards $6 million to border counties for border security. http://governor.state.tx.us/news/press-release/2558/ (accessed May 31. 2013).
29. State of Texas. 2013. Gov. Perry: Emergency response starts with individual preparedness. http://governor.state.tx.us/news/press-release/18598/ (accessed May 31, 2013).
30. Steen, Margaret. April 20, 2011. Game-changing trends beg evolution of emergency management planning. *Emergency Management: Strategy and Leadership in Critical Times.* http://www.emergencymgmt.com/disaster/Population-Shifts-Aging-Infrastructure-Emergency-Planning-041911.html?utm_source=embedded&utm_medium=direct&utm_campaign=Game-Changing-Trends-Beg-Evolution-of-Emergency-Management-Planning (accessed May 17, 2013).
31. Terrorism Liaison Officer Information Network. 2013. What is a terrorism liaison officer? http://www.tlo.org/what_is_tlo.html (accessed May 31, 2013).
32. Top 10 costliest disasters. 2013. *Bankrate.* http://www.bankrate.com/finance/insurance/top-10-costliest-natural-disasters-4.aspx (accessed May 31, 2013).
33. Waugh, William. 2003. The future of emergency management. Georgia State University. http://search.yahoo.com/r/_ylt=A0oG7thSEKlR3i0AJrFXNyoA;_ylu=X3oDMTEydjVrZDFlBHNlYwNzcgRwb3MDMQRjb 2xvA2FjMgR2dGlkA0RGRDVfODg-/SIG=15n7p44fi/EXP=1370063058/**https%3a//training.fema.gov/ EMIWeb/edu/docs/emfuture/ Future%2520of%2520EM%2520-%2520the%2520Future%2520of%2520Homeland% 2520Security%2520and%2 520EM%2520-%2520Waug.doc (accessed May 30, 2013).

34. Wing, Nick. 2013. Nidal Hassan, Fort Hood shooting suspect, has received $278,00 in salary since arrest," *The Huffington Post*. http://www.huffingtonpost.com/2013/05/21/nidal-malik-hasan-salary_n_3313519.html (accessed May 31, 2013).
35. Wood, Colin. 2013. The case for drones. *Emergency Management: Strategy and Leadership in Critical Times*. http://www.emergencymgmt.com/safety/Case-for-Drones.html (accessed May 31, 2013).

www.ingramcontent.com/pod-product-compliance
Ingram Content Group UK Ltd.
Pitfield, Milton Keynes, MK11 3LW, UK
UKHW051248180426
11947UKWH00020B/1601